Protecting Traditional Knowledge

This is the first comprehensive review of the Intergovernmental Committee (IGC) of the World Intellectual Property Organization (WIPO) established in 2000. It provides an in-depth consideration of the key thematic areas within WIPO discussions – genetic resources (GRs), traditional knowledge (TK) and traditional cultural expressions (TCEs) through the perspectives of a broad range of experts and stakeholders, including indigenous peoples and local communities. It also looks at how these areas have been treated in a number of forums and settings (including national systems and experiences, and also in trade agreements) and the interface with WIPO discussions.

Furthermore, the book analyses the process and the negotiation dynamics since the IGC received a mandate from WIPO members, in 2009, to undertake formal text-based negotiations towards legal instruments for the protection of GR, TK and TCEs. While there has been some progress in these negotiations, important disagreements persist. If these are to be resolved, the adoption of these legal instruments would be a significant development towards resolving key gaps in the modern intellectual property system. In this regard, the book considers the future of the IGC and suggests options which could contribute towards achieving a consensual outcome.

Daniel F. Robinson is Research Fellow at the International Centre for Trade and Sustainable Development (ICTSD) Geneva, Switzerland and Associate Professor in Environmental Studies at The University of New South Wales, Australia.

Ahmed Abdel-Latif is Special Assistant for Policy and Programme in the Office of the Director General at the International Renewable Energy Agency (IRENA), Abu Dhabi, United Arab Emirates.

Pedro Roffe is Senior Associate in Innovation, Technology and Intellectual Property at ICTSD, Geneva, Switzerland.

Routledge Research in International Environmental Law

International Environmental Law-making and Diplomacy
Insights and Overviews
Edited by Tuomas Kuokkanen, Ed Couzens, Tuula Honkonen and Melissa Lewis

Responsibilities and Liabilities for Commercial Activity in the Arctic
The Example of Greenland
Edited by Vibe Ulfbeck, Anders Møllmann and Bent Ole Gram Mortensen

Ecological Integrity and Global Governance
Science, Ethics and the Law
Laura Westra

Environmental Law and the Ecosystem Approach
Maintaining Ecological Integrity through Consistency in Law
Froukje Maria Platjouw

Stratospheric Ozone Damage and Legal Liability
US Public Policy and Tort Litigation to Protect the Ozone Layer
Lisa Elges

Ecological Restoration in International Environmental Law
Anastasia Telesetsky, An Cliquet and Afshin Akhtar-Khavari

Sustainable Development Principles in the Decisions of International Courts and Tribunals
1992–2012
Marie-Claire Cordonier Segger and H.E. Judge C.G. Weeramantry

Culture and International Economic Law
Valentia Vadi and Bruno de Witte

Natural Resources Law, Investment and Sustainability
Shawkat Alam, Jahid Hossain Bhuiyan and Jona Razzaque

Protecting Traditional Knowledge

The WIPO Intergovernmental Committee on Intellectual Property and Genetic Resources, Traditional Knowledge and Folklore

Edited by Daniel F. Robinson, Ahmed Abdel-Latif and Pedro Roffe

Routledge
Taylor & Francis Group
LONDON AND NEW YORK

earthscan
from Routledge

ICTSD
International Centre for Trade and Sustainable Development

First published 2017
by Routledge

2 Park Square, Milton Park, Abingdon, Oxfordshire OX14 4RN
52 Vanderbilt Avenue, New York, NY 10017

Routledge is an imprint of the Taylor & Francis Group, an informa business

First issued in paperback 2019

British Library Cataloguing-in-Publication Data
A catalogue record for this book is available from the British Library

Library of Congress Cataloging in Publication Data
Names: Robinson, Daniel F., editor. | Abdel-Latif, Ahmed, editor. | Roffe, Pedro, editor.
Title: Protecting traditional knowledge: the WIPO Intergovernmental Committee on Intellectual Property and Genetic Resources, Traditional Knowledge and Folklore / edited by Daniel F. Robinson, Ahmed Abdel-Latif and Pedro Roffe.
Description: London; New York : Routledge, 2017. |
Series: Routledge research in international environmental law | Includes bibliographical references and index.
Identifiers: LCCN 2016052547 | ISBN 9781138955448 (hbk) | ISBN 9781315666358 (ebk)
Subjects: LCSH: World Intellectual Property Organization. Intergovernmental Committee on Intellectual Property and Genetic Resources, Traditional Knowledge and Folklore. | Intellectual property. | Indigenous peoples—Legal status, laws, etc. | Traditional ecological knowledge—Law and legislation | Germplasm resources—Law and legislation.
Classification: LCC K1401.5 .P76 2017 | DDC 346.04/8089—dc23
LC record available at https://lccn.loc.gov/2016052547

ISBN: 978-1-138-95544-8 (hbk)
ISBN: 978-0-367-33523-6 (pbk)

Typeset in Times
by Keystroke, Neville Lodge, Tettenhall, Wolverhampton

Contents

Acronyms and abbreviations

ABS	access and benefit-sharing
ADR	alternative dispute resolution
ARIPO	African Regional Intellectual Property Organization
BD Act	Biological Diversity Act
BL	Biodiversity Law
CAFTA-DR	Dominican Republic–Central America Free Trade Agreement
CARIFORUM	Caribbean Forum of the African, Caribbean and Pacific Group of States
CBD	Convention on Biological Diversity
CGIAR Centres	Consultative Group of International Agricultural Research's Centres
CONAGEBIO	National Commission for Management of Biodiversity
COP	Conference of the Parties to the CBD
DG	Director General
DNA	deoxyribonucleic acid
DOO	disclosure of origin patent requirement
EFTA	European Free Trade Association
EPO	European Patent Office
EU	European Union
FAO	United Nations Food and Agriculture Organization
FFM	fact finding mission
FTA	free-trade agreement
GA	WIPO General Assembly
GEF	Global Environment Facility
GIPID	WIPO Global Intellectual Property Issues Division
GR	genetic resource
GRAATK	genetic resources and associated traditional knowledge
GRULAC	Latin American and Caribbean Group
IDA	international depositary authority
IGC	Intergovernmental Committee on Intellectual Property and Genetic Resources, Traditional Knowledge and Folklore
ILO	International Labour Organization

IP	intellectual property
IPA	International Publishers Association
IPI	Swiss Federal Institute of Intellectual Property
IPLCs/ILCs	Indigenous peoples and local communities
IPRs	intellectual property rights
IWGs	inter-sessional working groups
JPO	Japanese Patent Office
MAT	mutually agreed terms
NAFTA	North American Free Trade Agreement
Nagoya Protocol	Nagoya Protocol on Access to Genetic Resources and the Fair and Equitable Sharing of Benefits Arising from their Utilization
NGOs	non-governmental organizations
OAPI	African Intellectual Property Organization
PCT	Patent Cooperation Treaty
PGRFA	Plant Genetic Resources for Food and Agriculture
PIC	prior informed consent
PIP Framework	Pandemic Influenza Preparedness Framework
Plant Treaty	International Treaty on Plant Genetic Resources for Food and Agriculture
PLT	Patent Law Treaty
PPVFRA	Protection of Plant Varieties and Farmers' Rights Act
PTAB	Patent Trial and Appeal Board
PTAs	preferential trade agreements
R&D	research and development
RNA	ribonucleic acid
RTA	regional trade agreements
SACU	Southern African Customs Union
SCCR	Standing Committee on Copyright and Related Rights
SCP	WIPO Standing Committee on the Law of Patents
SMTA	Standard Material Transfer Agreement
TCE	traditional cultural expressions
TCEs/EoF	traditional cultural expressions/expressions of folklore
TK	traditional knowledge
Tkarihwaié:ri Code	Tkarihwaié:ri Code of Ethical Conduct on Respect for the Cultural and Intellectual Heritage of Indigenous and Local Communities
TKDL	Traditional Knowledge Digital Library
TO	Technical Office of the CONAGEBIO
TPP	Trans-Pacific Partnership Agreement
TRIPS	Trade-Related Aspects of Intellectual Property Rights
UK	United Kingdom
UN	United Nations
UNCTAD	United Nations Conference on Trade and Development

UNDRIP	United Nations Declaration on the Rights of Indigenous Peoples
UNESCO	United Nations Educational, Scientific and Cultural Organization
UNU-IAS	United Nations University, Institute of Advanced Studies
UPOV	International Union for the Protection of New Varieties of Plants
USA/US	United States of America
USPTO	United States Patent and Trademark Office
USTR	United States Trade Representative
WHO	World Health Organization
WIPO	World Intellectual Property Organization
WPPT	WIPO Performances and Phonograms Treaty
WTO	World Trade Organization

Contributors

Ahmed Abdel-Latif, Special Assistant, Policy and Programme, Office of the Director General, International Renewable Energy Agency (IRENA). Former Senior Programme Manager, Innovation, Technology and Intellectual Property, International Centre for Trade and Sustainable Development (ICTSD).

Jens Bammel, former Secretary General, International Publishers Association, Geneva.

Margo A. Bagley, Hardy Cross Dillard Professor of Law and Joseph C. Carter, Jr. Research Professor of Law, University of Virginia School of Law; Asa Griggs Candler Professor of Law, Emory University School of Law, USA.

Georges Bauer, Policy Adviser for the Sustainable Development and International Cooperation unit at the Swiss Federal Institute of Intellectual Property (IPI) in Bern, Switzerland.

Cyrill Michael Berger, lic.iur., LL.M., former Legal Adviser, Sustainable Development and International Cooperation, Swiss Federal Institute of Intellectual Property (IPI) in Bern, Switzerland.

José Borghino, Secretary General, International Publishers Association, Geneva.

Jorge Cabrera Medaglia, Professor Environmental Law, University of Costa Rica. Lead Counsel Biosafety and Biodiversity Law, Centre for Sustainable Development Law, Montreal.

Claudio Chiarolla, Legal Officer, Traditional Knowledge Division, WIPO.

Biswajit Dhar, Professor, Centre for Economic Studies and Planning, School of Social Sciences, Jawaharlal Nehru University, New Delhi.

Graham Dutfield, Professor of International Governance and founding director of the LLM in Intellectual Property Law at the University of Leeds, UK.

Susy Frankel, Professor of Law and Director of the New Zealand Centre of International Economic Law, Victoria University of Wellington. President ATRIP 2015–2017.

Martin Girsberger, Head of the Sustainable Development and International Cooperation unit at the Swiss Federal Institute of Intellectual Property (IPI) in Bern, Switzerland.

Dominic Keating, Director of the Intellectual Property Attaché Program at the United States Patent and Trademark Office, and Adjunct Associate Professor at the American University Washington College of Law.

Jessica C. Lai, Senior Lecturer, School of Accounting and Commercial Law, Victoria University of Wellington.

Charles Lawson, Professor of Law, Griffith University, Queensland Australia, and is Admitted as a barrister and solicitor to the Supreme Court of Queensland, Supreme Court of Victoria and the High Court of Australia.

Ricardo Meléndez-Ortiz, Chief Executive Officer, International Centre for Trade and Sustainable Development (ICTSD).

Dominic Muyldermans, Independent attorney and senior legal advisor to ABS-int, Ghent, Belgium; and senior legal consultant to CropLife International, Brussels, Belgium.

Chidi Oguamanam, Professor of Law currently affiliated with the Open African Innovation Research (Open AIR), Access and Benefit Sharing Canada (ABS Canada) Projects and the Centre for Law, Technology and Society at the University of Ottawa.

Ruth L. Okediji, William L. Prosser Professor of Law and Solly Robins Distinguished Research Fellow at the University of Minnesota Law School, Minneapolis, USA.

Marc Perlman, Associate Professor in the Department of Music, Brown University, Providence, Rhode Island, USA.

Daniel F. Robinson, Research Fellow at the International Centre for Trade and Sustainable Development (ICTSD) Geneva, and Associate Professor in Environmental Studies, Arts and Social Sciences Faculty, the University of New South Wales, Australia.

Pedro Roffe, Senior Associate, Innovation, Technology and Intellectual Property, International Centre for Trade and Sustainable Development (ICTSD).

Manuel Ruiz Muller, Director and Principal Researcher, International Affairs and Biodiversity Program, Peruvian Society for Environmental Law.

Maui Solomon, Barrister (New Zealand) and Indigenous Peoples Rights Advocate.

Brendan Tobin, International Consultant on Environmental Law and Human Rights, Ashoka Fellow, Adjunct Lecturer Irish Center for Human Rights, National University of Ireland, Galway.

Wend Wendland, Director, Traditional Knowledge Division, WIPO and Secretary to the WIPO IGC.

Foreword

The WIPO Intergovernmental Committee on Intellectual Property and Genetic Resources, Traditional Knowledge and Folklore

Growing recognition of the economic value of genetic resources and associated traditional knowledge, as well as a trend of increasing the protection of biotechnological inventions, has over the last two decades prompted biodiversity-rich countries and traditional knowledge holders to place concerns about the misappropriation of genetic patrimony at the forefront of international trade and intellectual property frameworks. However, the relationship between intellectual property, traditional knowledge and genetic resources continues to remain one of the most complex and debated issues in multilateral forums.

One reason for this is that discussions have taken place simultaneously in a number of arenas ranging from the Convention on Biological Diversity (CBD) to the World Trade Organization (WTO) and the World Intellectual Property Organization (WIPO). Ensuring coherence between these processes in order to respond to different agendas and to a diversity of stakeholders has correspondingly proved challenging. Progress has been uneven in different settings and with varying implications.

At the WTO, stalemate in the 2001 Doha Round negotiations has resulted in little headway on these issues, while the adoption by the CBD of the Nagoya Protocol on Access and Benefit Sharing in 2010 was a significant milestone with a bearing on deliberations in other forums. At WIPO, meanwhile, the Intergovernmental Committee on Intellectual Property, Genetic Resources, Traditional Knowledge and Folklore (IGC) has since its creation in 2000 witnessed in-depth discussions involving different stakeholders – in particular indigenous peoples – which have contributed to a better understanding of the complex issues at hand. In 2009, the IGC was mandated by WIPO member states to accelerate its work towards developing an international instrument or instruments to protect genetic resources, traditional knowledge and traditional cultural expressions. However, the IGC has not succeeded yet in reaching concrete results in this respect.

Against this background, this book offers the first comprehensive overview of the IGC's work. It aims to analyse the processes and substantive issues which have been at the centre stage of the IGC's deliberations for almost two decades, as well as to identify possible solutions to overcome the gap in positions between different countries and stakeholders. Seeking to encompass a broad and diverse range of perspectives, the volume includes contributions from a number of distinguished scholars, policymakers, figures from industry, civil society groups and indigenous peoples' representatives.

The collaboration between the International Centre for Trade and Sustainable Development (ICTSD) and the authors in this volume builds on our longstanding efforts to help to achieve progress in various international processes tackling the interface of trade, intellectual property, genetic resources and traditional knowledge. ICTSD has been actively engaged in these areas starting from just after its establishment 20 years ago when the discussion crystallised on the review of the WTO's Agreement on Trade-Related Aspects of Intellectual Property Rights (TRIPS) Article 27.3(b) and the relationship between the TRIPS Agreement and the CBD. Since then, ICTSD has continued to pursue concrete responses to the challenges and opportunities posed by rapidly advancing bio-based technologies and emerging global policy frameworks, through multi-stakeholder dialogues and targeted policy-oriented research.

The backdrop to ICTSD's work in this area is the conviction that, in a knowledge-driven global economy, a better understanding of intellectual property related issues is the key to guaranteeing informed policymaking in virtually all areas of development. This is especially true in the case of genetic resources and associated traditional knowledge. Our focus has been on ensuring that international processes take into account a diversity of perspectives and strive towards designing appropriate policies and norms that are supportive of sustainable development objectives and international commitments.

We sincerely hope that you will find this publication to be a useful contribution – in the well-established tradition of ICTSD – to building bridges between different stakeholders and to advancing mutually acceptable solutions to complex issues.

Ricardo Meléndez-Ortiz
Chief Executive Officer

Part 1

Introduction

1 Introduction

Mapping the evolution, state-of-play and future of the WIPO IGC

Daniel F. Robinson, Pedro Roffe and Ahmed Abdel-Latif

This book is the first comprehensive review of the World Intellectual Property Organization (WIPO) Intergovernmental Committee on Intellectual Property and Genetic Resources, Traditional Knowledge and Folklore (IGC) process. The IGC was established in 2000 and its deliberations are still ongoing.

During its first years, the IGC was focused mostly on analysis and the exchange of national experiences with the aim of achieving a better understanding of the interface between intellectual property and genetic resources (GRs), traditional knowledge and traditional cultural expressions (TCEs). In 2009, the IGC received a new mandate from WIPO members to undertake formal text-based negotiations towards legal instruments for the protection of GRs, TK and TCEs. Since then delegates have made progress towards consolidated texts which include draft provisions addressing a range of issues such as scope of protection, term of protection, beneficiaries, rights granted and exceptions and limitations. However, important disagreements still persist in some of these key areas under discussion. If such disagreements were to be resolved, the adoption of these legal instruments would be a significant development towards resolving key gaps in the modern international intellectual property (IP) system. Having binding international agreements on GRs, TK and TCEs would be a landmark in international law and in IP law, and could potentially contribute to the prevention of their misappropriation.[1] Having an international IP agreement in relation to GRs could complement the Nagoya Protocol on Access to Genetic Resources and the Fair and Equitable Sharing of Benefits Arising from their Utilization to the Convention on Biological Diversity (CBD). The IGC might also ultimately produce new international norms on TCEs, which would complement the 2003 UNESCO Convention for the Safeguarding of the Intangible Cultural Heritage.

On the other hand, if the IGC continues to inch along at its current pace or fails to deliver, it may confirm the opinion of some critics of the process, that the

1 There is still debate in the IGC context on the appropriate use of terminology with several developed countries focusing on the prevention of erroneous patents, while many developing countries and indigenous peoples argue more broadly for the prevention of misappropriation (and some used the term 'biopiracy' – see Robinson, 2010 – but without it being seriously considered for the text).

negotiations were forum shifted into WIPO to maintain a status quo in the intellectual property system – managing developing country and Indigenous representative expectations against an ever higher ratchet of IP minimum standards through other forums or processes like regional trade agreements (RTAs) and bilateral free trade agreements (FTAs) (see Drahos, 2004; Helfer, 2009). Indeed, in previous years there has been some debate over whether the TRIPS Council of the WTO would be the most appropriate forum to deal with GRs and TK, given that the TRIPS Agreement requires a review of Article 27.3(b) which deals with patentability or non-patentability of plant and animal inventions, and the protection of plant varieties. Furthermore, Paragraph 19 of the 2001 WTO Doha Declaration has broadened the discussion in TRIPS to also look at the relationship between the TRIPS Agreement and the CBD, the protection of traditional knowledge and folklore. Many developing countries also see negotiations in TRIPS as preferable for dealing with issues like erroneous patents and misappropriations because the WTO carries the force of a dispute settlement mechanism and sanctions in cases of non-compliance. As Okediji has noted: 'WIPO is neither formally charged nor structurally designed to accomplish the kind of linkage bargains now associated with the WTO' (Okediji, 2008, p73).

At the WTO, substantial proposals have been made on a disclosure of origin patent requirement over the years and most recently in the aftermath of the adoption of the Nagoya Protocol. However, there has been little progress in WTO discussions in the context of a broader stalemate in the Doha round of trade negotiations. In the meantime the WIPO IGC received a mandate to work towards international legal instruments and the WIPO Secretariat has produced a wealth of documentation to support the negotiations. The work of the IGC has had a broader mandate, including TCEs. Indigenous peoples have also been able to take part in the IGC deliberations as observers. Over the years, the IGC saw some gradual narrowing of the texts on GRs, TK and TCEs, but still with no clear consensus on what the international instruments should look like, how negotiations could be taken forward, and persistent disagreements between different stakeholders.

Against this background, the first aim of the book is to analyse and understand the inception, evolution and ongoing role of the WIPO IGC process within the context of a number of other related international forums. The second aim is to provide an in-depth consideration of the key thematic areas within WIPO discussions – GRs, TK and TCEs – and how they have been considered in other forums and contexts (in national systems and experiences, and also in free trade agreements), as well as the challenges that arise in seeking to reach agreement on mechanisms and measures to protect them. These challenges include conceptual, legal and political divergences between various stakeholders over the need for and substance of protections over GRs, TK and TCEs, which has meant the IGC is still deliberating after 33 sessions at the time of writing. The final aim of the book is to consider the IGC's future and to suggest options which could contribute towards consensus building to overcome some of the significant hurdles facing the negotiators in reaching agreement.

Two important considerations have guided the preparation of the book. First, the inclusion of a diversity of views on the issues considered by the IGC, including national perspectives as well as perspectives from different stakeholders whether from industry or from indigenous peoples. Second, the involvement of authors who, in many cases, have been actively involved in the negotiations as delegates or observers or as experts on GRs/TK/TCEs, and to ground the analysis on their experiences, their interpretation of the ongoing development of IGC texts, and also based on interviews and meetings with delegates and different stakeholders. These considerations also make the originality and value added of this book as a contribution to the existing literature on these issues; this is in addition to being the first comprehensive overview of the IGC.

The book is structured in four parts reflecting the context in which the WIPO IGC was established and operates and the relationship between the work of the WIPO IGC and other forums (Part 1), the thematic areas under negotiation (Part 2), stakeholder perspectives (Part 3) and the future of the IGC (Part 4). The scope of analysis focuses on the legal, political and technical aspects of the negotiations and working texts, and their potential to protect GRs, TK and TCEs.

Part 1 provides an introduction to the WIPO IGC and includes the three following chapters starting with a discussion of the origins and establishment of the IGC by Ahmed Abdel-Latif (Chapter 2). The chapter shows that many of the issues the IGC has been grappling with throughout its existence have been present since the very beginning. Chapter 3 by Wend Wendland, examines the evolution of the IGC process until its renewed mandate and current status. Through an 'insider's perspective' the chapter provides a comprehensive overview of how the IGC's mandate and the format of its discussions have evolved over time. Chapter 4 by Charles Lawson considers the relationship between the WIPO IGC and other related forums. It helps to provide an explanation of the international legal scope and coverage that exists, from which the perceived gap on IP and GRs, TK and TCEs has emerged and that the IGC has sought to address.

Part 2 analyses the substantive thematic areas of the IGC starting with Chapter 5 by Margo A. Bagley on a disclosure of origin (DOO) patent requirement. The author examines the evidence regarding the potential impacts of a DOO requirement relating to GRs and TK, including national systems that have implemented a DOO, and existing legal provisions that perform a similar function to a DOO requirement in countries like the US. Chapter 6 by Daniel F. Robinson and Claudio Chiarolla analyses 'complementary measures' and/or alternatives to a DOO requirement such as databases, contracts and codes of conduct relating to GRs and TK. This chapter examines the current arguments posed in the IGC against evidence from national experiences, contractual mechanisms like Mutually Agreed Terms (MAT) under access and benefit-sharing (ABS), and various codes of conduct for researchers and professionals that work with GRs and TK. Chapter 7 by Manuel Ruiz Muller discusses the legal protection of widely shared and dispersed traditional knowledge. The chapter explores a number of different legal regimes and *sui generis* systems for TK protection at the international and national levels and also provides additional suggestions for the protection of widely shared or held TK. Chapter 8 by Ruth L. Okediji explores the concept of the 'public domain'. This chapter highlights that the concept has been problematic and needs to be directly engaged with in the IGC, since it influences every aspect of the negotiations. As she explains, the competing visions of what a possible multilateral framework for GRs, TK and TCEs should look like, and the principles that inform them, are linked squarely to assumptions about the nature and role of the public domain. Chapter 9 by Marc Perlman discusses the negotiations on TCEs and folklore in the context of national experiences, those of Indigenous peoples and minority communities, and includes considerations for the different types of TCEs (e.g. music, recordings, dance, written works, designs). Chapter 10 by Brendan Tobin explores the role of customary law in governing and regulating the TK and TCEs of Indigenous peoples and local communities, as well as GRs that they control. Customary law has been a consideration in the IGC and has appeared in various forms in the text but, as Tobin notes, does not feature prominently in the latest working texts. This highlights the concern for Indigenous peoples that the process is too nation-state driven and reflects the positions of states, not necessarily 'knowledge-holders', indigenous resource custodians or the creators and authors of TCEs.

Part 3 is a 'stakeholder perspectives' section including national, Indigenous and industry viewpoints. It includes:

- An Indigenous perspective on the WIPO IGC by Maui Solomon
- Genetic resources, traditional knowledge and disclosure obligations: some observations from the life science industry by Dominic Muyldermans
- Indigenous culture and Indigenous publishing: balancing protection, preservation and promotion, by Jens Bammel and José Borghino
- Disclosure requirements: Switzerland's perspective, by Georges Bauer, Cyrill Michael Berger and Martin Girsberger
- India's position in the Intergovernmental Committee for the Protection of Traditional Knowledge, by Biswajit Dhar
- The WIPO IGC: a US perspective, by Dominic Keating
- Access and benefit sharing, intellectual property rights and the protection of traditional knowledge in Costa Rica: participation in and possible impacts of the IGC's work, by Jorge Cabrera Medaglia
- New Zealand, *mātauranga Māori* and the IGC, by Jessica C. Lai.

Part 4 provides some analysis on the future of the WIPO IGC. Chapter 19 by Pedro Roffe examines how recent preferential trade agreements have addressed genetic resources and traditional knowledge. However, despite the proliferation of these RTAs and FTAs there are few that provide any new substantive mandatory provisions in relation to GRs, TK and TCE and they limit themselves to best endeavours clauses and references to general principles on 'access and benefit sharing'. Chapter 20 by Ahmed Abdel-Latif discusses the future of the IGC, focusing on challenges to IP treaty making in WIPO and beyond. It argues that the traditional knowledge 'conundrum' is the biggest policy and normative challenge that WIPO has ever faced and that the future of the global intellectual property system hinges upon its satisfactory resolution. Chapter 21 by Susy Frankel highlights the challenges of cross-border protections for traditional knowledge, relating to the earlier chapter by Ruiz Muller. Frankel's chapter highlights that it is important that the IGC process ensures that there are some minimum standards that would enable not only domestic protection, but also cross-border protection as a result of national treatment. Chapter 22 by Graham Dutfield provides a critical analysis of traditional knowledge protection as broadly or narrowly defined, and reflects on the future of the IGC from this viewpoint. Chapter 23 by Chidi Oguamanam looks at the ramifications of the WIPO IGC for IP and development. He notes the hardening of positions in the IGC and recent slow progress, but also underlines the potential role the IGC could have in advancing the development imperative.

The book therefore provides a compilation of research, analysis, expertise and stakeholder perspectives on the state of the art of the IGC forum and in the international law and politics of the relevant thematic areas of IP, GRs, TK and TCEs. Furthermore, it encourages the various stakeholders to take forward the current debates on these important areas. What we hope is that this book highlights some *future options and directions* for the WIPO IGC delegates, representatives and secretariat at a time critical for the momentum of the IGC. Divisions still remain and there is a need to make further progress before a formal diplomatic conference is convened as envisaged in the IGC's mandate. Suggestions have been made to convert the IGC to a permanent Standing Committee as per the existing WIPO Standing Committees on Patents, Copyright and Trademarks. However, this is unlikely to reduce the uncertainties surrounding the negotiations and to bridge the gap in different positions. Could soft laws be a solution? But then would these be effective means to protect GRs, TKs and TCEs against misappropriation?

In our discussions with delegations, procedural issues have been raised among the delegates of the IGC and are not easily solved. Indeed several chapters in the book point to the concerns of stakeholders about the IGC mandate, the political will of some delegations to see the process result in tangible outcomes, and issues of capacity and resourcing.

Aside from process, it is worth reflecting on the *substance* of the IGC negotiations, which are focused to a large extent on the knowledge, cultural heritage and cultural products of Indigenous peoples and local communities. At the 29th session of the IGC, the chair, Ian Goss from Australia, played an Indigenous Australian song to highlight that the IGC is indeed dealing with complex histories and ancient cultures that have brought about the traditional knowledge, traditional cultural expressions and medicinal or food plants and animals which have been stewarded, managed or conserved for generations. If one thing has emerged from the IGC's deliberations it is the extraordinary difficulty in trying to devise a protection regime for the knowledge and heritage of Indigenous peoples and local communities in a state-centric setting and the risks it entails. As Drahos (2014, p84) highlights, 'the WIPO drafts are a good example of how far states have moved away from recognising the implications of cosmological connectionism for state law-making'. Coombe (2009, p406; citing Handler 2003, pp361–63) explains that to the extent to which heritage preservation and cultural property initiatives tend to assume an objectifying approach, they may fundamentally transform the symbolic processes they seek to protect by focusing too narrowly on objects, sites and traditions to the detriment of the semiotic dimensions of culture. State-centric law-making, thus may be dividing up aspects of Indigenous cultural heritage into legal categories as 'GRs' and 'TK' which align against patent laws, and 'TCEs' and folklore against copyright protections, because these divisions reproduce the pre-existing legal categories from which IP rights are protectable (see also Groth, 2012). And while Indigenous people attending the IGC are recognised as 'observers', the negotiations are undertaken by states. Observers are invited to speak, but not included in informal sessions and in the substantive drafting of the texts. While a fund established by the WIPO Secretariat and funded by states has meant that Indigenous representatives are supported to take part in the deliberations, some of them are still critical of the nature of the process and the extent to which it effectively takes on board their views.

Enormous amounts of time and effort have been spent trying to conclude international legal instruments such as the Nagoya Protocol and negotiating new norms at the WIPO IGC. As Tobin (2014, p178) has noted, attention is now turning to the actions Indigenous peoples can take unilaterally to protect their interests, including the development of biocultural or community protocols (which the Nagoya Protocol 'encourages') to protect cultural heritage, including TK and TCEs. Perhaps it is from a groundswell of biocultural protocol development and from Indigenous customary laws that the WIPO IGC delegates can also learn to shape more holistic international instruments that better reflect Indigenous concerns? Perhaps, as Dutfield (2013, p263) argues, policy-makers in WIPO and within states need to cede political space to allow the Indigenous peoples themselves to establish the rules of engagement. He states that (Dutfield, 2013, p263): 'legally and conceptually speaking, there needs to be a shift from imposing our legal "solutions" to accepting theirs: their own customary practices, norms and laws.' But given that the issues in question – misappropriations, erroneous patents, culturally offensive uses – are often transnational in nature, surely a global system needs to be established with the consent of international actors. Herein lies a massive challenge, pointed out by Antony Taubman who highlights the need to revisit the boundaries of intellectual property and the public domain (Taubman, 2005, see also the chapter by Okediji in this volume) and also underlines the heterogeneous qualities of TK (and these also relate to GRs and TCEs) the global protection of which leads to a paradox: 'to globalise diversity holistically' (Taubman, 2005, p525).

Sixteen years after the IGC's inception, it is still pushing ahead to determine what international architecture would do justice to the diversity of TK and related subject matter, given the heterogeneous peoples, cultures and environments involved. Even though advances have been made, this paradox still appears to be some way from being resolved.

References

Coombe, R. J. (2009). The expanding purview of cultural properties and their politics. *Annual Review of Law and Social Science*, 5(3): 393–412.

Drahos, P. (2004). Intellectual property and pharmaceutical markets: a nodal governance approach. *Temple Law Review*, 77: 401–424.

Drahos, P. (2014). *Intellectual Property, Indigenous People and Their Knowledge*. Cambridge University Press, Cambridge.

Dutfield, G. (2013). Transboundary resources, consent and customary law. *Law, Environment and Development Journal*, 9(2): 259–263.

Groth, S. (2012). *Negotiating Tradition: The Pragmatics of International Deliberations on Cultural Property*. Göttingen University Press, Göttingen.

Handler, R. (2003). Cultural property and cultural theory. *Journal of Social Archaeology*, 3(3): 353–65.

Helfer, L. R. (2009). Regime shifting in the international intellectual property system. *Perspectives on Politics*, 7(1): 39–44.

Okediji, R. L. (2008). WIPO-WTO relations and the future of global intellectual property norms. *Netherlands Yearbook of International Law*, 39(1): 69–125.

Robinson, D. (2010). *Confronting Biopiracy: Cases, Challenges and International Debates*. Routledge/Earthscan, London.

Taubman, A. (2005). Saving the village: conserving jurisprudential diversity in the international protection of traditional knowledge, in Maskus, K. E. and Reichman, J. H. (Eds.) *International Public Goods and Transfer of Technology: Under a Globalized Intellectual Property Regime*. Cambridge University Press, Cambridge, pp521–564.

Tobin, B. (2014). *Indigenous Peoples, Customary Law and Human Rights: Why Living Law Matters*. Routledge/Earthscan, Oxon.

2 Revisiting the creation of the IGC

The limits of constructive ambiguity?

Ahmed Abdel-Latif[1]

In 2000, the WIPO General Assembly established the Intergovernmental Committee on Intellectual Property and Genetic Resources, Traditional Knowledge and Folklore (IGC). Seventeen years later, the IGC's deliberations are still ongoing. The prospect of reaching consensus in these deliberations seems unlikely, at least in the near future. Yet, many of the divisive questions facing the IGC have remained virtually unchanged since its creation. What should be the ultimate outcome of the IGC's discussions? Should norm-setting be the Committee's main objective? Should genetic resources, traditional knowledge and folklore/traditional cultural expressions be bundled together and addressed by the same body? Should the IGC become a standing committee on equal footing with other WIPO standing committees?

The recurrence of these questions makes it timely to revisit the IGC's creation, which has seldom been the subject of an in-depth consideration on its own. Most of the accounts of the IGC's birth are rather descriptive as they tend to merely provide some background about the growing importance of the interface between genetic resources and traditional knowledge and intellectual property as well as relevant developments at WIPO and other international forums which then 'led' to the establishment of the IGC.[1] However, such accounts do not generally dwell sufficiently on the expectations and motivations of the different countries and actors which supported or accepted the decision at the time, nor do they dwell on the profound ambiguities and grey areas surrounding it and which have continued to 'haunt' the IGC's deliberations until today.

In this regard, the chapter seeks to address this lacuna. It starts by providing a necessary background about the IGC's origins. It then goes on to examine in detail the decision which formally established it and the dynamics behind it. Finally, the chapter looks at the IGC's creation in retrospect, reflecting on the extent to which these dynamics have changed. One lingering question that emerges is the

1 This chapter is written by the author in his personal capacity and does not reflect the views of any institution he might be affiliated with. The author was an Egyptian delegate to WIPO (2000–2004) and a member of the Egyptian delegation to the 26th WIPO General Assembly which adopted the decision to establish the IGC in 2000.

following: Was the IGC's creation just the result of a 'constructive ambiguity' and the minimum common denominator that could be reached among WIPO member countries at the time? Does it still have a bearing on the stalemate which its deliberations continue to face until today?

The origins of the IGC: a favourable global landscape and a crisis

The IGC's origins can be traced back to a number of developments both outside WIPO and within it. A number of parallel and inter-related dynamics at the national as well as international level in different multilateral forums fashioned a global landscape that was propitious for the establishment of the IGC. However, such a favourable global landscape was not in itself sufficient. In spring of 2000, disagreement concerning a proposal on genetic resources supported by developing countries threatened to derail the Diplomatic Conference convened to adopt the *Patent Law Treaty* (PLT). Developing countries agreed to withdraw their proposal, enabling a successful adoption of the PLT, in exchange for a commitment from developed countries that discussions regarding the issues they had raised would continue at WIPO in the framework of an especially dedicated body.

When looking at the origins of the IGC, it is also important to bear in mind that it brought together issues which had different international trajectories and backgrounds. Folklore was the oldest one to have been formally raised at the international level, as it emerged in the agenda of WIPO and UNESCO in the 1960s (Bannerman, 2015). The relationship between intellectual property, genetic resources and traditional knowledge was more recent, gaining momentum following the adoption of the Convention on Biological Diversity (1992) and of the Agreement on Trade Related Aspects of Intellectual Property Rights (1995). (See also Chapter 4 by Lawson on the relationship between the IGC and some of these forums.) This section provides an overview of the origins of the IGC in relation to these developments and issues.

The protection of folklore

Demands for the protection of folklore were first made at WIPO and UNESCO in the 1960s.[2] At a meeting for African Countries convened by WIPO in Brazzaville, in 1963, countries present requested the protection of folklore to be examined by the organization, particularly in the context of the revision of the *Berne Convention on Literary and Artistic Works*. However, the revision of the Berne Convention (1967) did not yield significant results in this regard.

In 1978, discussions on the protection of folklore took place at the WIPO Governing Bodies. At their request, WIPO and UNESCO convened several meetings of a Committee of Governmental Experts, which led to the adoption, in 1982, of the *Model Provisions for National Laws on the Protection of Expressions of Folklore against Illicit Exploitation and Other Prejudicial Actions*. In 1984, the two organizations jointly convened another group of experts which examined a draft treaty based on the Model Provisions. However, a majority of the participants considered it was premature to establish an international treaty.

Starting in the mid-1990s, there was renewed interest in the international protection of folklore, and in further action at the international level.

The *WIPO Performances and Phonograms Treaty* (WPPT), adopted in 1996, made an explicit reference to 'expressions of folklore'. Article 2 of the Treaty, on Definitions, considers 'performers' as 'actors, singers, musicians, dancers, and other persons who act, sing, deliver, declaim, play in, interpret, or otherwise perform literary or artistic works or expressions of folklore'.

In April 1997, a UNESCO/WIPO World Forum on the Protection of Folklore, held in Phuket, adopted a Plan of Action in which participants expressed the view 'that at present there is no international standard protection for folklore and that the copyright regime is not adequate to ensure such protection'. The Plan of Action suggested that 'Regional consultative fora should take place' and that a Committee of Experts should be set up, and 'should complete the drafting of a new international agreement on the *sui generis* protection of folklore . . . in view of the possible convocation of a Diplomatic Conference'.[3]

Pursuant to this recommendation, WIPO organized four regional consultations on the protection of expressions of folklore in 1999.[4] The consultations called on WIPO and UNESCO to intensify their work in this area with a focus on the development of a *sui generis* form of legal protection at the international level. Three of the four regional consultations endorsed the recommendation for the establishment of a WIPO Standing Committee to facilitate future work in this area (see Chapter 9 by Perlman for further analysis of TCEs and folklore in the IGC).[5]

The interface between intellectual property, genetic resources and traditional knowledge

The relationship between intellectual property, genetic resources and traditional knowledge became the subject of increased international interest in the 1990s against the backdrop of growing concerns about 'biopiracy'. Several cases of misappropriation of genetic resources and associated traditional knowledge involving patents had come to public attention and were the subject of significant media coverage during this time period (Dutfield, 2004 and Robinson, 2010).

For developing countries, the CBD's adoption heralded a new global regime regulating genetic resources which was based on a number of key principles: sovereign control over genetic resources by countries which in turn granted access to such resources based on prior informed consent (PIC), mutually agreed terms (MAT) and equitable sharing of the benefits arising from their utilization, for example in the context of Research and Development (R&D) activities.

The CBD also contained ground breaking provisions on IPRs,[6] which the United States considered to be detrimental to the interests of its biotechnology industry and played an important role in it ultimately not ratifying the Convention.

In addition to genetic resources, the CBD contained a provision – Article 8(j) – on traditional knowledge which stated that:

> Each contracting Party shall, as far as possible and as appropriate, subject to national legislation, respect, preserve and maintain knowledge, innovations and practices of indigenous and local communities embodying traditional lifestyles relevant for the conservation and sustainable use of biological diversity and promote their wider application with the approval and involvement of the holders of such knowledge, innovations and practices and encourage the equitable sharing of the benefits arising from the utilization of such knowledge, innovations and practices.

Shortly after the adoption of the CBD, the TRIPS Agreement came into existence as part of the new agreements concluded in the context of the Uruguay round of negotiations which resulted in the creation of the World Trade Organization (WTO).

The TRIPS agreement established new minimum standards of intellectual property protection that were binding on all WTO members. In the area of patents, Article 27(1) of the TRIPS agreement required that patents be 'available for any inventions, whether products or processes, in all fields of technology, provided that they are new, involve an inventive step and are capable of industrial application'. Many developing countries saw a potential conflict between these new patent rules and CBD provisions. They began to raise this issue in a forceful manner at WTO meetings.

In the preparatory meetings for the WTO Seattle Ministerial Conference (1999), several groups of developing countries submitted proposals for the revision of Article 27.3(b) of the TRIPS Agreement,[7] and for the introduction of a new negotiation mandate in the WTO that would allow *sui generis* protection for the traditional knowledge of local or indigenous communities.[8]

The growing global interest in intellectual property issues associated with genetic resources and traditional knowledge did not go unnoticed at WIPO.

In 1998, the WIPO Secretariat undertook nine fact-finding missions (FFMs) on traditional knowledge, innovations and creativity in different regions: South Pacific, South Asia, Southern and Eastern Africa, North America, West Africa, Arab countries, South America and Central America and the Caribbean. Their objective was 'to identify and explore the intellectual property needs and expectations of new beneficiaries, including the holders of indigenous knowledge and innovations' in order to 'promote the contribution of the intellectual property system to their social, cultural and economic development'.

The report resulting from the FFMs 'filled several important gaps in the international context in which the traditional knowledge discourse had unfolded within and outside WIPO' (Okediji, forthcoming 2017). It raised WIPO's profile, and gave it credibility and leadership showing that it could make an important substantive contribution to international discussions on these issues. Ultimately, the report provided the intellectual foundation of much of the subsequent analytical work that would be undertaken in the IGC context (Okediji, forthcoming 2017).

Issues relating to intellectual property, genetic resources and traditional knowledge also started to be raised in a number of WIPO bodies.

At the third session of the *Standing Committee on the Law of Patents* (SCP), in September 1999, Colombia proposed a new provision in the text of the draft Patent Law Treaty which was being discussed by the SCP[9] and which aimed to harmonize and streamline formal procedures relating to the filing of patent applications.

The provision suggested by Colombia stated the following:

1 All industrial property protection shall guarantee the protection of the country's biological and genetic heritage. Consequently, the grant of patents or registrations that relate to elements of that heritage shall be subject to their having been acquired legally.
2 Every document shall specify the registration number of the contract affording access to genetic resources and a copy thereof where the goods or services for which protection is sought have been manufactured or developed from genetic resources, or products thereof, of which one of the member countries is the country of origin.

According to Colombia, 'the purpose of its proposal, which was based on an initiative by the Andean Community, was to provide for the consideration, by the Standing Committee, of the procedural or formal aspects of the protection of genetic and biological resources, as well as their legal acquisition'. Many developing countries[10] supported the proposal. However, Germany, supported by the United States, Japan, France, the Republic of Korea, Romania and Finland (speaking on behalf of the European Community) considered that it 'related to issues of substantive law and was therefore not appropriate for inclusion in the draft Treaty'.[11]

In view of the lack of agreement on the Colombian proposal, the SCP invited the International Bureau to include the issue of protection of biological and genetic resources on the agenda of a *Working Group on Biotechnological Inventions*, to be convened, at WIPO, in November 1999. The SCP also invited the International Bureau to convene a separate meeting early in 2000 to consider this issue.[12]

The *Working Group on Biotechnological Inventions* met on 8–9 November that year, to identify issues related to biotechnology and intellectual property rights, which might be included in the WIPO work program in the 2000–1 biennium. It established a questionnaire to gather information about the protection of biotechnological inventions, in WIPO Member States.[13] It also considered a number of matters for its future work including, the 'review application of certain legal standards to early stage and certain other biotechnology inventions under patent standards of WIPO Member States'.[14]

In response to the invitation by the SCP, the WIPO Secretariat also organized a *Meeting on Intellectual Property and Genetic Resources* on 17 and 18 April, 2000. The Chairman's Conclusions stated that the exchange of views at the meeting produced a clear consensus in favour of 'setting up of an appropriate forum within WIPO for future work' on these issues.

Around the same time period, in May 2000, the Conference of the Parties (COP) of the CBD adopted Decision V/16 which established that the program of work for the implementation of Article 8(j) should 'be carried out in collaboration with other relevant organizations, including the World Intellectual Property Organization (WIPO)'. The decision also recognized the importance of *sui generis* and other appropriate systems for the protection of traditional knowledge of indigenous and local communities and the equitable sharing of benefits arising from its use.

Another COP Decision – Decision V/26 – invited relevant international organizations, including WIPO, to analyse 'issues of intellectual property rights as they relate to access to genetic resources and benefit sharing including the provision of information on the origin of genetic resources, if known, when submitting applications for intellectual property rights, including patents'.

In May 2000, the Diplomatic Conference to adopt the Patent Law Treaty (11 May to 2 June) began its work. At the conference, Colombia tabled the proposal that it had previously put forward at the SCP with the support of many developing countries. Industrialized countries opposed it again arguing that it was a matter of substantive patent law while the PLT focused on formalities and procedural matters. As previously mentioned, the disagreement erupted into a crisis that threatened to derail the entire diplomatic conference.

To avoid this, WIPO's Director General undertook consultations on how discussions concerning genetic resources could continue at WIPO. A statement was ultimately agreed upon and read out by the Director General, which stated:

> Member State discussions concerning genetic resources will continue at WIPO. The format of such discussions will be left to the Director General's discretion, in consultation with WIPO Member States.[15]

After the Conference, consultations continued and resulted in a proposal by the Director General that 'a distinct body should be established within WIPO to facilitate such discussions and that, in addition to the issue of genetic resources, the discussions should also include the results of WIPO's previous work on the related fields of traditional knowledge and expressions of folklore'.[16]

The WIPO General Assembly decision to establish the IGC

Following the outcome of the consultations above mentioned, the agenda item *Matters Concerning Intellectual Property and Genetic Resources, Traditional Knowledge and Folklore* was added to the agenda of the WIPO General Assembly, which was held from 25 September to 3 October 2000.

Two sets of documents were considered under this agenda item: A Note by the Dominican Republic on behalf of the Group of Countries of Latin America and the Caribbean (GRULAC) which contained two documents *Traditional Knowledge and the Need to Give it Adequate Intellectual Property Protection* and *WIPO Committee on the Relationship between Intellectual Property, Genetic Resources and Traditional Knowledge*[17] (WO/GA/26/9) and a Secretariat document entitled *Matters Concerning Intellectual Property and Genetic Resources, Traditional Knowledge and Folklore*[18] (WO/GA/26/6).

The first GRULAC document, *Traditional Knowledge and the Need to Give it Adequate Intellectual Property Protection*, outlined the objectives of 'an intellectual property system for traditional knowledge and associated innovations and practices'. It pointed to some possible *sui generis* solutions for the protection of traditional knowledge noting that 'there has been an increase throughout the world in the adoption of *sui generis* systems for the protection of the intellectual property rights in various kinds of subject matter'. Finally, the document requested the 'creation of a Standing Committee on access to the genetic resources and traditional knowledge of local and indigenous communities. The work of that Standing Committee would have to be directed towards defining internationally recognized practical methods of securing adequate protection for the intellectual property rights in traditional knowledge.'

The second GRULAC document, *WIPO Committee on the Relationship between Intellectual Property, Genetic Resources and Traditional Knowledge* further fleshed out the objectives and functions of the Committee to be established. It clearly specified that the Committee should have a norm-setting function as it would:

> (c) devise and draft proposals for legislative texts, including draft recommendations, resolutions, treaties and other instruments, for the regulation and standardization of principles and standards at the national or the international level on intellectual property aspects of the use and exploitation of genetic resources and biodiversity, and of the traditional knowledge referred to above;

The document also mentioned that the Committee would have to 'coordinate its action with that of other relevant intergovernmental organizations'.

Finally, it envisaged that the new Committee could conduct its work jointly with other WIPO bodies such as the SCP and the Standing Committee on Trademarks, Industrial Designs and Geographical Indications. For instance, the document suggested that the Committee could join forces with the SCP to examine issues such as: 'products of nature, absolute novelty, legitimacy of use and exploitation of genetic resources and traditional knowledge and simplification of revocation and claim procedures'. It considered this necessary to deal with the situation in the patent system where 'some of the problems of lack of protection and "biopiracy" reported by the possessors of biological and genetic material and traditional knowledge in connection with its use seem to be the result of a regulatory system that is not capable of resolving certain situations that arise from the use and exploitation of biological or genetic material existing in nature, or of putting an end to certain practices that produce results prejudicial to those possessors'.

Several developing countries expressed their support to the proposals made by GRULAC during the General Assembly deliberations. The African Group went on to formally associate itself with the GRULAC proposals.[19] Bulgaria, speaking on behalf of the Central European and Baltic States Group, also welcomed them, suggesting that they be included among the papers to be considered at the first session of the new Committee.[20]

The Secretariat document *Matters Concerning Intellectual Property and Genetic Resources, Traditional Knowledge and Folklore* focused, in part II, on the Proposal for the establishment of the new WIPO body. Paragraph 13 stated the following:

> Following the recommendations of the regional consultations on folklore, the consensus achieved at the Meeting on Intellectual Property and Genetic Resources, the commitment reached in the context of the Diplomatic Conference for the Adoption of the Patent Law Treaty and subsequent consultations with Member States, the Member States might wish to consider the establishment of an Intergovernmental Committee on Intellectual Property and Genetic Resources, Traditional Knowledge and Folklore for the purpose of discussions on these subjects.

The document elaborated that each one of the three themes to be addressed by the IGC 'cuts across the conventional branches of intellectual property law and does therefore not fit into existing WIPO bodies, . . . and that the three themes are closely interrelated, and none can be addressed effectively without considering aspects of the other.'[21]

The document proposed that the Intergovernmental Committee 'report any recommendations for action that it might formulate to the WIPO General Assembly'.[22]

Finally, it invited 'the WIPO General Assembly to approve the establishment of the WIPO Intergovernmental Committee on Intellectual Property and Genetic Resources, Traditional Knowledge and Folklore',[23] which the General Assembly did, based on the elements mentioned above.

Mostly developing countries took the floor expressing their strong support for the establishment of the IGC. Many emphasized that the issues it would address were of 'paramount importance to developing countries'.[24] At the same time, several delegations, such as Malaysia, speaking on behalf of Asian countries, stressed that this 'was not an issue involving only, and confined to, developing countries, but an issue of concern and interest to all'.[25] Nigeria noted that the question should not be one of a North–South divide.[26] The delegation went on to state the following:

> . . . the present attempt to draw a fine line between intellectual property and traditional knowledge would lead nowhere. Referring to an earlier statement about all human activities being knowledge-based, it stated that whether one came from the North or the South, traditional knowledge predated other forms of knowledge. It suggested that the issue of traditional knowledge be examined with a common objective, namely to recognize its existence, to assign it a role, and to devise a scheme to reward its contribution to humanity.

Colombia emphasized the need to undertake 'work which would enable a very precise definition of the concept of each of the subjects tabled'.[27] Egypt agreed with 'the Director General's plea that these matters should not be politicized'.[28] India was also of the view that 'discussions in the proposed Committee should be held in a technical and non-emotive manner so that the Member States could move rapidly forward towards a consensus'. It suggested that the Committee 'might constitute specialized expert subgroups to deal with specific aspects of operationalizing the protection of traditional knowledge systems'.[29]

In contrast, only two developed countries spoke during the consideration of this agenda item: Japan on behalf of Group B (industrialized countries) and France on behalf of the European Union. Japan voiced the belief that these issues 'should be discussed without prejudging the outcome of the work of the Committee and proposed that the work commence with a sharing of experiences among all Member States to define and delimit the scope of the issues'.[30] France highlighted that 'each and every one of the issues to be considered by the Committee covered a different branch of intellectual property and that none of the existing bodies within WIPO could take all these matters up together' and thus there was a 'need for a new forum which would have a specific mandate and would be accountable to the General Assembly for its work, with a view to arriving at very specific results'.[31]

Regarding how the Committee's work should proceed, Bulgaria, on behalf of the Central European and Baltic States Group, emphasized the 'importance of a step-by-step approach, pending the outcome of the first meeting of the proposed Committee and subsequently determining the most appropriate framework for discussions on the three related, but separate, topics'. It advocated 'a realistic approach' and 'urged the Member States not to over-estimate the possible solutions'.[32]

While the statements made by countries during the Twenty-Sixth WIPO General Assembly, in which the proposal for the IGC's creation was approved, showed broad support for the proposal, the contrast between the way it was expressed by developed and developing countries was glaring. Developed countries were not very enthusiastic to say the least.

The creation of the IGC: motivations, expectations and perceptions

The motivations and considerations at play in the genesis of the IGC were varied and differed according to the countries and actors involved. Chief among them were WIPO's effort to reassert its prominence in global intellectual property governance after the adoption of the TRIPS Agreement. In effect, the conclusion of the TRIPS Agreement, bringing intellectual property under the realm of the WTO and its newly established dispute settlement machinery, posed a formidable challenge to WIPO. Many interrogations emerged about whether the organization would still be relevant in the post-TRIPS world. The leadership of WIPO's Secretariat was acutely aware of this and sought to respond to this challenge in different ways. For instance, it rapidly moved to conclude an agreement with the WTO in 1996 by virtue of which WIPO became the main multilateral technical assistance provider for the implementation of the TRIPS agreement in developing countries. WIPO also sought to address, through its norm-setting activities, new and emerging issues which had not been tackled by the TRIPS Agreement. This was reflected in the adoption of the WIPO Internet Treaties in 1996. WIPO's resurgence was fuelled by the growing resources of the organization as a result of the steady increase in its revenues from the registration systems it administered, first and foremost the Patent Cooperation Treaty (PCT) (Okediji, 2008).

Kamil Idris, WIPO's first Director General from a developing country (Sudan), elected in 1997, pursued this drive towards a stronger engagement by WIPO with global policy issues, particularly *vis à vis* the WTO and other United Nations agencies. He clearly signalled this ambition in his acceptance speech following his election, which included a reference to traditional knowledge and genetic resources in the context of outlining WIPO's growing activism in the multilateral setting, after a long period of cautious isolationism. In that speech, he stated:

> It is apparent to all that we live in an age of increasing interdependence, which affects all the dimensions of our existence, including the Organization's relations with other international organizations. In such an age, I consider, therefore, that . . . WIPO should reach out to its partners in the United Nations family of organizations and to the World Trade Organization to ensure that the expertise and knowledge that WIPO possesses is also placed at the service of its sister organizations. This would seem to be particularly necessary as intellectual property assumes increasing importance and thus becomes a factor in a range of fields as diverse as indigenous technology, intangible heritage, and folklore, the exchange of meteorological data and trade relations. I look forward to exploring such global issues and to making available WIPO's expertise, in an open and constructive manner, to other international organizations to provide assistance for the protection of intellectual property in the post-TRIPS era.[33]

In this context, the key role the WIPO Secretariat played in the establishment of the IGC has often been overlooked but is increasingly underlined (Okediji, forthcoming 2017). In reality, the formal creation of the IGC and its mandate resulted from a proposal by the WIPO Director General on the basis of consultations he undertook after the PLT Diplomatic Conference. As previously mentioned, the report of the FFMs will provide the intellectual foundation for the IGC's deliberations at least in its early years.

At the same time, the growing international engagement pursued by WIPO's Secretariat on matters relating to traditional knowledge, genetic resources and folklore was supported, to a large extent, by many developing countries. Several reasons explain this. First, WIPO being a UN agency, it was considered by many developing countries, at the time, as a potentially more friendly forum to their interests than the WTO which was outside the UN system and was witnessing a strong polarization between developed and developing countries as reflected in the collapse of the Seattle Ministerial Conference in 1999. WIPO was also perceived as a more 'expert', 'technical' and 'neutral' forum which was not 'politicized' in the way the WTO had become. Finally, the election of Kamil Idris had raised expectations among developing countries that their priorities would be given greater prominence in the work of the organization, in the context of an international intellectual property system which was perceived to have been largely shaped by the interests of industrialized countries as well as those of intellectual property rights holders and their trade associations.

Such perceptions were reflected in some of the interventions by developing countries during the deliberations regarding the IGC's creation at the WIPO General Assembly. For instance, Kenya, after recalling that it had submitted proposals to the WTO's Ministerial Conference in Seattle, expressed its appreciation 'that this matter had now been brought to the right forum where . . . the political tone would be less' and underlined 'the importance of seeing the particular aspects handled professionally and legally'.[34] Brazil, on its part, stated:

> In the international fora it has been argued that WIPO should play a meaningful role in international debates on intellectual property and genetic resources, traditional knowledge and folklore. Given its expertise and resources, it believes that WIPO, within its mandate, could bring a significant contribution to the international debate.[35]

In addition to these considerations, many developing countries saw in the creation of a distinct body within WIPO a clear recognition of the importance of these issues, which would now stand firmly at the top of the organization's agenda and at the heart of its work programme. They had huge expectations that the IGC would establish a new international framework for the protection of genetic resources, traditional knowledge and folklore. From their perspective, this would result in a significant re-balancing of the international intellectual property system which would become truly universal and legitimate as it would have proven its ability to also protect the intangible intellectual assets of developing countries, and not only those of developed countries.

Tanzania and Cote d'Ivoire voiced this forcefully at the WIPO General Assembly. Tanzania highlighted the huge potential social and economic benefits for developing and least-developed countries if proper intellectual property protection was accorded to genetic resources, traditional knowledge and expressions of folklore. It stressed that 'such measures would enhance the understanding of intellectual property concepts among the population of these countries and would make intellectual property more meaningful for the benefit of mankind.'[36] In the same vein, Cote d'Ivoire considered that this undertaking would 'open up intellectual property to millions of human beings who had until now been living on the margins of the globalization.' It referred particularly to African populations in rural environments, with illiteracy rates of more than 70 per cent, but who were full of creativity and innovation.[37]

For developing countries, the establishment of the IGC was a symbolic and significant step that in itself superseded any other considerations, such as the fact that the new body did not have a clear mandate. Developing countries as well as other countries were acutely aware of this. The Dominican Republic, speaking on behalf of GRULAC, stressed that it 'would have liked to see broader terms of reference and more clearly defined responsibilities for the Committee'.[38] Bulgaria, on behalf of the Central European and Baltic States Group, requested clarification on four key questions: 'What would be the relationship of the proposed Committee with the other bodies in WIPO's overall structure? What would be its competencies? What sorts of end-results were to be expected? What would be the main difference between the Intergovernmental Committee and a Standing Committee?'[39]

As the General Assembly deliberations show, the issue of whether the IGC should be a Standing Committee has been raised since its inception. The recommendations emerging from some of the regional consultations organized by WIPO prior to the IGC's creation clearly called for the establishment of a Standing Committee. So did the GRULAC proposals. However, this was opposed by industrialized countries, as Standing Committees had a clear norm-setting mandate. WIPO Standing Committees are also permanent bodies. In contrast, the IGC has a two-year life span which countries need to agree to renew every two years. On a more substantive level, it might imply that traditional knowledge was considered a separate category of intellectual property rights on equal standing with patents, copyright and trademarks. Developing countries had the expectation that, despite this concession on their part, the IGC's deliberations would evolve in a way that would eventually lead towards the IGC becoming a Standing Committee with a clear norm-setting mandate.

Apart from the mandate and the legal status of the IGC within WIPO, the relationship between the ICG's deliberations and discussions in other intergovernmental organizations were unclear. For many developing countries, discussions at the IGC were not a substitute for deliberations in other forums. The Dominican Republic, speaking on behalf of GRULAC, clearly stated its hope that the proposals of the Director General 'were not an attempt to replace consideration of these questions in any other multilateral fora where they might be analysed, negotiated upon, or discussed'.[40] Bangladesh made the more ambitious proposal 'that WIPO should work closely with other international agencies such as the WTO, the United Nations Conference on Trade and Development (UNCTAD), the United Nations Food and Agriculture Organization (FAO) and the Secretariat of the Convention on Biological Diversity in evolving a new global regime in this regard'.[41] At the minimum, the expectation of many developing countries was that the IGC's deliberations would help achieve progress in the discussions on their proposals at the WTO, with regard to, for instance, the definition of traditional knowledge.

Japan made a reference to discussions at other relevant international forums as it underlined, speaking on behalf of Group B, 'the importance of maintaining close relations with other intergovernmental bodies which also deal with this issue'.[42] However, this rather general statement did not describe the manner in which developed countries perceived the nature of the interface between the discussions at the IGC and in other international forums, and how it should be articulated and evolve.

With regard to the relationship between the IGC and other WIPO bodies, there was also a lack of clarity and differing views. The GRULAC Note envisioned that the new body could conduct its work 'jointly' with other WIPO bodies such as the Standing Committee on Patents. Japan, on its part, underlined that the new Committee should study all aspects of this issue, but when it 'recognized that particular issues, such as patents, required specialized expertise, the corresponding WIPO Standing Committees should be responsible for addressing these issues.'[43] It was, ultimately, not clear, in practice, how this relationship would unfold.

Regarding the substantive work to be undertaken by the Committee, several developing countries underlined that it should focus on the creation of *sui generis* systems. Malaysia, on behalf of the Asian Group, 'highlighted the possibility of creating a *sui generis* system for the protection of traditional knowledge and a system that would avoid unscrupulous exploitation of such knowledge.'[44] China emphasized that 'traditional knowledge represented the achievements made over time by a nation during its long history' and that 'it embodied the distinctive characteristics of the nation's culture'. It suggested that, as the rights holders of traditional knowledge could not be identified, 'Governments should actively participate in this work so as to make it a part of the nation's cultural policy'.[45] A number of countries made reference to biopiracy in their statements. Peru 'expressed its concern that this wealth was endangered . . . by improper and illicit use'.[46] Egypt 'suggested that WIPO had a major role to play in this area by providing protection against piracy'. Sudan proposed that 'the mandate of the Intergovernmental Committee should include the discussion of ways and means to fight against piracy'.[47]

The high expectations of developing countries manifested at the creation of the IGC were met with circumspection by industrialized countries which sounded a cautionary note, particularly regarding the body's expected outcome. Japan, on behalf of countries of Group B, voiced 'the belief of the Group that these issues should be discussed without prejudging the outcome of the work'.[48] France, on behalf of the EU noted 'that these were very complex issues and for the moment it seemed difficult to predict the outcome of future discussions'.[49]

For the sake of achieving a consensus on the IGC's creation, developing countries accepted a mandate which did not include an explicit reference to norm-setting, which was unclear or silent regarding a number of key issues, such as the relationship with discussions at other intergovernmental bodies, the relationship with other WIPO bodies and, more importantly, which did not specify what the nature of the Committee's outcome would be.

The creation of the IGC in retrospect

In retrospect, the creation of the IGC was no doubt an important procedural step. However, the 'constructive ambiguity' which enabled it to happen took place against the backdrop of a deep divide among countries on the intellectual and conceptual framing of the issues at hand and the very essence of what the new body should aim to accomplish. To give just one example: the term 'protection' was not retained in the IGC's mandate despite attempts by developing countries to include it. The very notion that genetic resources, traditional knowledge and folklore required 'protection' was not acceptable to a number of industrialized countries at the time. The Committee was just to discuss the relationship between 'intellectual property and genetic resources, traditional knowledge and folklore'. It is also not clear why the issue of 'folklore' was bundled together with genetic resources and traditional knowledge, despite having a very different international trajectory and background.

This primacy of process over substance would continue to mark the IGC throughout its existence, as if procedural advances could help achieve progress in the substantive deliberations. In 2009, the IGC's mandate changed for the first time significantly in comparison to the original mandate adopted at its creation. This new mandate required the IGC to 'undertake text-based negotiations with the objective of reaching agreement on a text of an international legal instrument (or instruments) which will ensure the effective protection of GRs, TK and TCEs'. The term 'protection' featured in the new mandate which also stipulated that the Committee's work was not just to 'discuss' the issues at hand but to 'negotiate' new international instruments. Despite the change in mandate, the IGC's discussions would remain inconclusive during the subsequent years as a result of persistent disagreements over the nature of the protection to be established and of the legal instruments to be negotiated. So despite the a new mandate which envisaged norm setting, the need for new legal instruments continued to be questioned with calls made for more analysis and studies to better understand the issues under consideration and efforts to focus the work of the committee more on practical measures such as contracts and databases. A sense of *déjà vu* has permeated the IGC's discussions throughout its existence.

Since the very beginning, it seemed as if many industrialized countries were supportive of the IGC as a 'process without an outcome': a process of deliberations which kept the issues alive on WIPO's agenda but that was not necessarily meant to reach any concrete normative outcomes.

It is important also to underline that the IGC's establishment took place in the context of 'misperceptions' about WIPO's orientations, particularly among many developing countries. This is seldomly mentioned when the IGC's creation is evoked. In effect, the IGC's establishment predated by four years the launch of the WIPO Development Agenda, an initiative by Brazil and Argentina in 2004, which sought to have WIPO adopt a more balanced approach to intellectual property protection that was supportive of development and public policy objectives.[50] As previously mentioned, in the 1990s developing countries tended to perceive WIPO primarily as a UN agency which was generally friendly to their interests, in contrast to the more 'hostile' WTO recently established. Few developing countries, particularly among their diplomatic representations in Geneva, had a clear understanding of the orientations and institutional dynamics of WIPO, including for instance the fact that it derived nearly 90 per cent of its revenues from the PCT, a registration system used mostly by patent holders from industrialized countries, or the fact that by virtue of its history it tended to be aligned with the interests of intellectual property right holders groups. However, this 'misperception' was not just on the part of developing countries. There was a true knowledge gap in terms of the study of WIPO and the understanding of its orientations. There were virtually no academic studies on the organization itself with the exception of a few specialized works on the key treaties it administered. Public interest non-governmental organizations (NGOs) were almost entirely absent from its deliberations. However, this situation began to change with the release of the report of the Commission on Intellectual Property Rights (CIPR) in 2002 which contained strong criticism of WIPO's approach to promoting intellectual property.[51]

Soon after the IGC's creation, many developing countries started to realize that they had brought their demands for measures to address 'biopiracy' and the misappropriation of genetic resources and traditional knowledge by the patent system to a forum which was a staunch supporter of that system and derived most of its revenues from patent fees. This meant in practice that the IGC's discussions and their outcome could not be detrimental to the interests of patent holders in any significant manner and that the framing of the discussions would need to cater to their views. It is interesting to note, in this connection, that the Secretariat document containing the proposal to establish the IGC does not mention the term misappropriation even once. Since the very first session of the IGC, industrialized countries expressed reservations about the reference to 'biopiracy' or even 'misappropriation' to characterize cases where patents had been wrongfully granted over genetic resources and associated traditional knowledge. The more neutral and technical term of 'erroneous patents' was privileged. Ultimately, few cases of 'biopiracy' were brought forward to the attention of the Committee and extensively discussed. Over time the creation of a positive system of protection of traditional knowledge and traditional cultural expressions became the issue of greater focus in the IGC's deliberations and absorbed most of the Committee's time and energy. While many developing countries are supportive of efforts in this direction, it is nevertheless revealing of the paradigmatic shift brought by the IGC's work to the framing of these issues at the international level. Many developing countries have pointed throughout the IGC's existence to the irony stemming from the fact that the issues which were at the root cause of the IGC's creation – misappropriation of genetic resources and the possible measures in patent law to address it – have witnessed very limited progress, if any.

The frustrations of developing countries on this front grew further when in 2001, the same year the IGC's first session was held, the WIPO Secretariat launched the *WIPO Patent Agenda* which aimed, among other objectives, to conclude a new treaty to harmonize substantive patent law (Correa, 2002). During discussions on a draft Substantive Patent Law Treaty (SPLT) at the WIPO SCP, proposals by developing countries to include issues relating to genetic resources and traditional knowledge in the scope and provisions of the treaty were rejected by industrialized countries which argued that the IGC was now the right body to deal with them. Similarly, when developing countries raised these matters at the TRIPS Council of the WTO similar arguments were made about the IGC being the appropriate forum to discuss them. Yet, at the IGC, industrialized countries showed no political will to tackle these issues in a meaningful manner.

Quickly, developing countries started to realize that a key motivation for developed countries in agreeing to the IGC's creation was to use its existence to 'sanitize' and 'quarantine' genetic resources and patent related issues and the tensions they raised from the mainstream of patent and intellectual property discussions. From their perspective, the IGC discussions could help defuse these tensions while not addressing the substance of the matters under consideration. This was the exact opposite of what developing countries had aspired to at the IGC's creation as the proposal of GRULAC clearly stated, particularly with regard to interaction with other WIPO bodies. When the IGC's mandate was renewed in 2003, developing countries insisted that a mention should be added stating that its work was 'without prejudice' to deliberations in other forums. Yet this would be of limited effectiveness.

In this regard, the relationship between the IGC and other WIPO bodies as well as the WTO and the CBD would continue to be an important dynamic throughout the IGC's existence. For instance, industrialized countries would be particularly keen to renew the IGC's mandate at the approach of WTO ministerial conferences as a non-renewal could put more onus on the WTO to take up these issues.[52] Developing countries would strive to capitalize on outcomes at the CBD such as the conclusion of the Nagoya Protocol (2010) to advance their positions and demands at the IGC and the TRIPS Council. In fact, the evolving relationship between these forums in the context of the 'forum-shifting' strategies pursued by different countries is essential to understand the dynamics at play regarding these issues at the international level (Okediji, 2008 and Abdel-Latif, 2013). This is not surprising as 'forum-shifting' has become an important feature of global IP governance (Helfer, 2009; Sell, 2011) which poses formidable challenges in terms of achieving coherent outcomes across different forums and tends to be most problematic for weaker and smaller countries whose capacities to coordinate their positions across these forums tends to be more limited (Abdel-Latif, 2005).

Looking at the IGC's creation in retrospect, it is also interesting to observe some important changes in the dynamics of both developing and developed country participation.

With regards to developing countries, the role of GRULAC which was instrumental in the IGC's creation has become less prominent in the Committee's deliberations. This can be attributed to several factors. First, a general sense of *fatigue* among delegations regarding multilateral discussions given how long the IGC's deliberations have spanned with no concrete normative outcome achieved. Second, several countries from the region (such as Peru, Colombia and Central American countries) concluded Free Trade Agreements with the United States and the European Union in which provisions relating to genetic resources and traditional knowledge were included. While these provisions are not strongly worded, they give some level of recognition to the demands these countries have advanced for many years on these issues (see Chapter 19 by Pedro Roffe in this volume).

Most importantly, several key Latin American countries started to have serious misgivings about the desirability of a positive system of protection for traditional knowledge as the IGC's deliberations unfolded, in view of the risks that could result in the commodification of traditional knowledge. This was the case for Brazil, in particular, which was confronted with a lack of agreement among its domestic stakeholders, including indigenous groups, on this matter. As a result, Brazil will be most active throughout the IGC's deliberations in advancing the case of defensive protection and the introduction of a mandatory disclosure requirement in patent applications, and less so on the positive protection side. It is also interesting to note that a Like Minded Group of Countries (LMG) consisting of a number of developing countries[53] from different regions was established in 2009 with a view to push forward the IGC's text-based deliberations and has made a number of submissions to the Committee.[54] However, throughout recent years, the African Group of countries has emerged as the most active and vocal grouping of developing countries in the IGC setting, particularly in terms of championing a positive system of protection for traditional knowledge and traditional cultural expressions.

On the side of developed countries, they are less today of the monolithic group that made up at IGC's creation. EU countries and Switzerland have accepted the principle of the disclosure requirement in patent applications and have introduced it in their domestic law. They have also shown a willingness to engage constructively in the Committee's text-based negotiations. In contrast, the United States, Canada, Japan and Korea have remained strongly opposed to the principle of the disclosure requirement and have recurrently sought to stall the Committee's text based deliberations, and move the focus from binding normative outcomes in favour of more analysis, evidence-gathering and defensive measures such as databases. The positions of these countries tend to espouse those of the biotechnology industry whose stated views have virtually remained unchanged since the IGC's creation in terms of opposition to the disclosure requirement or to efforts to devise a positive system of protection (see Chapter 12 by Muyldermans in this volume).

Finally, the participation by indigenous people has become an important feature of the IGC's dynamics. At the time of the IGC's creation, it was not clear how this participation would take shape and how it would impact the Committee's work. In retrospect, it has certainly contributed to legitimize the IGC process in the face of other forums discussing these issues – such as the TRIPS Council – in which it is absent. Yet, indigenous groups have also expressed over the years their dissatisfaction with their participation in the IGC as mere 'observers' and have demanded an enhanced status which would enable them to play a more pro-active role in the deliberations, for instance, by submitting written proposals. They have also manifested their frustration at the extent to which countries have taken on board their concerns and views in the deliberations. At the same time, they have expressed a very broad range of views and positions which show, in some cases, considerable differences in opinion regarding the role of intellectual property with regard to traditional knowledge and traditional cultural expressions. After the adoption of the United Nations Declaration of the Rights of Indigenous Peoples (UNDRIP) in 2007, many indigenous groups sought to 'import' some of its wording and provisions in the IGC's deliberations, a move which was met with the opposition of a number of countries, in particular those which had not signed it or had reservations towards it (see Chapter 11 by Solomon in this volume).

Conclusion

The IGC's creation remains an important moment in time for IP and GRs, TK and TCEs. An understanding of why and how the IGC was established is critical to grasp the dynamics characterizing its deliberations throughout its existence and the reasons why it has been unable to reach binding normative outcomes until today. In retrospect, it is clear that countries had different expectations and perceptions not only about the ultimate goal of the IGC's deliberations, but also about their intellectual framework and conceptual basis. The implications of having these deliberations under the aegis of an organization whose orientations and institutional dynamics were also not well understood by many countries and actors at the time of the IGC's creation. Thus, for many years, the IGC remained a process without an outcome. While 'constructive ambiguity' helped create the IGC as a process, it has not yet enabled it to reach consensus on an outcome. However, the absence of an outcome does not mean that nothing has happened since the IGC's creation. The exchange of national experiences and the considerable body of analysis which has emerged over the years have paved the way for a better understanding of the issues at hand. A number of practical measures have been adopted, such as the IPC's expansion to enhance treatment of traditional knowledge as prior art, the incorporation of traditional knowledge into materials required by the Patent Cooperation Treaty, and the establishment of technical standards for documenting traditional knowledge. Negotiations since 2009 have resulted over the years in draft normative texts which, while still reflecting important differences, have reached a level of granularity that is unprecedented. Never before have such in-depth discussions on these issues taken place in an intergovernmental setting. It remains to be seen whether countries are able to overcome their differences to bring the IGC process to a fruitful conclusion.

Notes

1 An exception to this is a paper by Okediji (forthcoming 2017).
2 For an overview of WIPO's role in relation to the protection of folklore and traditional knowledge, see Bannerman (2015).
3 The Phuket Plan of Action records that '[t]he participants from the Governments of the United States of America and the United Kingdom expressly stated that they could not associate themselves with the plan of action'. See, *UNESCO-WIPO World Forum on the Protection of Folklore, Phuket, Thailand, April 8 to 10, 1997*, WIPO Publication No. 758, 1998, p. 235, ftp://ftp.wipo.int/pub/library/ebooks/wipopublications/wipo_pub_758efs.pdf (accessed 19 December 2016).
4 These consultations were held for African countries in South Africa (March 1999); for countries of Asia and the Pacific region in Vietnam (April 1999); for Arab countries in Tunisia (May 1999); and for countries of Latin America and the Caribbean in Ecuador (June 1999).

5 Recommendation of consultations for Arab region, Asia/Pacific region and Latin America and Caribbean region.

6 Most notably Article 16(5) which stated: The Contracting Parties, recognizing that patents and other intellectual property rights may have an influence on the implementation of this Convention, shall cooperate in this regard subject to national legislation and international law in order to ensure that such rights are supportive of and do not run counter to its objectives.

7 Article 27.3(b) of the TRIPS Agreement states that Members may also exclude from patentability:

(b) plants and animals other than micro-organisms, and essentially biological processes for the production of plants or animals other than non-biological and microbiological processes. However, Members shall provide for the protection of plant varieties either by patents or by an effective *sui generis* system or by any combination thereof. The provisions of this subparagraph shall be reviewed four years after the date of entry into force of the WTO Agreement.

8 Proposals by Kenya on behalf of the African Group (WT/GC/W/302), by Bolivia, Colombia, Ecuador, Nicaragua and Peru (IPJW/165) and by Cuba, Honduras, Paraguay and Venezuela (WTO WT/GC/W/329).

9 Proposal by the Delegation of Colombia, *Protection of Biological and Genetic Resources*, Standing Committee on the Law of Patents, 6–14 September 1999, SCP/3/10, www.wipo.int/edocs/mdocs/scp/en/scp_3/scp_3_10.pdf (accessed 19 December 2016).

10 Bolivia, Paraguay, China, Namibia, Cameroon, Mexico, South Africa, Chile, Cuba, India, Kenya, Costa Rica and Barbados, Report of the Third Session of the Standing Committee on Law of Patents, Geneva, September 6 to 14, 1999, para.204, SCP/3/11, para. 203, www.wipo.int/edocs/mdocs/scp/en/scp_3/scp_3_11.pdf (accessed 19 December 2016).

11 Ibid., para.204

12 Ibid., para.208

13 The WIPO Secretariat subsequently sent a questionnaire to the Member States and compiled information from the responses in document WIPO/GRTKF/IC/1/6, www.wipo.int/edocs/mdocs/tk/en/wipo_grtkf_ic_1/wipo_grtkf_ic_1_6.pdf (accessed 19 December 2016).

14 Working Group on Biotechnology, 8–9 November, 1999, WIPO/BIOT/WG/99, www.wipo.int/meetings/en/details.jsp?meeting_id=3906 (accessed 19 December 2016).

15 WIPO General Assembly, *Matters Concerning Intellectual Property and Genetic Resources, Traditional Knowledge and Folklore*, Document prepared by the Secretariat, WO/GA/26/6, para.9, www.wipo.int/edocs/mdocs/govbody/en/wo_ga_26/wo_ga_26_6.pdf (accessed 19 December 2016).

16 Ibid., para.10

17 WIPO General Assembly, *Traditional Knowledge and the Need to Give it Adequate Intellectual Property Protection* and *WIPO Committee on the Relationship between Intellectual Property, Genetic Resources and Traditional Knowledge*, Documents prepared by the Group of Countries of Latin America and the Caribbean (GRULAC), WO/GA/26/9, www.wipo.int/edocs/mdocs/govbody/en/wo_ga_26/wo_ga_26_9.pdf (accessed 19 December 2016).

18 WIPO General Assembly, *Matters Concerning Intellectual Property and Genetic Resources, Traditional Knowledge and Folklore*, Document prepared by the Secretariat, WO/GA/26/6, www.wipo.int/edocs/mdocs/govbody/en/wo_ga_26/wo_ga_26_6.pdf (accessed 19 December 2016).

19 Report of the 26th Session, WIPO General Assembly, Geneva, 25 September to 3 October, 2000, WO/GA/26/10, www.wipo.int/edocs/mdocs/govbody/en/wo_ga_26/wo_ga_26_10.pdf, Statement by the Delegation of Uganda on behalf of the African Group, para.31 (accessed 19 December 2016).

20 Ibid., Statement by the Delegation of Bulgaria on behalf of the Group of Central European and Baltic States Group, para.33

21 WO/GA/26/6, para.15
22 Ibid., para.16
23 Ibid., para.19
24 Supra, 15, Statement by the Delegation of Jordan, para.21
25 Ibid., Statement by the Delegation of Malaysia on behalf of the Asian Group, para.32
26 Ibid., Statement by the Delegation of Nigeria, para.52
27 Ibid., Statement by the Delegation of Colombia, para.66
28 Ibid., Statement by the Delegation of Egypt, para.38
29 Ibid., Statement by the Delegation of India, para.37
30 Ibid., Statement by the Delegation of Japan on behalf of Group B, para.34
31 Ibid., Statement by the Delegation of France on behalf of the European Union, para.41
32 Ibid., Statement by the Delegation of Bulgaria on behalf of the Group of Central European and Baltic States Group, para.33
33 Statement by the Director General of WIPO, WIPO General Assembly, 22 September–1 October 1997, Report, WO/GA/XXI/13, para.8, www.wipo.int/edocs/mdocs/govbody/en/wo_ga_xxi/wo_ga_xxi_13.pdf (accessed 19 December 2016)
34 Supra, 15, Statement by the Delegation of Kenya, para.46
35 Ibid., Statement by the Delegation of Brazil, para.47
36 Ibid., Statement by the Delegation of Tanzania, para.45
37 Ibid., Statement by the Delegation of Cote d'Ivoire, para.53
38 Ibid., Statement by the Delegation of the Dominican Republic on behalf of GRULAC, para.30
39 Ibid., Statement by the Delegation of Bulgaria on behalf of the Group of Central European and Baltic States Group, para.33
40 Ibid., Statement by the Delegation of the Dominican Republic on behalf of GRULAC, para.30
41 Ibid., Statement by the Delegation of Bangladesh, para.54
42 Ibid., Statement by the Delegation of Japan, on behalf of Group B, para.34
43 Ibid.
44 Ibid., Statement by the Delegation of Malaysia on behalf of the Asian Group, para.32
45 Ibid., Statement by the Delegation of China, para.35
46 Ibid., Statement by the Delegation of Peru, para.36
47 Ibid., Statement by the Delegation of Egypt, para.38
48 Ibid., Statement by the Delegation of Japan, on behalf of Group B, para.34
49 Ibid., Statement by the Delegation of France on behalf of the European Union, para.41
50 Proposal by Argentina and Brazil for the Establishment of a Development Agenda for WIPO, 2004 (WO/GA/31/11). One of the recommendations comprising the Development Agenda adopted in 2007 – Recommendation 18 – called on WIPO 'to accelerate the process on the protection of genetic resources, traditional knowledge and folklore, without prejudice to any outcome, including the possible development of an international instrument or instruments'.
51 Report of the Commission on Intellectual Property Rights (CIPR), *Integrating Intellectual Property Rights and Development Policy*, 2002. The report stated that 'WIPO should give more explicit recognition to the fact that IP protection has both benefits and costs, and give greater emphasis to the need for IP regimes to be appropriately tailored to the individual circumstances of developing countries', p.158, http://www.iprcommission.org/papers/pdfs/final_report/ciprfullfinal.pdf
52 It has come to public knowledge that the United States periodically made demands over the years in the capitals of many developing countries enciting them to approve the renewals of the IGC's mandate.
53 The Group includes countries such as Algeria, Angola, Bangladesh, Colombia, Egypt, India, Indonesia, Malaysia, Myanmar, Namibia, Pakistan, Peru, South Africa, Tanzania, Thailand and Zimbabwe.
54 For instance, see *Recommendations of the Second Session of Like Minded Countries Meeting on the protection of genetic resources, traditional knowledge and folklore*,

WIPO/GRTKF/IC/19/8, www.wipo.int/edocs/mdocs/tk/en/wipo_grtkf_ic_19/wipo_grtkf_ic_19_8.pdf (accessed 19 December 2016).

References

Abdel-Latif, A. (2005). *Developing Country Coordination in International Intellectual Property Standard-Setting,* South Centre, T.R.A.D.E. Working Paper 24, South Centre.

Abdel-Latif, A. (2013). Genetic Resources, Patents and Benefit Sharing: State of Play and Challenges Facing Multilateral Discussions, in *Intellectual Property in the Pharmaceutical Industry*, Jacques De Werra (ed.), Schulthess.

Bannerman, S. (2015). The World Intellectual Property Organization and Traditional Knowledge, in *Indigenous Intellectual Property: A Handbook of Contemporary Research*, Research Handbooks in Intellectual Property series, Edward Elgar Publishing.

Correa, C. (2002). *The WIPO Patent Agenda: The Risks for Developing Countries*, South Centre, T.R.A.D.E. Working Paper 12, South Centre.

Dutfield, G. (2004). *Intellectual Property, Biogenetic Resources and Traditional Knowledge*. Earthscan/Routledge, London.

Helfer, L. R. (2009). Regime shifting in the international intellectual property system, *Perspectives on Politics*, 7(1): 39–44.

Okediji, R. L. (2008). WIPO-WTO relations and the future of global intellectual property norms, *Netherlands Yearbook of International Law*, 39(1): 69–125.

Okediji, R. L. (forthcoming 2017). *When Should We Invent International Intellectual Property Rights?*

Robinson, D. F. (2010). *Confronting Biopiracy: Challenges, Cases and International Debates*, Earthscan/Routledge, London.

Sell, S. K. (2011). TRIPS was never enough: Vertical forum shifting FTAS, ACTA, and TPP, *Journal of Intelectual. Property Law*, 447: 18.

3 The evolution of the IGC from 2001 to 2016

An insider's perspective[1]

Wend Wendland

Introduction

After marathon negotiations, and as dawn broke, WIPO delegates to the 2015 WIPO General Assembly (GA) agreed on renewing the mandate of the IGC for 2016 and 2017. One year earlier, in September 2014, delegates had not agreed on an IGC work plan for 2015 and, after nearly 30 sessions since the IGC first met in 2001, it did not meet in 2015. Had there not been agreement at the 2015 GA on renewal of the mandate, the IGC process may well have been over, at least in its present form.

This chapter describes what intervened between 2001 and 2016 – particularly what led to the suspension of work in 2014 and the renewal of the mandate in October 2015.

In 2000, the establishment of the IGC had been in recognition by all Member States of the need for a forum such as the IGC to address the linkages between IP and GRs, TK and TCEs.[1]

However, the IGC's story is one of diverging expectations among countries as to its precise purposes and goals. These differences have never been truly resolved, and they lie at the heart of the tensions that have marked the IGC's work since its establishment.

On the one hand, 'demandeurs' have consistently called for the adoption of binding international instruments providing for the 'effective protection of GRs, TK and TCEs'. On the other hand, although there may be wide support for some form of a disclosure requirement in patent law, 'non-demandeurs' have remained unconvinced, so far, of the need for new international IP norms on TK and TCEs (disclosure is discussed in Chapter 5 by Bagley).

The high ambitions of demandeurs led relatively soon to the IGC focusing on texts of international legal instruments before, at least from the perspective of non-demandeurs, there was common understanding on core questions such as the

1 Wend Wendland, Director, Traditional Knowledge Division WIPO and Secretary to the WIPO IGC. Any views expressed are those of the author alone and do not necessarily constitute the views of the WIPO Secretariat or any of its Member States. The author thanks Ed Harris, Simon Legrand, Fei Jiao and Alice Manero for their comments and assistance. All errors are his sole responsibility.

objectives of protection, the scope of protectable subject matter, the extent of the rights to be granted and the nature of the beneficiaries.

Demandeurs' frustrations would grow as they sensed a lack of political will on the part of non-demandeurs and were unable to generate any meaningful leverage. Equally, demandeurs would battle to seize the initiative by presenting non-demandeurs with specific textual proposals and engaging them in a true negotiation.

Most often, the site of the struggle would be the biennial renewal of the mandate of the IGC, during which decisions on convening a Diplomatic Conference and the meaning of terms such as 'legal instrument' and 'text-based negotiations' would be especially divisive. On each occasion, the IGC or GA would eventually agree a new mandate for the IGC, but the over-negotiated mandates would be ambiguous and provide an unstable foundation for continued work.

The IGC would wrangle over the design of an adequately inclusive yet efficient working methodology. The ideal would be a methodology that facilitates the reaching of widely acceptable outcomes that are flexible and pragmatic yet consequential.

Given the sizable trust-deficit among delegations, a key difficulty would be empowering smaller working groups to make progress on specific technical or policy issues. In the absence of a strong consensus on the purposes and directions of the IGC, countries would disagree heatedly over issues such as the number of sessions each year and their length, the status of certain documents, and the logistical arrangements for informal expert work.

Working so far on a package-deal concept, in which all three themes of the IGC are treated separately but together, coupled with the consensus principle, in which a single delegation can potentially block any decision, the IGC battles to this day to work out an effective working methodology. After 15 years, and as fatigue with the process mounts, many consider that some re-thinking is needed as to the objectives, expected outcomes and working methodologies of the IGC.

Despite these difficulties, however, the IGC's mandate has been renewed six times since 2000, and it was renewed again for 2016 to 2017. This supports the view that countries agree on the need for a body such as the IGC. Additionally, despite the complexity of the IGC's agenda, the process has produced some high quality substantive materials, launched and guided policy and legislative developments at the national and regional levels and enabled new stakeholders, such as indigenous peoples, to participate for the first time in international IP policy making. What is more, IGC participants have collectively developed draft texts on GRs, TK and TCEs that are work in progress but are perhaps the bases for what could eventually become broadly acceptable international instruments.

This chapter is in three parts. The first part identifies key moments in the IGC's evolution between 2001 and 2016, during which the IGC met 29 times. There were also three meetings of inter-sessional working groups (IWGs). In the second part, this period is divided into eight phases, and a more detailed session-by-session account is presented. The last part is a brief telling of the renewal of the IGC's mandate for 2016–2017 and the first session under the new mandate, IGC 29 in February 2016.

Key moments in the IGC's evolution between 2001 and 2016

Particular moments

The first real crisis came at IGC 5 in July 2003. For the first time, the IGC had to decide on renewal of its mandate. In what would become a pattern in subsequent years, the IGC couldn't agree and the WIPO GA had to step in.

In 2004, as demandeurs' frustration rose at the lack of progress on the development of legal instruments, a significant moment was the publication of Draft Articles on TK and TCEs. Non-demandeurs were displeased, and work on 'texts' was effectively frozen for almost five years. Further, little progress was being made on GRs.

The ensuing stalemate led unexpectedly to a dramatically empowered IGC mandate for the 2010–2011 biennium. This mandate, referring for the first time to 'text-based negotiations' and to the possible convening of a Diplomatic Conference, would have a far-reaching effect. On the one hand, it raised expectations among demandeurs; on the other, it drew non-demandeurs into normative work before they believed it was time. The constructive ambiguity of that mandate eased its adoption, but a continuing mismatch between the expectations of Member States would form an insecure platform for the IGC's future work.

The establishment of the IWGs in 2010 was a breakthrough. For the first time, technical work could be undertaken by smaller, less formal expert working groups. Considerable substantive progress was made and the three IWG sessions launched what was perhaps the most productive period in the IGC's work.

By 2012, however, the IGC plenary had re-asserted itself at the expense of smaller working groups. Progress slowed and the IGC struggled to undertake genuine 'text-based negotiations'. There was much back and forth, with, at times, more 'back' than 'forth'. Many perceived a significant chasm between the ambition reflected in the mandate and the maturity of the substantive texts.

The tireless efforts of IGC Chair Ambassador McCook (Jamaica) from 2012 to 2014 marked another defining phase in the IGC's life. He instilled procedural rigour and would repeatedly implore demandeurs and non-demandeurs to engage with each other, focus on what they have in common rather than what separates them and truly 'negotiate' in a spirt of partnership.[2]

The breakdown in late 2014 that led to the suspension of the IGC's work during 2015 was a pivotal moment, with potentially significant implications for the future of the process. The IGC would be revived one year later, with a renewed mandate for 2016–2017. However, the first session under the new mandate would disappoint many.

Recurring issues

Several contentious issues have coursed through the IGC's life.

There has been, for example, a persistent tension between 'non-normative' and 'normative work'.

In the early years of the IGC, greater attention was paid to fact-gathering, case-studies, general policy discussion and practical tasks. Over time, the focus would move to work on 'articles' and the development of an international legal instrument/s. The shift was gradual and propelled mainly by developing countries which from the outset saw the IGC primarily as a norm-setting forum. The mandate agreed on in 2009, which launched 'text-based negotiations', was, for many, a logical culmination of this shift. For non-demandeurs, however, the commencement of formal negotiations was too soon. They believed that foundational work – and common understanding on – 'objectives and principles' were needed before work could begin on substantive 'articles'.

The process has also seen a recurring tension between 'substance' and 'process'. The IGC has struggled to resolve a triangle of related questions, namely, the substance or content of the IGC's work, the legal character of its outcome(s) and the design of the IGC process itself.

One manifestation of this difficulty was the resurgence, at regular intervals, of whether the outcome/s of the IGC should be 'binding' or 'non-binding'. This question ignited a divisive debate whenever it arose. Many developing countries, seeking signs of good faith from non-demandeurs, desired assurances that future outcomes would be 'binding'. Many developed countries were unwilling to agree to be bound by instruments whose content was not yet clear, preferring 'soft norms' at most.

This question was one of the main causes of the breakdown in IGC 5.[3] It reappeared in the negotiation of the mandate for the 2010–2011 biennium. At that time, the matter was temporarily settled by the carefully constructed 'international legal instrument' and the reference to a possible 'Diplomatic Conference' (the word 'binding' was notably absent but many believed 'legal instrument' and 'Diplomatic Conference' were clear enough). The issue detonated again during IGC 27 in 2014. The new 2016–2017 mandate now explicitly states, however, that the nature of the outcome is open.[4]

Initially, the IGC was able to focus more on substance, but as the optimism that had characterized the first years evaporated and the IGC battled to initiate genuine 'text-based negotiations' as from 2010, attention turned to finding more creative working methodologies, such as the IWGs, 'high-level segments', 'thematic' and 'cross-cutting' sessions, expert drafting groups and 'informal informals'. In designing these tools, the goal was to strike a balance between inclusiveness and efficiency.

As the IGC would spend more and more time on procedural questions, most developed countries seemed to remain unpersuaded as to the need for and benefits of normative outcomes. Carrying the burden of proof, yet lacking meaningful leverage, developing countries seemed to struggle to translate their political ambition into specific textual proposals, engage those with whom they disagreed and convince them of the need for change.

The crises that frequently accompanied the annual renewals of the IGC mandate and/or workplans demonstrated the mismatch of expectations and understandings of the delegations. Unnerved by what appeared to be a lack of conviction on the part of many developed countries, developing countries would revisit the mandate question every year. The often bitter disputes that ensued only further deepened distrust between delegations. In several sessions, substance would be pushed to the margins as only the IGC's mandate would be discussed, and the yearly obligation of the GA to 'decide on convening a Diplomatic Conference' was too frequent and placed an artificial strain on the process.

The relationship between the three themes also arose regularly: should they be dealt with separately, but in parallel? Should two or more themes be merged into a single text? Should a theme that showed particular progress be 'fast-tracked'? The IGC had moved over time from treating the themes separately (but equally and in parallel) to, in 2014, discussing 'cross-cutting issues'. The relative priority accorded by countries to the issues has also shifted over time.

Another issue that would appear routinely was the tension between an 'intellectual property' approach and approaches 'beyond intellectual property' (such as a human rights approach or a 'safeguarding/preservation' approach), and between 'conventional IP' and '*sui generis* IP' approaches.

Finally, always close to the surface were questions about the relationship between the work of the IGC and other processes and instruments, both within WIPO (such as the WIPO Development Agenda and the SCP) and external to WIPO (such as the CBD, the Nagoya Protocol, the UNDRIP, the FAO's Plant Treaty). In its early years, the IGC responded to requests for technical studies from the CBD's COP. Later, there was cooperation between the Secretariats of WIPO and of UNESCO and the Permanent Forum. As could be expected, some countries also engaged in 'forum shopping'. At different times and on GRs and associated TK in particular, delegations would give greater or lesser importance to the IGC depending on developments outside of WIPO, such as in the lead up to the adoption of the Nagoya Protocol in 2010. Discussions in the WTO also influenced the IGC.

Session-by-session accounts

The first four sessions: analysis, exchange of practical experiences, policy debate and practical outcomes; frustration emerges at slow progress on an international legal instrument(s)

The IGC met for the first time in April 2001.[5]

A challenge, given the vastness of the IGC's mandate, was to organize the work in a sound and coherent manner, prioritizing where possible. The Secretariat prepared an important document[6] which identified a mix of practical and policy 'tasks', covering both 'defensive' and 'positive' approaches. GRULAC submitted a helpful paper for the session.[7]

However, before substantive work could begin, there was a delay as the IGC could not agree on its Chair. After almost two days, a solution was found – Ambassador Marchi (Canada) would share the chairmanship of the first session with Ambassador Chak Mun See (Singapore).

So began the inaugural session of the IGC – an inauspicious beginning, some would say.

In what remained of the first session, delegations began to feel their way forward. The task-based approach proposed by the Secretariat was accepted. Interventions were mostly general. Already then, however, the African Group clearly envisaged an international binding outcome.[8] The different expectations of some developed countries were also evident: they believed that the development of a new uniform global IP-like regime was not the best fit for IPLCs, and that the IGC should clarify concepts, focus on non-binding guidance and gather national experiences.[9]

The following three sessions (IGC 2,[10] IGC 3[11] and IGC 4[12]) continued, relatively calmly, to discuss the tasks agreed at IGC 1. Following a multi-disciplinary approach, the IGC's work combined fact-finding, analysis, exchanges of practical experience and policy debate. There was more focus, however, on 'non-normative' work. For example, proposals at IGCs 3, 4 and 5 which might have tended towards 'normative' work on TCEs (even the modest exercise of updating the non-binding 1982 UNESCO-WIPO Model Provisions) did not receive enough support. Similarly, attempts to launch work on *sui generis* elements of TK protection[13] were deflected.

The IGC's mandate for 2001 and 2003 made no mention, of course, of 'international instrument(s)'. This change would only come in the mandate for the 2004–2005 biennium, and 'text-based negotiations' would only commence formally in 2010.

These early years of the IGC led to a number of practically useful outcomes.

For example, steps to strengthen the defensive protection of TK and associated GRs,[14] particularly the enhanced recognition of TK as prior art, included a revision of the IPC so as to expand its coverage of TK[15] and incorporation of TK periodicals in the PCT's minimum documentation,[16] as well as the establishment of technical standards for TK documentation.[17]

At this time, work also began on developing capacity-building tools, such as a 'toolkit' on TK documentation,[18] a practical guide on the protection of TCEs[19] and a database of contractual provisions concerning the IP aspects of access to GRs.[20]

Another output at this time was a technical study on patent disclosure requirements relating to TK and GRs, responding to an invitation from the CBD's COP.[21] It was at the impetus, therefore, of the CBD that this normative question – a possible mandatory disclosure requirement – began to be tackled by the IGC.

However, impatience would mount as to lack of progress on 'legal instruments(s)' and, eventually, demandeurs would seek to shift the focus away from so-called 'practical' initiatives to 'normative' work.

The Committee had swathes of empirical information about national and regional approaches available to it,[22] gathered through Member State submissions, responses to questionnaires issued by the Secretariat[23] and panels organized during IGCs 4[24] and 5.[25] Further, the Secretariat was prolific,[26] publishing, in addition to the many formal documents for IGC sessions, several volumes of case-studies.[27] Over time, however, the value of empirical information would be questioned within the IGC, and at no time has a rigorous cost-benefit analysis or impact assessment been undertaken by the IGC.

By IGC 3 in June 2002, an extended seven day session, the workload of the IGC was enormous. The various tracks of the discussions were becoming complicated to follow, fuelling perhaps the impatience of many developing countries at what they perceived as an overly 'technical' approach. The 'task-based' methodology was helping to structure the IGC's expansive work programme, but it may also have complicated the discussions.[28] While technically of a high quality, some Secretariat documents were wordy. At one point, the Secretariat was even accused by an NGO of 'analysis overkill'.[29]

It became difficult for delegations to follow the work and detect where progress was being or could be made.

At IGC 3, the African Group restated its objective of a 'legally binding international instrument that recognizes, protects and rewards traditional knowledge and innovations'.[30]

IGC 5, July 2003: the first crisis and a new mandate for 2004–2005

Substantive work on the three themes continued at IGC 5, held in July 2003, but the session would be remembered for another reason. The IGC's mandate for 2001–2003[31] was drawing to an end, and IGC 5 had to make a recommendation on a new mandate to the GA meeting in September 2003.

By now, the IGC had met on four previous occasions. The IGC's agenda was vast and complex and, by July 2003, it had considered almost 80 documents.

The enormity of the IGC's task was now clearer: the IGC was embarking on nothing less than a profound re-imagining of the entire IP system, and all this in the absence of solid legislative experience at the national or regional levels, in the context of a complex web of inter-relationships with other international instruments such as the CBD, demandeur countries with diverse objectives and agendas at the helm, and some non-demandeur countries not yet fully convinced as to the need for the IGC.

The excitement and optimism that had infused the IGC in its early years were dissipating as demandeurs' expectations as to quick normative outcomes soured. Other countries, recalling that the IGC was merely a 'forum for discussion', believed further conceptual understandings were needed before embarking on any 'normative' work.

Shortly before the fifth session, the Islamic Republic of Iran organized an inter-regional seminar in Isfahan,[32] which led to the 'Isfahan Declaration', submitted to IGC 5 in order to try to shape the directions of the session.[33]

The discussion of the mandate at IGC 5 was heated and no agreement could be reached. There were essentially two main issues: the form and mandate of 'future work'. 'Form' referred to the mechanism through which the work would be conducted and 'mandate' referred to the directions and nature of future outcomes. Consideration was at that time already being given to the possible conversion of the IGC into a standing committee.[34] The African Group would formally propose this 12 years later in 2015.

For the African Group, supported by the Asian Group, Venezuela, Brazil and others, it was clear that, although the IGC 'had undertaken important information gathering and policy discussion, it had not emphasized the need to negotiate a legally-binding international instrument to protect GR, TK, and folklore. . . [the IGC's work] had thus. . . continually been used as an excuse to exclude or delay work in other fora'.[35]

Many developed countries had a different view. While Group B[36] and the USA[37] welcomed the practical and fact-gathering work of the IGC, Australia[38] and Canada[39] believed that commencing negotiations on binding language and instruments would be premature.

In the margins, IGC Chair Olsson (Sweden) attempted to broker an agreement. He presented a series of non-papers for consideration[40] but none were accepted.

In the end, the differences could not be bridged. The wording of the formal decision taken by the IGC suggests that it was the African Group's insistence on its position that led to the breakdown.[41]

IGC 5 was a low-point for the IGC – its first real crisis.

It was left to the GA of 2003 to resolve the matter. The WTO trade talks in Cancun[42] had just collapsed, and WIPO Director General Idris had, in May that year, been re-appointed for a second term. There was considerable pressure for the Assemblies to go smoothly, including a successful renewal of the IGC's mandate.

It was only after informal consultations led by GA Chair Ambassador Kessedjian (France) that the deadlock was broken and a mandate for the 2004–2005 biennium was agreed.

The mandate included new references to a 'focus, in particular, on a consideration of the international dimension of [the questions being addressed by the IGC], without prejudice to the work pursued in other fora' and that 'no outcome of its work is excluded, including the possible development of an international instrument or instruments' (at the last minute, the word 'focus' replaced 'cover' and the reference to 'or instruments' was added).[43] For the first time, a reference to 'international instrument or instruments' appeared in the IGC mandate, as did the concept 'the international dimension'. Further, the mandate required the IGC to 'accelerate' its work. The precedent was also established that the IGC would report back to the GA after just one year.

The shift towards normative work had now explicitly begun.

The new reference to 'without prejudice to the work pursued in other fora' was seen as a particularly important achievement for developing countries, which, observers believed, could use such language to prevent developed countries from blocking discussion of GRs and TK issues in other fora such as the TRIPS Council.[44]

Thus, the IGC's work entered a second phase. In a pattern to be repeated later, this pragmatic yet ambiguous new mandate was agreed despite the higher level of ambition of many developing countries and misgivings by several developed States. Venezuela disassociated itself from the decision after its adoption.[45]

This same mandate was extended, more or less without change, by future GAs (in 2005 and 2007) to cover the period 2006 to 2009. It was only later in 2009 that the mandate was dramatically re-set, in what would be a momentous turning point in the IGC's evolution.

However, while the new mandate for 2004–2005 largely reflected developing country proposals, it did not provide the clarity and specificity that developing countries had sought. Lettington and Nnadozie[46] pointed out that, in view of the mandate's ambiguity, clear and specific goals for the IGC would have been needed quickly, which would, in turn, have required a unity of purpose among the majority of developing countries. As the IGC looked forward to IGC 6, it was essential, from a developing country viewpoint, that well-justified and specific substantive proposals be put forward by them to focus the discussion. At that stage, however, a common position and focus were still missing and there was still no Member State textual proposal.[47]

The strengthened 2004–2005 mandate: IGCs 6 to 8; efforts to 'accelerate' the IGC's work; first appearance of Draft Articles on TK and TCEs

Initially designed as a forum for discussion, the new mandate reflected the changing expectations of countries, in particular, the strong interest by many countries in binding, normative outcomes. Expectations were very high.

At the same time, as scepticism among developing countries as to the IGC's effectiveness grew, the IGC risked becoming irrelevant. The IGC was, therefore, at a critical point.

The challenge was how to give effect to the GA's direction to 'accelerate' the work. Progress was needed on: (i) clarifying the substance (or 'content') of the IGC's work; (ii) building political consensus; and (iii) designing a process that could deliver agreed outcomes. Each of these needed to be considered in parallel. This trifecta of issues continues, up to 2016, to remain a challenge.

A sensitive question that began to make its appearance at this time was whether the IGC should agree first on the legal nature of its future outcome/s or, rather, on their substance (content). This question would detonate periodically in the IGC's evolution.

A related question was whether the IGC should proceed directly to work on the text (articles) of a 'legally binding international instrument', or whether delegations should first try to reach agreement on the 'objectives and principles' of such an instrument.

These questions would provoke the next crisis in the IGC – the introduction, at IGC 8 in June 2005, of Draft Articles on TK and TCEs.

The evolution of the Draft Articles had been incremental. At IGC 6[48] the documents on TK and TCEs had comprised 'legal and policy options' and, in IGC 7,[49] 'policy objectives and core principles' in a short, instrument-like format.

By IGC 8[50] the documents contained not only draft 'objectives and principles' but also draft 'articles' ('substantive provisions').[51]

The thinking within the Secretariat at that time was that draft 'texts' would respond to the required 'acceleration' and help sharpen the focus of the discussions: draft texts in the form of concise articles, rather than longer discussion documents, might help to illuminate the possible elements of an international consensus, while also illustrating the precise points of difference.

The Secretariat, however, came under withering criticism from many developed countries, which accused it of having overstepped its mandate and for having tilted the playing field in favour of those delegations seeking a legally binding international instrument(s).[52] Most developing countries welcomed the revised documents. The session could not agree on further work on the two documents, on which direct work would not resume for several years.

On the 'international dimension', the Secretariat produced a significant document[53] but it was not fully discussed. Like several substantive documents prepared for the IGC, it seemed to pass through the IGC without being properly digested. IGC 8 also saw the EU file an important proposal on disclosure requirements.[54]

IGC 8 was the last session of the IGC in the 2004–2005 biennium, and in a weary stalemate, the session agreed upon the extension of the mandate for 2006–2007, which the GA of 2005 endorsed.

IGCs 9 to 14 (2006 to 2009): the IGC treads water; frustration grows; attention turns to improving the methodology of the IGC; IGC 14 fails to agree on a mandate for 2010–2011

The IGC divide is not exclusively 'north-south'. For example, at IGC 8, Norway had been supportive of continued work on TK and TCEs. In February 2006, in advance of IGC 9,[55] Norway organized an informal discussion among key IGC delegations, after which the country made a submission to IGC 9.[56]

In 2006 and 2007, the IGC met three times in extended sessions (IGCs 9,[57] 10[58] and 11[59]), under the Chairmanship of Ambassador I Gusti Agung Wesaka Puja (Indonesia).

IGC 10 developed a 'List of Key Issues' on TK and TCEs. Subsequently,[60] comments on the key issues were collated, tabulated and 'factually extracted'. The IGC was now working 'around' the texts but not 'on' them. The texts themselves were effectively frozen.

By now, the volume of paperwork before the IGC was staggering. Documents commissioned by the IGC relevant to TK alone now numbered 27.[61] IGC 12 then requested the preparation of 'gap analyses', which were duly prepared by the Secretariat for consideration at IGC 13.[62] These were comprehensive but not extensively debated by the IGC.

Again, frustration began mounting at the lack of progress on normative work. Member States tried to bolster the process through meetings out of Geneva, such as an Asian-African Forum, held in Bandung[63] (leading to a Declaration at IGC 11).[64] However, most developed countries still seemed sceptical as to the precise objectives of the IGC.

At around this time, the UNDRIP was adopted[65] and it added further incentives for the IGC.

The 'practical' tasks of the IGC, which had borne some tangible benefits in the early years and had not been fully completed, were by now all but forgotten.

By IGC 11[66] it was once again time to reconsider the mandate of the IGC for 2008–2009. After extensive discussion, the IGC reached agreement on a recommendation to the 2007 GA and the GA renewed the IGC's mandate. The mandate for 2008–2009 was largely reflective of the preceding mandate.

As the IGC entered its eighth year, there was a distinct lull. Little progress was being made on what mattered most to developing countries, namely the adoption of an internationally legally binding instrument. For 'non-demandeurs', for whom the status quo was adequate, this seemed satisfactory.

With a stalemate on substance, greater attention began to be paid to working methods. Creative thinking was needed. At IGC 12[67] the possibility of inter-sessional work was raised, and countries were invited to make proposals at IGC 13,[68] and the African Group did so.[69]

In September 2008, WIPO elected a new Director General, Francis Gurry (Australia), who, up to that point, had directly overseen the work of the TK Division.

IGC 14 took place from 29 June to 3 July, 2009. By now, substantive work on TK and TCEs had virtually ground to a halt, as demandeurs insisted on working on the 'substantive provisions', while others wished to discuss 'objectives and principles'. The IGC was treading water on GRs too.

The focus at IGC 14 was on modalities for inter-sessional work and the mandate for 2009–2010.

Despite that, the IGC Chair, Ambassador Gauto Vielman (Paraguay) had conducted informals before the session.[70] 'Future Work' was difficult and consumed almost the entire session.

Once again, the African Group was prominent. It submitted a proposal on the mandate, which it amended a few days later, adding the words 'internationally legally binding' before 'instrument/instruments'.[71] This proposal received support from a number of developing countries. A key part of the proposal, however, was a new reference to 'text-based negotiations'. The EU suggested that the IGC prepare a declaration and, in parallel, conduct 'outcome oriented deliberations'.[72] The EU's proposal was only submitted on the session's last day, however, and the African proposal remained the sole basis of the Committee's deliberations.

As time for the IGC seemed to be running out, there was some interest among even demandeurs for a non-binding declaration as a first step. Indeed, it may be recalled that the WIPO Program and Budget for 1998 and 1999 had explicitly articulated the benefits of 'soft law' instruments, at least as interim steps.[73] This would, therefore, have been the moment to reassure demandeurs that a declaration would not be empty rhetoric or close the door on future treaty negotiations – rather, it could have laid a firm political basis for future work and manifest what the substance of a future binding instrument might look like. However, the most vocal demandeurs were determined to press ahead and seek agreement on a binding instrument, and further discussion of interim outcomes was prevented.

Key sensitivities related to the mandate renewal were the relationship between the three subjects, inter-sessionals and the terms 'legally binding', 'instrument' and 'text-based negotiations'. There was no agreement on future work at the session.[74]

It was with this unfortunate end to IGC 14, that Member States and observers braced themselves for the September 2009 GA.

The 2009 General Assembly, the new IGC mandate and the launch of 'text-based negotiations'

The Assembly would end dramatically, late into the night, on 1 October 2009, with the adoption of a robust and ambitious new IGC mandate. This was a significant moment in the life of the IGC. Given the despair among delegations after IGC 14, the adoption of this mandate at the GA was unexpected.

The African Group proposal from IGC 14 was still on the table and the EU[75] and the USA[76] responded with proposals of their own. Australia filed a proposal[77] similar to that of the African Group.

GA Chair Ambassador Dumont (Argentina) played an active role in the negotiations that took place in the margins. He produced a draft text which became the focus of the negotiations. The text evolved: while earlier drafts[78] included references to the EU's 'declaration' proposal, and to the African Group's wish for 'binding', these ideas were eventually dropped. Finally, after agreement in informals, the text was rushed down to the plenary, the Assembly (which had already completed its other business) was hastily reconvened, and the new 2010–2011 mandate was gavelled. It was near midnight. Demandeurs were generally delighted. Indonesia believed that the new mandate would 'breathe life and vigor' into the IGC[79] while Tunisia described it as a 'watershed'.[80] Other demandeurs believed the new mandate was a logical progression from the consensus that had existed at the IGC's creation, and that, after eight years, it was time for the IGC to move more deliberately towards the development of international legal instrument(s).[81] Others, however, such as the EU and the USA, were more guarded.[82]

And so, from 2010, 'text-based negotiations' became formally part of the IGC's mandate. The shift to normative work, begun many years earlier, was now complete.

These negotiations would be aimed at 'reaching agreement on a text of an international legal instrument (or instruments) which will ensure the effective protection of GRs, TK and TCEs'. The three existing IGC texts were to 'constitute the basis of the Committee's work'. The GA would, a mere one year later, be required to 'decide on convening a Diplomatic Conference'. The mandate also endorsed the organizing of IWGs.

With the adoption of this mandate, expectations among demandeur countries were high that a Diplomatic Conference would be convened soon.

However, there were still important divergences among Member States as to the exact goals and substance of the IGC's work, and several States had doubts about the ambitious (and ambiguous) new mandate. As they saw it, work on international instruments would begin prematurely, in the absence of agreement on certain fundamental questions, a cost-benefit analysis and effective national systems that international instruments could be modelled on.

Senegal insightfully described the consensus that had been reached as follows: 'consensus was the fact that one was not satisfied with everything but accepted everything'.[83]

The IGC's most productive phase (IGCs 15 to 19): the inter-sessional working groups and other procedural innovations enable substantive progress; a mandate for 2012–2013 agreed

As the IGC was not making significant progress on substance, the new methodological tool, the IWGs, was keenly awaited. However, arrangements for the IWGs could only be agreed on following protracted informal consultations. In the interim, IGC 15[84] took place in December 2009. The breakthrough on the arrangements for the IWGs came in May 2010, at the first session (IGC 16[85]) chaired by Ambassador Owade (Kenya).[86]

The aim was that the IWGs be effective, transparent and representative.[87] In their design, there were essentially three issues that presented difficulty: (1) the precise mandate of the IWGs (and their relationship with the IGC plenary), (2) the methods of work, and (3) who could participate in the IWGs.[88]

On substance, IGC 16 requested documents on the term 'public domain' and on the various forms in which TK may be found (see Chapter 8 by Okediji, this volume, for further discussion of the public domain).[89] The second document triggered an important differentiation between different forms of TK (for example, 'secret' and 'widely available') that would begin, helpfully, to inform the work of the Committee later at IGC 27 in 2014.

As from IGC 16, the IGC entered what was perhaps its most productive few years, propelled by the new mandate and armed with a new tool, the IWGs.

The IWGs were organized thematically. The IGC texts, first introduced by the Secretariat in 2005, had become complicated as a result of layer upon layer of comments and proposals collected in tables and other ancillary documents. IWG 1 on TCEs[90] produced revised articles which were not only shorter and simpler but developed solely by IGC participants, rather than the Secretariat. Countries were thus beginning to take ownership of the texts. IWG 2[91] also produced a more streamlined TK text.

The IWGs were effective because the participants were experts and small in number and the sessions were less formal. They enabled frank and technical exchanges, and were constructive, dynamic and participatory.[92] Importantly, the IWGs uncorked work on the draft texts, effectively blocked since 2005.

IWG 3[93] on GRs was more complicated, however. The Nagoya Protocol had been adopted in late 2010 and IGC 17[94] had given detailed instructions to IWG 3 on how it should address its work. As commenting on the GRs documents had reached saturation point, the IWG had to make progress in narrowing the materials, and a working methodology and work plan for the session were devised to advance this goal.[95] IWG 3 came out with two concise documents which were transmitted to the IGC.

Would the IWG texts be accepted, however, by the IGC plenary as the focus of future negotiations? Yes, as IGC 17 readily agreed to accept the TCEs text of IWG 1, and the same happened to the TK and GRs texts at IGC 18.[96]

The IGC's acceptance of the results of the IWGs legitimized the utility of smaller expert groups. After the lengthy prior consultations and preparations, the IWGs were seen as having been legitimately created, and their results, therefore, legitimate.

Once the IWGs had completed their work, it soon became apparent, however, that the IGC did not have a clear methodology for how to 'negotiate' the new texts.

As some other countries had done to support the process, the Russian Federation organized an International Symposium[97] just before IGC 17, which led to a submission by the Russian Federation to that session.[98]

By the end of IGC 18, in May 2011, there was one text only on each of TK and TCEs and two on GRs. This was an improvement on what had been, until the IWGs, a blizzard of documents.

IGC 19[99] was, again, largely preoccupied with the mandate question, as the 2010–2011 mandate was coming to an end. The 'Future Work' agenda item commenced early on in the plenary, and discussions were informed by a proposal from like-minded countries (LMCs).[100]

IGC 19 managed to reach agreement on a new mandate for the IGC for 2012–2013, and it was adopted by the GA. On substance, IGC 19 made some progress. In relation to TK and TCEs, the IGC agreed to focus, for the first time, on certain 'key issues', namely definition of subject matter, beneficiaries, scope of rights, and exceptions and limitations. This useful step allowed the IGC to narrow its enquiry to these four core and interdependent issues (see Chapter 7 by Ruiz Muller on TK and Chapter 9 by Perlman on TCEs).

The four issues did not include the 'objectives and principles' part of the TK and TCEs texts, however – these parts would, therefore, not be discussed meaningfully in many subsequent sessions. Other issues not regarded at that juncture as 'key' (such as term of protection, international protection and formalities) were not fully addressed either. The same four issues remained the focus of the IGC's work until its suspension in late 2014. This means that no or little work has been done on the other issues since mid-2011.

The texts emanating from IGC 18 were relatively concise and focused.[101] The LMCs also produced substantive inputs for IGC 19, in the form of almost complete textual proposals on each of the IGC's themes.[102] This was perhaps the first time – 10 years after the IGC's first session – that a group of demandeurs tabled complete treaty-language proposals.

Procedurally, this session broke new ground in using 'facilitators', two or three delegates who assisted in drafting text in line with the plenary's directions. The 'facilitators' tool worked and the articles addressing the four key issues were further simplified.

The end of IGC 19 was also the end of Ambassador Owade's term as IGC Chair. IGC 19 was one of the most successful sessions of the IGC. It managed to reach agreement on a new mandate for the next biennium and make progress on substance. As the DG of WIPO stated, 'Ambassador Owade presided over the most productive years of the IGC'.[103] Ambassador Owade capped off his term by penning valuable documents on 'pending issues' to guide subsequent IGC sessions.[104]

2012–2013: a new mandate; Ambassador McCook becomes IGC Chair; attempts to begin real 'text-based negotiations'; further procedural innovations

IGC 20 was the first session to take place under the newly minted 2012–2013 mandate, which contained a few new elements, in particular that the IGC should 'expedite' its text-based negotiations and hold thematic sessions. The mandate stipulated, again, that only one year later the GA should 'decide on convening a Diplomatic Conference'. This would place considerable pressure on the IGC in 2012.

IGC 20 was also the first session under a new Chair, Ambassador McCook (Jamaica), who would remain Chair of the IGC until 2014 (IGCs 20 to 28, inclusive).

IGC 20, an eight-day session, addressed only GRs – this would be the only time that the IGC would cover GRs before the GA would decide on convening a Diplomatic Conference. Propelled by Ambassador McCook, the IGC began to consider more rigorously how to conduct 'text-based negotiations'.

By IGC 20, the number of documents on GRs began to increase again and it was clear that the IGC could not begin 'text-based negotiations' unless it could consolidate existing materials into a single document. The tension, as always, was between efficiency and inclusiveness. 'Facilitators'[105] were appointed to rationalize existing documents into a single text, which would be presented to the plenary for 'negotiation' and 'live drafting'. The text, as at the end of the session, would be noted, not 'adopted' and transmitted to the GA.

Somewhat complex rules for the 'live drafting' and for the treatment of various in-session revisions of the documents were developed along the way, causing some confusion at times.

Overall, IGC 20 was viewed as a success. The session consolidated all the documents into one and transmitted it to the 2012 GA. There was now, for the first time on GRs, a single document, albeit heavily bracketed. 'Objectives and principles' had also been considered together with the various substantive provisions (by contrast, the IGC would not as thoroughly address 'objectives and principles' on TK and TCEs). By now, GRs seemed to be emerging as the topic on which most progress was likely.

At the end of the session, and after much wrangling, there was agreement to include a 'Chairman's Note', which emphasized that the text was work-in-progress and without prejudice to the positions of participants.

At IGC 21,[106] on TK, the working methodology was refined further. Unlike with GRs, there was already a single text on TK, and reducing divergences in the text was a bigger challenge than had been the case with GRs. Once again, this was the first and last chance to deal with TK before the 2012 GA.

In preparatory consultations before the session, the core questions were identified as: whether to focus only on the four key issues mentioned earlier; how to deal with 'objectives and principles'; how to deal with the LMCs text;[107] and, the relationship between the TK and TCEs texts.[108] The goal was to retain no more than three distinct options under each substantive article. As some delegations had criticized the complexity of the working methodology at IGC 20, simpler and clearer rules of engagement were developed.

IGC 21 ended with a new version of the TK articles, which included comments by the facilitators[109] as to areas of convergence and issues on which further clarification were needed.

As with the GRs document, the session ended with disagreement on whether the document should include a 'Chairman's Note'. In the end, exasperated delegations agreed with a compromise brokered by the USA, which stated simply, in relevant part, '[The document] represents a work in progress'.

Overall, views as to the session's success were mixed. The African Group and other demandeurs were not satisfied that true 'text-based negotiations' were under way.[110]

At this stage, the IGC's schedule was unrelenting. In the 28 months since early 2010, the IGC and IWGs had met ten times, IGC 21 included, with sessions sometimes a mere seven or eight weeks apart. This placed a strain on delegations, the Chair and the Secretariat (WIPO's TK Division), and allowed little inter-sessional time for consultations, reflection and preparation.

Whereas, in the past, the TK Division had been able both to administer the IGC and meet the swelling demand from countries and IPLCs for complementary capacity-building and technical assistance, by now servicing the IGC had become an almost full-time task of the Division. The Division's role is purely facilitatory and administrative, and it consciously strives to provide a service that is neutral, professional and efficient. Inter-sessional preparations with the Chair intensified and sessions were preceded by extensive consultations with Member States and briefings for IGC participants. The TK Division's website[111] was revamped and an Indigenous Portal was added.[112] Greater attention was being paid to supporting IPLCs' effective participation. Since 2009, an 'Indigenous Fellow' had begun to work in the TK Division and, in April 2013, the Secretariat teamed up with the Secretariat of the Permanent Forum on Indigenous Issues to organize an Indigenous Expert Workshop which produced a submission reflecting indigenous people's perspectives on the texts.[113] Complementary capacity-building and technical assistance activities continued to the extent possible, however.

Following IGC 21, delegates were back a mere eight weeks later for IGC 22[114] on TCEs. However, a bruising procedural dispute delayed the session's start, concerning whether the agenda should include an item on 'Future Work'. A compromise was reached and the agenda was duly adopted[115] but by then it was already the second day of the session.

The mixed outcomes of IGC 21 resulted in changes to the working methodology for IGC 22. The challenge remained to design a process that would be transparent, fair, inclusive, efficient and orderly, disciplined and focused on 'text-based negotiations'.

For the first time, 'informal expert groups' were tried. They were intended to provide a safe informal space for a relatively small number of experts from all regions to resolve key policy and technical issues. The composition, working methodology and rules of engagement of these expert groups were wrestled over in consultations prior to IGC 22.

IGC 22, with Kim Connolly-Stone (New Zealand) as facilitator, ended with a cleaner TCEs text which was, as with the GRs and TK texts, noted and transmitted to the 2012 GA. Adding the informal expert group mechanism was working, at that stage.

Soon, however, the size of the group would grow, and eventually by IGC 24, delegations, anxious about transparency and inclusiveness, requested that a live audio-visual feed lead from the informal expert group to a larger room, so that all delegations including observers could follow the work being done in the informal process. This all but meant the informal expert groups lost their effectiveness. Furthermore, countries guarded the exclusive decision-making authority of the plenary, and gains made in the informals would often be rolled back by the plenary. The effectiveness of the informal expert groups would diminish over time.

The struggle now moved, again, to the 2012 GA. The IGC's report to the GA included the three texts produced at IGCs 20, 21 and 22.[116] Even though the IGC already had a mandate for the 2012–2013 biennium, the GA saw an attempted re-negotiation of some of the language of the mandate, with African countries in particular calling for a Diplomatic Conference in 2014 and the re-introduction of IWGs. In the end, neither of these ideas found their way into the agreed workplan.

The workplan for 2013 laid out a programme similar to 2012.[117] IGC 23 focused on GRs, having not discussed them since February 2012. There was immense pressure to make progress on all three themes, because the mandate would run out later in the year. Some Member States stepped up to support the process. Inspired by Indonesia's Bali meetings, India (February 2013), South Africa (April 2013) and Thailand (July 2013) organized pre-IGC preparatory meetings. These were generally seen as useful, although they lacked a formal linkage with the IGC. The 'IGC Retreat' in Thailand resulted in a comprehensive report to IGC 25.[118]

In 2013, the IGC would meet three times. The methodology that had been tried in 2012 was maintained, although refinements and adjustments were made. 'Ad hoc informal drafting groups' (a small number of countries tasked with working on language on one specific issue or article) were also experimented with. In 2013, Ambassador McCook began to provide informal 'issues papers' in advance of each session to help focus and orientate the IGC's work.

Each session that year followed more or less the methodology of IGCs 20, 21 and 22, and would note and transmit the next versions of the texts to the 2013 GA. Progress was made in particular on the GRs text. Progress on the TK and TCE texts was less evident. Despite other proposals re-submitted, especially on GRs at IGC 23,[119] the only documents formally submitted to the 2013 GA were the 'texts' on GRs, TK and TCEs.[120]

Progress on the TCEs text was especially elusive as delegates were preoccupied by the mandate renewal question. South Africa referred to the TCEs text that emerged at the end of that part of IGC 25 as 'bloated, long and unclear'.[121] The achievements of the IWGs were being rolled back in plenary.

The discussions on the mandate were difficult and failed to produce an agreed recommendation to the GA. Initial statements by delegations reflected the well-known divergences in expectations and perceptions. The African Group believed, for example, that 'with goodwill and strong commitment and engagement from Member States, agreements could easily be reached on them'.[122] On the other hand, the USA 'deplored that a detailed discussion of facts and a meaningful technical discussion [was] lacking'.[123] It seemed that perspectives and expectations had changed little since the first sessions of the IGC.

A facilitator, Ian Goss (Australia), attempted to piece together a compromise draft mandate, but in vain.[124] So ended IGC 25, one of the low points in the IGC's life.

A new mandate for the IGC for the 2014–2015 biennium was agreed upon at the GA of 2013, however. The consultations that led to the agreement – facilitated again by Goss – were far from easy. After the adoption of the mandate, some interventions demonstrated that the 'consensus' reflected an uneasy compromise.[125]

2014–2015: an amended mandate; IGCs 26 to 28; some progress on GRs but little on TK and TCEs; the IGC does not meet in 2015

The new mandate featured a few novel aspects, including an Ambassadorial/Senior Capital-based Officials meeting; ten day-long back-to-back TK and TCE sessions, including 'consideration of cross-cutting TK/TCE issues'; the new phrase 'With a view to finalizing the text(s) within the biennium'; and, a 'cross-cutting GR/TK/TCE session and a stocktaking and making a recommendation to the 2014 GA'. For the first time, the mandate referred to the possibility of studies and examples, noting that they should not delay progress in the text-based negotiations.

It may be said that the discrepancy between the ambition and sense of urgency reflected in the language of the mandate and the actual state of the substantive negotiations was now very wide. Furthermore, the means and timeframe allowed for the substantive work to make headway seemed inadequate.

The first session under the 2014–2015 mandate was IGC 26[126] focused on GRs.

As before, work continued on the text, and some further 'cleaning' was achieved. The text as at the end of the session was transmitted to the 2014 GA. However, this time there was one difference – the texts that would emanate from IGC 26 (GRs) and 27 (TK and TCEs) would be transmitted to the GA 'subject to any agreed modifications on cross-cutting issues the IGC may make at its July 2014 session'.

The 2014 GA would have to decide only on the IGC workplan for 2015 (as there was already a mandate for 2014–2015). There was, at that stage, little indication of the breakdown that would occur at the 2014 GA.

At IGC 26, Ambassador McCook, re-elected for 2014–2015, presided over the first Ambassadorial/Senior Capital-based Officials meeting, which was intended 'to share views on the key policy issues relating to the negotiations and to further inform/guide the process'.

After extensive consultation,[127] the Secretariat proposed guidelines[128] for the conduct of the meeting, and suggested that delegations address questions such as, in respect of each theme, what was the policy issue that needed to be resolved as a priority and why; what should be dealt with in an international legal instrument and what could be dealt with at the national level; and, what suggestions were there for common ground on the issues that needed to be resolved internationally? Regarding the process as a whole, what new negotiating pathways and modalities might there be to make further progress?

At the conclusion of this high-level segment, the Chair delivered, on his own authority, an oral summary.[129] It was widely felt that this segment did not add much new which could 'further inform/guide the process'.

Next came IGC 27.[130] Again, much preparation went into the design of the session and extensive consultations were necessary. To assist delegations, the Chair suggested four cross-cutting TK/TCE issues, namely (a) the meaning of 'traditional', (b) the beneficiaries of protection, in particular, the role of States or 'national entities', (c) the nature of the rights, including the meanings of 'misappropriation' and 'misuse', and, (d) the treatment of publicly available and/or widely diffused TK and TCEs.[131]

By now it was clear that the expert drafting group method had become less efficient, and delegates were becoming bogged down in detail. It was during IGC 27 that the binding/non-binding question detonated again,[132] leading to a somewhat tense exchange between the Chair and the Delegation of Sweden. The Chair queried whether it was possible for the negotiations to continue if there were delegations which *a priori* could not accept an outcome that could have treaty-like effect.[133]

Right before IGC 27, Indonesia had organized a meeting in Bali[134] at which a 'rights pyramid' (or 'tiered approach') had been proposed. This approach posited that different levels of rights might be accorded to different forms of TK depending on the TK's degree of diffusion – for example, 'secret' TK might have stronger protection than 'publicly available' TK. The idea gained some traction and it was then discussed in IGC 27. Subject to its further refinement, such an approach showed great promise. However, the IGC would soon be suspended.

IGC 28,[135] which met in July 2014, was a short session, comprising a review of the cross-cutting elements contained within the GRs, TK and TCEs text(s) from IGCs 26 and 27, a stocktaking of the Committee's progress and the making of a recommendation to the GA.

To assist delegations, Ambassador McCook published an informal paper reflecting his views on possible cross-cutting issues.[136]

The texts were transmitted, without change, to the GA. After extensive consultations, there was no agreement, however, on a recommendation to the GA on a workplan for the next year.

Once again, morale in the IGC was at a low. Frustration among demandeurs was especially high as the 2014 GA commenced. The 2014 GA had to agree on the IGC's workplan for 2015 and 'decide on convening a Diplomatic Conference'.[137]

While preliminary agreement on a workplan was achieved quite easily, discussions foundered on the familiar question of convening a Diplomatic Conference. The African Group insisted that the GA convene a Diplomatic Conference in either 2015 or 2016. Other regional groups seemed more flexible and pragmatic. Goss, as facilitator, tried various formulations to satisfy both the African Group and the rest, issuing some 12 draft decisions for consideration.[138] Even Goss's eleventh hour minimalist draft aimed at simply keeping the process alive by ensuring that the IGC would meet at least once in 2015[139] was not accepted by the African Group which held its position that it could not agree to any 2015 IGC workplan unless the GA also agreed to 'convene a Diplomatic Conference in 2016 to adopt a legally binding instrument or instruments'.[140] This position did not receive support from other regional groups. No decision could, therefore, be agreed on, including in respect of a workplan.[141]

After 13 years, and 28 sessions and three IWGs, the IGC process came to a halt. The IGC did not meet in 2015.

Views on the wisdom of this suspension were divided. Many argued that it would have been wiser to keep the process alive. While slow, there had been progress in the substantive work of the IGC. The GRs text was approaching a certain level of maturity and existing gaps were more political than technical. The recently introduced pyramid approach to TK and TCEs needed refinement but showed signs of illuminating a way forward. There was some momentum, therefore, that risked being lost. After a 'gap year', would the IGC be revived?

Renewal of the IGC's mandate 2016

One year later in October 2015, as weary delegates emerged from WIPO's conference hall after agreement on a new IGC mandate for 2016–2017, there was relief that the IGC had been resuscitated and that the investment made in the process up to 2014 would not have been in vain. The IGC would meet again as from early 2016. However, as before, the new mandate is a delicately constructed compromise. Its development was facilitated by Goss, who drew on three formal proposals[142] and a suite of other suggestions.[143]

The text of the new mandate includes a number of wording changes that States considered significant. For example, it was considered that the addition of the word 'including' before 'text-based negotiations' would broaden the IGC's work beyond solely text-based negotiations. The mandate now explicitly recognizes that the IGC's negotiations do not prejudge the nature of the outcome(s).

The new mandate clarifies that the work of the IGC relates to the IP dimension of the protection of GRs, TK and TCEs and that such protection should be 'balanced'.

The IGC should now reach 'a common understanding' on certain core issues identified in the mandate. The term 'evidence-based approach' is also new, as is a call upon the WIPO Secretariat to organize inter-sessional seminars and workshops, building on successful WIPO Seminars held in 2015.[144]

One especially helpful change is that the GA has granted the IGC two years within which to work before the GA should take stock of progress made, and decide on whether to convene a Diplomatic Conference or continue negotiations. The next significant GA session for this work takes place in September/October 2017, therefore.

The Committee may also consider the conversion of the Committee into a Standing Committee. Another novelty is that each theme will be addressed twice, in two successive IGC sessions.

Conclusion

The first session of the IGC under the new mandate took place in February 2016.[145] The session focused on GRs, with several working documents, including a 'Consolidated Document Relating to Intellectual Property and Genetic Resources',[146] the text last worked on at IGC 26 in February 2014. The IGC elected a new Bureau for the biennium.[147] The plenary had a high-quality dialogue on a list of 'core issues' proposed by the Chair, but ran into difficulties when trying to work on text. Based on the plenary dialogue, facilitators[148] presented revised text but the plenary could not agree on its status. In the end, the pre-existing Consolidated Text was transmitted to the next IGC session, with one relatively small change. In addition, as required, the IGC transmitted to the next session an 'Indicative List of Outstanding/Pending Issues to be Tackled/Solved at the Next Session'.[149]

The challenges experienced at IGC 29 were familiar to followers of the IGC. Difficulties continue to stem from what many see as an over-negotiated and ambiguous mandate, persistent divergences among countries on objectives and core policy issues and the absence, so far, of an effective and lasting working methodology. With three further sessions of the IGC in 2016, the immediate priority for the IGC should be to devise a methodology that enables the gaps to be narrowed and, eventually, the reaching of a flexible and balanced yet sufficiently consequential outcome.

More broadly, after 15 years, at least some are calling for a re-thinking of the IGC's objectives and expected outcomes. An envisagable outcome could take the form of a succinct and pragmatic text (or texts) on one or more themes of the IGC, with an emphasis on high-level principles and provisions suited to an international instrument. However, how and when should such a text/s be developed?

It is possible that an intergovernmental body such as the IGC cannot narrow the gaps much further. While the IGC's current texts are already a significant achievement, some consider that alternative or complementary routes are now needed to complete the work. Informal Member State-led initiatives to first develop texts outside the IGC are not yet delivering on the promise they once showed. One idea would be for the IGC Chair to be mandated to prepare a text for the Member States to consider, negotiate and finalize. As the critical 2017 GA approaches, there will certainly be other suggestions too.

There is still reason to be hopeful. The renewal of the mandate for 2016–2017 evidences that countries still believe, as they did in 2000, that the issues require discussion and resolution. Further, there has already been a significant technical and political investment by IGC participants, and there is a wealth of substantive materials and draft texts to draw on.

Despite a somewhat discouraging start at IGC 29, the new mandate may yet carve new pathways for the IGC and enable it to make genuine progress in 2016 and beyond.

Notes

1 The term initially used in the IGC was 'expressions of folklore'.
2 See generally 'The Negotiating Strategy of UNCLOS III: Developing and Developed Countries as Partners – A Pattern for Future Multilateral International Conferences?', Beesley A., *Law and Contemporary Problems*, Vol. 46, No. 2, 1983.
3 July 7 to 15, 2003.
4 IGC mandate for 2016–2017, available at www.wipo.int/tk/en/igc/ (accessed 19 December 2016) or www.wipo.int/export/sites/www/tk/en/igc/pdf/igc_mandate_1617.pdf (accessed 19 December 2016).
5 April 30 to May 3, 2001.
6 WIPO/GRTKF/IC/1/3.
7 WIPO/GRTKF/IC/1/5. The African Group (WIPO/GRTKF/IC/1/10) and Zambia (WIPO/GRTKF/IC/1/12) also made submissions.
8 WIPO/GRTKF/IC/1/10, para. 3.3.
9 See, for example, the statement by the USA at WIPO/GRTKF/IC/1/13, para. 49.
10 December 10 to 14, 2001.
11 June 13 to 21, 2002.
12 December 9 to 17, 2002.
13 See WIPO/GRTKF/IC/3/8 and WIPO/GRTKF/IC/4/8.
14 See WIPO/GRTKF/IC/5/6.
15 IPC/CE/30/11, paras 47 to 53 and IPC/CE/32/12, paras 83 to 91.
16 PCT/MIA/9/4 and PCT/MIA/7/5.
17 WIPO/GRTKF/IC/4/14 (Asian Group submission) and WIPO/GRTKF/IC/5/6.
18 WIPO/GRTKF/IC/5/5.
19 WIPO/GRTKF/IC/3/10 and WIPO/GRTKF/IC/5/3.
20 WIPO/GRTKF/IC/5/9 ; see also Chapter 6 by Robinson and Chiarolla.
21 WIPO/GRTKF/IC/5/10.
22 For a comprehensive overview of the work of the IGC during its first four sessions, see WIPO/GRTKF/IC/5/12, especially from paras 61 to 72.
23 The unit within the Secretariat that organizes and facilitates the IGC process and conducts the Secretariat's work on GRs, TK and TCEs is the Traditional Knowledge Division (TK Division), formerly the Global Issues Division.
24 December 9 to 17, 2002.
25 July 7 to 15, 2003.
26 The author pays warm tribute to two colleagues at the time, Tony Taubman and Shakeel Bhatti, who contributed enormously in the formative years of the IGC.
27 WIPO, *National Experiences with the Protection of Expressions of Folklore/Traditional Cultural Expressions: India, Indonesia and the Philippines*, written by Valsala Kutty (2002); WIPO, *Minding Culture: Case Studies on Intellectual Property and Traditional Cultural Expressions*, written by Terri Janke (2003); WIPO-UNEP, *The Role of Intellectual Property Rights in the Sharing of Benefits Arising from the Use of Biological Resources and Associated Traditional Knowledge*, written by Anil Gupta (2004).
28 For example, on TK, four main 'tasks' had initially been proposed at IGC 1; by IGC 3, each of these tasks comprised up to four or five 'activities' and, in respect of each activity, there were various proposed 'options'. See, for example, WIPO/GRTKF/IC/3/6.

29 Grain, *The Great Protection Racket: Imposing IPRS on Traditional Knowledge*, Seedling (January 20, 2004), page 3.
30 WIPO/GRTKF/IC/3/15, at page 5.
31 Because the IGC had been established in the middle of the 2000–2001 biennium, the first mandate of the IGC was for a three-year period. From then on, the IGC fell in step with the biennial cycles of the WIPO Program and Budget.
32 June 16 to 18, 2003.
33 WIPO/GRTKF/IC/5/14.
34 Internal notes on file with the author.
35 WIPO/GRTKF/IC/5/15, para. 123.
36 WIPO/GRTKF/IC/5/15, para. 125.
37 WIPO/GRTKF/IC/5/15, para. 138.
38 WIPO/GRTKF/IC/5/15, para. 131.
39 WIPO/GRTKF/IC/5/15, para. 155.
40 Dated July 10, July 12 and July 13, 2003, on file with the author.
41 WIPO/GRTKF/IC/5/15, paras 174 to 176. See also Grain, *op. cit.*
42 September 10 to 14, 2003.
43 Notes on file with the author.
44 'Cancun casts shadow over WIPO Assemblies', *Bridges,* Vol. 3, No. 17 (October 3, 2003).
45 WO/GA/30/8, para. 96.
46 Lettington, R., and Nnadozie, K., *A review of the Intergovernmental Committee on Genetic Resources, Traditional Knowledge and Folklore at WIPO*, T.R.A.D.E. Occasional Papers 12, South Centre, pages 21–22 (December 2003).
47 The African Group submitted a document to IGC 6 on objectives, principles and elements of an international instrument/s (WIPO/GRTKF/IC/6/12). However, it did not constitute a fully developed draft instrument/s.
48 March 15 to 19, 2004.
49 November 1 to 5, 2004.
50 June 6 to 10, 2005.
51 For TCEs, see WIPO/GRTKF/IC/8/4 and for TK, see WIPO/GRTKF/IC/8/5.
52 WIPO/GRTKF/IC/8/15 Prov. 2, paras 21 and 93; WIPO/GRTKF/IC/8/15 Prov. 2, paras 24 and 88; WIPO/GRTKF/IC/8/15 Prov. 2, para. 89; WIPO/GRTKF/IC/8/15 Prov. 2, para. 86; WIPO/GRTKF/IC/8/15 Prov. 2, para. 87.
53 WIPO/GRTKF/IC/6/6.
54 WIPO/GRTKF/IC/8/11.
55 April 24 to 28, 2006.
56 WIPO/GRTKF/IC/9/12. Notes on the Norway meeting on file with the author.
57 April 24 to 28, 2006.
58 November 30 to December 8, 2006.
59 July 3 to 12, 2007.
60 February 25 to 29, 2008.
61 WIPO/GRTKF/IC/12/5(a).
62 October 13 to 17, 2008. The analyses were in WIPO/GRTKF/IC/13/4(b) Rev (TCEs) and 13/5(b) Rev (TK).
63 June 18 to June 20, 2007.
64 WIPO/GRTKF/IC/11/12 (submission by Indonesia).
65 WIPO/GRTKF/IC/12/INF/6.
66 July 3 to 12, 2007.
67 February 25 to 29, 2008.
68 October 13 to 17, 2008.
69 WIPO/GRTKF/IC/13/10.
70 WIPO/GRTKF/IC/14/INF/6.
71 WIPO/GRTKF/IC/14/8 Rev.

72 WIPO/GRTKF/IC/14/11.
73 WIPO Program and Budget 1998 and 1999, page viii.
74 WIPO/GRTKF/IC/14/12. For a detailed account of IGC 14, see 'WIPO stalemate: No agreement reached on genetic resources, traditional knowledge and folklore', *Knowledge Ecology International*, available online at http://keionline.org/node/235 (dated July 7, 2009) (accessed 19 December 2016).
75 WO/GA/38/17.
76 WO/GA/38/18.
77 WO/GA/38/19.
78 On file with the author.
79 WO/GA/38/20, para. 231.
80 WO/GA/38/20, para. 225.
81 See also *'Turning Point' at WIPO Pulls Traditional Knowledge Debate Out at Eleventh Hour*, IP Watch, October 3, 2009 and *Genetic Resources and Traditional Knowledge: Can the New Mandate Deliver on Biopiracy*, Musungo, S., Iqsensato, October 15, 2009.
82 WO/GA/38/20, paras 226 and 232.
83 WO/GA/38/20, para. 221.
84 December 7 to 11, 2009. IGC 15 was chaired by Ambassador Juan José Ignacio Gomez Camacho (Mexico).
85 May 3 to 7, 2010.
86 WIPO/GRTKF/IC/16/8, para. 255.
87 Notes on file with the author.
88 Notes on file with the author.
89 These documents were published for IGC 17 as WIPO/GRTKF/IC/17/INF/8 and INF/9, respectively. Also a glossary was published as WIPO/GRTKF/IC/17/INF/13.
90 July 19 to 23, 2010.
91 February 21 to 25, 2011.
92 See, for example, *WIPO sees First Real Progress on Text for Protection of Folklore in 10 Years*, IP Watch (July 26, 2010).
93 February 28 to March 4, 2011.
94 December 6 to 10, 2010.
95 Internal notes on file with author.
96 May 9 to 13, 2011.
97 International Symposium on Intellectual Property and Traditional Knowledge, Traditional Cultural Expressions and Genetic Resources: Towards Sustainable Development for Indigenous Communities, October 31 to November 4, 2010, Saint Petersburg, Russian Federation.
98 WIPO/GRTKF/IC/17/INF/14.
99 July 18 to 22, 2011.
100 WIPO/GRTKF/IC/19/8 (submitted by Indonesia on behalf of the Like Minded Countries).
101 WIPO/GRTKF/IC/19/4, WIPO/GRTKF/IC/19/5 and WIPO/GRTKF/IC/19/6.
102 WIPO/GRTKF/IC/19/9, WIPO/GRTKF/IC/19/10 and WIPO/GRTKF/IC/19/11.
103 WIPO/GRTKF/IC/19/12, para. 10.
104 Such as WIPO/GRTKF/IC/20/INF/4.
105 Goss (Australia), Suchanandan (South Africa), Raina and Dhar (India).
106 April 16 to 20, 2012.
107 WIPO/GRTKF/IC/21/5.
108 Notes on file with the author.
109 Facilitators were Lesieur (Canada), Bonnet Lopez (Colombia) and Tala (Egypt).
110 WIPO/GRTKF/IC/21/7 Prov. 2, para. 273.
111 www.wipo.int/tk/en/ (accessed 19 December 2016).
112 www.wipo.int/tk/en/indigenous/ (accessed 19 December 2016).

113 WIPO/GRTKF/IC/25/INF/9.
114 July 9 to 13, 2012.
115 WIPO/GRTKF/IC/22/1 Prov. 4.
116 WIPO/GA/41/15.
117 WIPO/WO/GA/41/18.
118 July 15 to 24, 2013.
119 Such as *Joint Recommendation on Genetic Resources and Associated Traditional Knowledge* (WIPO/GRTKF/IC/23/5); *Proposal for the Terms of Reference for the Study by the WIPO Secretariat on Measures Related to the Avoidance of the Erroneous Grant of Patents and Compliance with Existing Access and Benefit-Sharing Systems* (WIPO/GRTKF/IC/23/6); and, *Joint Recommendation on the Use of Databases for the Defensive Protection of Genetic Resources and Traditional Knowledge Associated with Genetic Resources* (WIPO/GRTKF/IC/23/7). These documents were simply noted.
120 WIPO/WO/GA/43/14.
121 WIPO/GRTKF/IC/25/8 Prov. 2, para. 355.
122 WIPO/GRTKF/IC/25/8 Prov. 2, para. 267.
123 WIPO/GRTKF/IC/25/8 Prov. 2, para. 275.
124 WIPO/GRTKF/IC/25/8 Prov. 2, para. 423.
125 See, for example, statement by the USA – WO/GA/43/22, para. 182.
126 February 3 to 7, 2015.
127 Internal notes on file with the author.
128 Circular, dated January 22, 2014.
129 WIPO/GRTKF/IC/26/8 Prov. 2, para. 63.
130 March 24 to April 4, 2014.
131 Chair's non-paper on file with the author.
132 WIPO/GRTKF/IC/27/10, paras 69 to 77.
133 WIPO/GRTKF/IC/27/10, paras 69 to 77.
134 *Consultative Meeting on the Work of the Intergovernmental Committee ('IGC') on Intellectual Property and Genetic Resources, Traditional Knowledge and Folklore*, March 11 and 12, 2014.
135 July 7 to 9, 2014.
136 On file with the author.
137 Document WO/GA/46/6. Ambassador Kairamo (Finland) chaired the 2014 GA.
138 On file with the author.
139 WO/GA/46/12, para. 152.
140 WO/GA/46/12, para. 153.
141 WO/GA/46/12, para. 173.
142 WO/47/16, WO/GA/47/17 and WO/GA/47/18.
143 On file with the author.
144 See www.wipo.int/tk/en/#meetings (accessed 19 December 2016).
145 IGC 29 (February 15 to 19, 2016).
146 WIPO/GRTKF/IC/29/4.
147 Ian Goss (Australia) as Chair and Jukka Liedes (Finland) and Ambassador Michael Tene (Indonesia) as Vice-Chairs.
148 Margo Bagley (Mozambique) and Emelia Hernández Priego (Mexico).
149 Annexed to the decisions of the session, available at www.wipo.int/meetings/en/doc_details.jsp?doc_id=330356 (accessed 19 December 2016).

4 Relationships between the work of the WIPO IGC and other forums

*Charles Lawson**

Introduction

Traditional knowledge (TK), genetic resources (GRs) and traditional cultural expressions/expressions of folklore (TCEs) have been the subject of discussion at the World Intellectual Property Organization (WIPO) since October 2000.[1] The evolving concern about these matters is recognition that TK, GRs (and their associated TK) and TCEs (including TK) have developed increased economic, scientific, cultural and commercial values, and that intellectual property is increasingly relevant and important to their protection and exploitation.[2] WIPO's first tentative step in this area was to convene the Intergovernmental Committee on Intellectual Property and Genetic Resources, Traditional Knowledge and Folklore (IGC) in 2001.[3] Since then there have been deliberations over thirty-two sessions with the first twelve sessions until 2008 reaching no consensus about any of the identified issues.[4]

A critical dimension of the IGC's work was the early recognition that any developments in its text-based negotiations must be 'without prejudice to the work pursued in other fora'[5] and the 'international dimension' of the IGC's work.[6] As a consequence these other forums have been central to the IGC's negotiations.[7] In its early stages the IGC was uncertain about the kinds of outcome that were being negotiated in the IGC forum – a binding or non-binding normative international instrument (or instruments), 'authoritative or persuasive interpretations or elaborations of existing legal instruments', 'a high level political resolution, declaration or decision . . . stating a norm against misappropriation and misuse', 'guidelines or model laws', or coordinated 'national legislative developments'.[8] Since the 2011 renewed mandate from the WIPO General Assembly,[9] the fifteenth IGC sessions onwards have focused on text-based negotiations, and importantly, these negotiations have been influenced by developments in other forums. The purpose of this chapter is to review these other forums and their progress. The forums considered in turn are: the United Nations' *Convention on Biological Diversity* (CBD), the Food and Agriculture Organization of the United Nations' *International Treaty on Plant Genetic Resources for Food and Agriculture* (Plant Treaty), the World Trade Organization's *Agreement on Trade-related Aspects of Intellectual Property Rights* (TRIPS), the *United Nations Declaration on the Rights of Indigenous*

Peoples (UNDRIP), and the United Nations Educational, Scientific and Cultural Organization (UNESCO) agreements with a focus on the *Convention for the Safeguarding of the Intangible Cultural Heritage*. The chapter then concludes with some commentary about these developments. The conclusion is that these various forums have remained distinctly separate with limited acknowledgement in the submissions that similar, and perhaps often the same, proposals are being presented in other forums. Despite this, the CBD developments provide useful insights, the Plant Treaty is at an early stage and any insights may be limited, TRIPS appears to be awaiting the outcomes of the IGC, UNDRIP has provoked very new thinking, and UNESCO agreements have provided some useful direction. While these other forums and their dealings with aspects of the IGC developing GR, TK and TCE instruments are important, none of them provide a comprehensive solution or alternative to the IGC.

United Nations' *Convention on Biological Diversity* (CBD)

The Preamble to the CBD provides:

> *Conscious* of the intrinsic value of biological diversity and of the ecological, genetic, social, economic, scientific, educational, cultural, recreational and aesthetic values of biological diversity and its components . . .
>
> *Recognizing* the close and traditional dependence of many indigenous and local communities embodying traditional lifestyles on biological resources, and the desirability of sharing equitably benefits arising from the use of traditional knowledge, innovations and practices relevant to the conservation of biological diversity and the sustainable use of its components.

The CBD Article 8(j) then provides:

> Each Contracting Party shall, as far as possible and as appropriate . . .
>
> (j) Subject to its national legislation, respect, preserve and maintain knowledge, innovations and practices of indigenous and local communities embodying traditional lifestyles relevant for the conservation and sustainable use of biological diversity and promote their wider application with the approval and involvement of the holders of such knowledge, innovations and practices and encourage the equitable sharing of the benefits arising from the utilization of such knowledge, innovations and practices.

The first meeting of the Conference of the Parties to the CBD (COP) in 1994 scheduled the first consideration of implementing these obligations at the third COP.[10] And at the third COP in 1996 the matters were considered as components: 'questions related to the respect, preservation and maintenance of the knowledge, innovations and practices of indigenous and local communities', 'questions related to promoting the wider application of such knowledge, innovations and practices with the approval of the holders' and 'questions related to the encouragement of the equitable sharing of the benefits arising from their use'.[11] The outcome of discussions was to establish an inter-sessional process including a workshop and a report on the likely meanings of the CBD obligations.[12] The fourth COP in 1998 established an Ad Hoc Open-ended Inter-sessional Working Group to Address the Implementation of Article 8(j) and Related Provisions,[13] and the fifth COP in 2000 extended the mandate of the working group and set out a work programme, including:[14]

(a) Taking measures, mechanisms and initiatives to foster and promote indigenous and local community participation in decision making about TK, innovations and practices (Tasks 1, 2 and 4).[15]
(b) To prepare guidelines for mechanisms, legislation and other initiatives to ensure equitable benefit sharing, prior informed consent and the obligations of countries of origin of GRs (Task 7).[16]
(c) To develop guidelines or recommendations for the conduct of cultural, environmental and social impact assessments for sacred sites, land and waters occupied by indigenous and local communities (Task 9).[17]
(d) To develop standards and guidelines for the reporting and prevention of unlawful appropriation of TK and related GRs (Task 10).[18]
(e) To access existing laws, 'particularly' intellectual property instruments and their implications for TK, innovations and practices (Task 11).[19]
(f) To develop guidelines to assist in the development of legislation or other mechanisms, including *sui generis* systems, to implement the Article 8(j) obligations (Task 12).[20]
(g) To develop guidelines about the repatriation of information, including cultural property, to facilitate the recovery of TK of biological diversity (Task 15).[21]

The sixth COP in 2002 called for reporting on actions and developments implementing TK obligations,[22] continued the development of guidelines for the conduct of cultural, environmental and social impact assessments,[23] continued the development of participatory mechanisms for indigenous and local communities to participate,[24] and started the Ad Hoc Open-ended Inter-sessional Working Group to Address the Implementation of Article 8(j) and Related Provisions considering a *sui generis* system for protecting TK, in alignment with the WTO and WIPO, and taking into account national practices, case studies and other relevant experiences.[25] The sixth COP also engaged the Ad Hoc Open-ended Working Group on Access and Benefit-sharing to consider the disclosure, certificates of origin, customary law, and so on, dimensions to protecting knowledge, information and practices in access and benefit-sharing arrangements.[26] Perhaps most significantly the sixth COP adopted the voluntary *Bonn Guidelines on Access to Genetic Resources and Fair and Equitable Sharing of the Benefits Arising out of their Utilization.*[27] These guidelines were conceived to be useful 'as inputs when developing and drafting legislative, administrative or policy measures on access and benefit-sharing' as it related to Article 8(j).[28] Notably, the guidelines included a soft disclosure requirement:

> Contracting Parties with users of [GRs] under their jurisdiction should take appropriate legal, administrative, or policy measures, as appropriate, to support compliance with prior informed consent of the Contracting Party providing such resources and mutually agreed terms on which access was granted. These countries could consider, *inter alia*, the following measures . . . to encourage the disclosure of the country of origin of the [GRs] and of the origin of traditional knowledge, innovations and practices of indigenous and local communities in applications for intellectual property rights.[29]

The seventh COP in 2004 continued reporting on indigenous and local community knowledge, information and practices relevant to conservation and sustainable development, including the elements to develop a plan of action,[30] endorsed the *Akwé: Kon Guidelines* (fulfilling Task 9),[31] continued to develop mechanisms for indigenous and local communities to participate,[32] and continued to identify elements of *sui generis* systems for the protection of TK, innovations and practices.[33] The seventh COP also tasked the Ad Hoc Open-ended Working Group on Access and Benefit-sharing with developing an international regime on access and benefit-sharing within the framework of the CBD and including within its scope '[t]raditional knowledge, innovations and practices in accordance with Article 8(j)'.[34]

The eighth COP in 2006 continued reporting on indigenous and local community knowledge, information and practices relevant to conservation and sustainable development,[35] started to consider 'the possibility of developing technical guidelines for recording and documenting [TK], innovations and practices',[36] and sought collaboration between the Ad Hoc Open-ended Inter-sessional Working Group to Address the Implementation of Article 8(j) and Related Provisions and Ad Hoc Open-ended Working Group on Access and Benefit-sharing in the elaboration and negotiation of an international regime on access and benefit-sharing relevant to TK, innovations and practices associated with GRs.[37] And the Ad Hoc Open-ended Working Group on Access and Benefit-sharing continued elaborating and negotiating an international regime as a means of effectively implementing Article 8(j).[38]

The ninth COP in 2008 continued reporting on indigenous and local community knowledge, information and practices relevant to conservation and sustainable development,[39] considered a composite report on the status and trends regarding the knowledge, innovations and practices of indigenous and local communities, relevant to the conservation and sustainable use of biodiversity,[40] started developing guidelines for documenting TK,[41] developing a plan of action for the retention of TK, including measures and mechanisms to address the underlying causes for the decline of TK,[42] continued developing the elements of a *sui generis* system to protect TK, innovations and practices,[43] and continued to develop elements of an ethical code of conduct 'to ensure respect for the cultural and intellectual heritage of indigenous and local communities relevant for the conservation and sustainable use of biological diversity'.[44]

The tenth COP in 2010 adopted the *Nagoya Protocol on Access to Genetic Resources and the Fair and Equitable Sharing of Benefits Arising from their Utilization* (*Nagoya Protocol*).[45] The tenth COP also established the Open-ended Ad Hoc Intergovernmental Committee for the Nagoya Protocol on Access to Genetic Resources and the Fair and Equitable Sharing of Benefits Arising from their Utilization,[46] continued developing the elements of a *sui generis* system to protect TK, innovations and practices,[47] and adopted the voluntary *Tkarihwaié:ri Code of Ethical Conduct on Respect for the Cultural and Intellectual Heritage of Indigenous and Local Communities* (*Tkarihwaié:ri Code*).[48] Significantly the tenth COP revised the multi-year programme on the implementation of Article 8(j) and related provisions, recommitting to: fostering and promoting indigenous and local community participation in decision making about TK, innovations and practices (Tasks 1, 2 and 4);[49] preparing guidelines for initiatives to ensure equitable benefit sharing, prior informed consent and the obligations of countries of origin of GRs (Task 7);[50] developing standards and guidelines for the reporting and prevention of unlawful appropriation of TK and related GRs (Task 10);[51] developing *sui generis* systems to implement the Article 8(j) obligations (Task 12);[52] and developing guidelines about the repatriation of information, including cultural property, to facilitate the recovery of TK of biological diversity (Task 15).[53]

The eleventh and twelfth COPs in 2012 and 2014, respectively, continued these tasks: reporting on progress implementing the programme of work on Article 8(j) and related provisions,[54] developing mechanisms to promote the effective participation of indigenous and local communities in the work of the CBD;[55] developing guidelines for ensuring equitable benefit sharing, prior informed consent, the obligations of countries of origin of GRs, preventing unlawful appropriation and implementing Article 8(j);[56] developing best-practice guidelines for the repatriation of TK relevant to the conservation and sustainable use of biological diversity;[57] and developing *sui generis* systems for the protection, preservation and promotion of TK, innovations and practices.[58]

The major outcomes from the CBD's deliberations have been the *Nagoya Protocol* addressing TK in the context of GRs. The COPs consider and adopt protocols (Arts 23(4)(c) and 28(2)), and Contracting Parties to the CBD (Art 32(1)) become bound by a protocol when they become a Part to the protocol (Art 34(1)). The effect of the *Nagoya Protocol* is to formally recognise TKs associated with GRs and provide the machinery for dealing with these TKs. By way of summary, and using the clustering of issues that have crystallised at WIPO, the *Nagoya Protocol* addresses:

(a) *Defensive protection of GRs to prevent the grant of a patent where there is a misappropriation (or misuse) of GRs ('biopiracy')* – The *Nagoya Protocol* requires Parties to:

   (i) 'establish mechanisms' to inform potential users of TK about their obligations, including the Access and Benefit-sharing Clearing-House.[59]

(ii) 'endeavour to support' the development of 'community protocols' about TKs, minimum requirements for mutually agreed terms and model contract clauses.[60]
(iii) designate a 'national focal point' for information about accessing GRs and associated TK.[61]
(iv) establish an 'Access and Benefit-sharing Clearing-House' to provide information about relevant 'legislative, administrative and policy measures', 'permits or their equivalent' as evidence of 'prior informed consent' and 'mutually agreed terms', and information about 'model contract clauses', 'methods and tools' and 'codes of conduct and best practice'.[62]
(v) establish 'checkpoints' to 'monitor and to enhance transparency about the utilization of [TK]' such as collecting relevant information, requiring certificates of compliance, and so on.[63]
(vi) encourage the 'development, update and use of voluntary codes of conduct, guidelines and best practices and/or standards' for access and benefit-sharing GRs and associated TK.[64]
(vii) take measures to promote awareness of GRs and their associated TK, including education and training, information dissemination, and so on.[65]
(viii) cooperate in general 'capacity-building, capacity development and strengthening of human resources and institutional capacities' to implement the agreed obligations.[66]

(b) *Disclosure requirements in patent applications* – The *Nagoya Protocol* requires Parties to:

(i) 'take appropriate, effective and proportionate legislative, administrative or policy measures' to regulate meeting standards of 'prior informed consent' and 'mutually agreed terms',[67] including establish an 'Access and Benefit-sharing Clearing-House' to provide information about relevant 'legislative, administrative and policy measures', 'permits or their equivalent' as evidence of 'prior informed consent' and 'mutually agreed terms'.[68]
(ii) establish 'checkpoints' to 'monitor and to enhance transparency about the utilization of [TK]' such as collecting relevant information, requiring certificates of compliance, and so on.[69]

(c) *Mutually agreed terms for fair and equitable benefit-sharing* – The *Nagoya Protocol* requires Parties to:

(i) 'take legislative, administrative and policy measures, as appropriate' to share the benefits from exploiting GRs and their associated TK 'with indigenous and local communities holding such knowledge'.[70]
(ii) 'take measures, as appropriate' to ensure the 'prior informed consent' of 'indigenous and local communities' when granting access where the 'prior informed consent' requires 'legislative, administrative and

policy measures, as appropriate' that are certain, clear (including rules and appeals), transparent, fair, non-arbitrary, disclosed, determined by a 'competent national authority' in writing, and notify the Access and Benefit-sharing Clearing House.[71]

(iii) consider a need for a 'global multilateral benefit-sharing mechanism' to 'support the conservation of biological diversity and the sustainable use of its components globally'.[72]

(iv) 'endeavour to support' the development of 'community protocols' about TK, minimum requirements for mutually agreed terms and model contract clauses.[73]

(v) establish 'checkpoints' to 'monitor and to enhance transparency about the utilization of [TK]' such as collecting relevant information, requiring certificates of compliance, and so on.[74]

(vi) encourage the 'development, update and use of voluntary codes of conduct, guidelines and best practices and/or standards' for access and benefit-sharing GRs and associated TK.[75]

(vii) take measures to promote awareness of GRs and their associated TK, including education and training, information dissemination, and so on.[76]

(viii) cooperate in general 'capacity-building, capacity development and strengthening of human resources and institutional capacities' to implement the agreed obligations.[77]

The text of the *Nagoya Protocol* expressly recognises the inter-relationship between GRs and TK,[78] and applies to GRs covered by the CBD and TK associated with those GRs.[79] Any benefits (both monetary and non-monetary)[80] are then to be shared on 'mutually agreed terms' with the indigenous and local communities holding such GRs and TK.[81] Further, in accessing GRs, 'prior informed consent' or 'approval and involvement' of indigenous and local communities is required 'where they have the established right to grant access to such resources',[82] and in accessing TK associated with GRs, 'prior informed consent' or 'approval and involvement' of indigenous and local communities with 'mutually agreed terms' is required.[83]

The subject matter of the TK relevant to the *Nagoya Protocol* is TK associated with the GRs being 'indigenous and local communities' customary laws, community protocols and procedures'.[84] Some care, however, was taken not to restrict 'the customary use and exchange of GRs and associated TK within and amongst indigenous and local communities in accordance with the objectives of the [CBD]' (presumably being conservation and sustainable uses).[85] Then the *Nagoya Protocol* addresses enforcement mechanisms for TK associated with GRs, including 'situations of non-compliance' and 'co-operat[ion] in cases of alleged violation':[86]

> Each Party shall take appropriate, effective and proportionate legislative, administrative or policy measures, as appropriate, to provide that [TK]

> associated with [GRs] utilized within their jurisdiction has been accessed in accordance with prior informed consent or approval and involvement of indigenous and local communities and that mutually agreed terms have been established, as required by domestic access and benefit-sharing legislation or regulatory requirements of the other Party where such indigenous and local communities are located.[87]

Importantly, the *Nagoya Protocol* was intended to complement other international agreements addressing GRs and TK and will not apply where some other 'specialised international access and benefit-sharing instrument applies that is consistent with, and does not run counter to the objectives of the [CBD] and this [*Nagoya*] *Protocol*'.[88] This is likely to cover access and benefit-sharing regimes like those set out in the Plant Treaty and the Pandemic Influenza Preparedness Framework (PIP Framework).

Unlike the binding effect of the *Nagoya Protocol*, the other measures addressing TK in the CBD are voluntary. Thus, the *Akwé: Kon Guidelines*[89] set out a decision making framework, providing that decision makers and managers should:

> Take into account the traditional knowledge, innovations and practices of indigenous and local communities as part of environmental, social and cultural impact-assessment processes, with due regard to the ownership of and the need for the protection and safeguarding of traditional knowledge, innovations and practices.[90]

And the *Tkarihwaié:ri Code* aims to promote respect, preservation and the maintenance of TK, innovations and practices relevant for the conservation and the sustainable use of biodiversity.[91] The ethical principles are intended to:

> promote respect for the rights of indigenous and local communities to enjoy, protect, and pass on to future generations, their cultural and intellectual heritage, including traditional knowledge, innovation and practices relevant for the conservation and sustainable use of biodiversity and it is according to these principles that others should engage with indigenous and local communities.[92]

The future work of the Ad Hoc Open-ended Inter-sessional Working Group to Address the Implementation of Article 8(j) and Related Provisions is likely to prove significant. In particular, the guidelines for ensuring equitable benefit sharing, prior informed consent, the obligations of countries of origin of GRs, and preventing unlawful appropriation,[93] the best-practice guidelines for the repatriation of TK relevant to the conservation and sustainable use of biological diversity,[94] and the outcomes of inquiries into *sui generis* systems for the protection, preservation and promotion of TK, innovations and practices,[95] are all likely to be directly relevant to WIPO's deliberations. Perhaps surprisingly, the delegates have cited the learnings from the CBD and *Nagoya Protocol* as: preserve policy space for countries at the national level to devise implementing and agreed principles and standards consistent with their national laws;[96] avoiding disagreement over a definition of 'traditional' by having no definition;[97] avoiding disagreement over the use of the term 'peoples', especially in the phrase 'indigenous and local communities';[98] and useful language, particularly for the preamble.[99] Maybe importantly, there was also express recognition that not defining TK was 'instrumental' in reaching agreement.[100] For the IGC this might not be so easy. Under the CBD and *Nagoya Protocol*, the contract between the resource holder and the bioprospector requires mutually agreed terms and it seems likely the parties will be able to define and describe the TK for the purposes of their particular agreement. For the IGC, the definitions of TK and TCEs are likely to be more important as the instruments appear to be formulated as a series of rights attaching to a subject matter – in effect requiring that subject matter to be defined, even if broadly defined.[101] A solution might be to leave this to national laws.[102]

Food and Agriculture Organization of the United Nations' *International Treaty on Plant Genetic Resources for Food and Agriculture* (Plant Treaty)

The Plant Treaty imposes a Standard Material Transfer Agreement (SMTA) on specific food crops and forages in the public domain (the 'Multilateral System') (Arts 10 and 11). Within this scheme the Plant Treaty provides:

> The Contracting Parties recognize the enormous contribution that the local and indigenous communities and farmers of all regions of the world, particularly those in the centres of origin and crop diversity, have made and will continue to make for the conservation and development of plant [GRs] which constitute the basis of food and agriculture production throughout the world (Art. 9).

These are known as 'Farmers' Rights' with the Plant Treaty requiring Contracting Parties to take 'responsibility for realizing Farmers' Rights, as they relate to plant GRs for food and agriculture' with the obligation that '[i]n accordance with their needs and priorities, each Contracting Party should, as appropriate, and subject to its national legislation, take measures to protect and promote Farmers' Rights' (Art. 9.2). The suggested content of Farmers' Rights might include: the protection of TK relevant to plant GRs for food and agriculture; equitable benefit sharing from using plant GRs for food and agriculture; participation in decision making about the conservation and sustainable use of plant GRs for food and agriculture; and 'any rights that farmers have to save, use, exchange and sell farm-saved seed/propagating material, subject to national law and as appropriate' (Arts 9.2 and 9.3). Some Contracting Parties have adopted schemes implementing the Plant Treaty obligations, although the content of Farmers' Rights actually required by the Plant Treaty remains unresolved.[103]

The Plant Treaty's SMTA makes no explicit recognition of Farmers' Rights, albeit the 'Funding Strategy' that distributes the amounts from payments made from commercialised products,[104] does potentially address allocating amounts to farmers (particularly farmers in developing countries) (Arts 13.3–13.5).[105] The Governing Body has adopted a resolution about implementing the PGRFA Treaty's Farmer's Rights provision.[106] At this stage, Farmers' Rights have not been considered as an area of compliance under the PGRFA Treaty[107] with the existing commitments merely being a 'pledge . . . to enhance the implementation of the [Plant] Treaty',[108] the collecting together of views from Contracting Parties and interested agencies (such as WIPO) about implementing Farmers' Rights, and the development of action plans to implement these commitments.[109]

Farmers' Rights have, however, gained increased importance in the context of intellectual property with the decision of the Consultative Group of International Agricultural Research's Centres (CGIAR Centres) to implement its *CGIAR Principles on the Management of Intellectual Assets*[110] that includes a consideration of these rights articulated in the *Implementation Guidelines for the CGIAR IA Principles on the Management of Intellectual Assets*.[111] These implementation guidelines require CGIAR Centres and those receiving funding from donors through the CGIAR Consortium and CGIAR Centres to comply with applicable national laws protecting and promoting Farmers' Rights in countries where they are working (such as access and benefit-sharing laws, plant variety protection laws, national seed laws, laws about land, resource and cultural rights of indigenous and or local peoples, national constitution, and so on), and CGIAR Centres 'should, where possible, seek to work in ways that promote' Farmers' Rights as they are articulated in the Plant Treaty.[112] The implementation guidelines then provide a list of 'practical actions' CGIAR Centres can take to satisfy these obligations, such as '[e]nsuring that research results . . . are shared with farmers from whom PGRFA or associated information was accessed'.[113] To date the experience with TK is limited, with only the International Potato Centre reporting experiences with TK and Farmers' Rights.[114]

World Trade Organization's *Agreement on Trade-related Aspects of Intellectual Property Rights* (TRIPS)

Built into the World Trade Organization's (WTO) *Agreement on Trade Related Aspects of Intellectual Property Rights* (TRIPs) was a requirement to review the exclusion of plants and animals (except 'micro-organisms') from patentability (Art. 27.3(b)). This review crystallised, in part, into an examination of the relationship between the CBD and TRIPS about the conservation and sustainable use of genetic material, and the relationship of TRIPS with concepts of TK and Farmers' Rights.[115] The Doha Ministerial Declaration provided further direction,[116] both in identifying the outstanding implementation-related issues (Paragraph 12),[117] and the TRIPS/CBD interactions and the protections of TK and folklore (Paragraph 19).[118] The approach of the TRIPS Council has been to identify the elements as separate agenda items and then often consider them together: review of the provisions of Article 27.3(b); relationship between the TRIPS agreement and the CBD; and protection of TK and folklore.[119] While there appears to have been a crystallisation of some of these issues there remains no agreement.[120] There have been many, many papers submitted to the TRIPS Council about TK.[121] These may be summarised as including:

(a) *International protection of TK is required* – That protection is required to enable indigenous peoples and communities to legitimately control their TK and derive the benefits from using their TK.[122] This is asserted to be necessary because of: a common economic interest in promoting the uses of TK, equity in sharing the benefits from exploiting TK, proper recognition of the contribution of farmers in conserving and using biological diversity (especially plant varieties) and promoting food security, sustaining cultures, recognising the link between the protection of TK and protection of the environment, promoting and fulfilling development objectives, conformity between TRIPS and other international agreements about TK, and protecting the trans-boundary uses of TK without permission (biopiracy).[123]
(b) *Domestic TK schemes may be sufficient* – That international protection may not be appropriate as domestic schemes are probably more effective because they are already in force and are well understood, there is no evidence that domestic schemes are not sufficient, sharing domestic experiences is necessary to determine areas of inadequacy, national systems can be adequately internationalised, and international systems need to be supported by widely implemented domestic schemes.[124]
(c) *Appropriate forums to address these matters* – That WIPO is the appropriate forum for developing an international scheme,[125] although this is not universally accepted with some considering discussions should continue in many forums in parallel.[126]
(d) *Patents grants covering TK* – That patents are being granted on TK that originates or is controlled by others, is already in the public domain or used without appropriate consents.[127] This is argued to be a result of improperly applying the patent threshold standards of prior art,[128] and that this might be resolved through databases of TK[129] and disclosure in patent applications.[130]
(e) *Access and benefit sharing* – That accessing and using TK without authorisation can be addressed through making the existing intellectual property schemes protect TK,[131] contracts with the appropriate TK holders,[132] disclosure requirements in patent applications,[133] or an appropriate *sui generis* scheme.[134]

The most significant recent development has been the advanced proposal to introduce an obligation to disclose the origin of GRs and TK in patent applications.[135] This proposal is to amend TRIPS with a new Article 29*bis* that provides, in part:

> Where the subject matter of a patent application concerns, is derived from or developed with biological resources and/or associated traditional knowledge, Members shall require applicants to disclose the country providing the

> resources and/or associated traditional knowledge, from whom in the providing country they were obtained, and, as known after reasonable inquiry, the country of origin. Members shall also require that applicants provide information including evidence of compliance with the applicable legal requirements in the providing country for prior informed consent for access and fair and equitable benefit-sharing arising from the commercial or other utilization of such resources and/or associated traditional knowledge.[136]

The nature of the issues in the TRIPS Council appears to be to undertake text-based negotiations 'to include a mandatory requirement for the disclosure of the country providing/source of [GRs], and/or associated [TK] . . . in patent applications'.[137] While these matters remain under ongoing discussions as a regular agenda item before the TRIPS Council these discussions do not appear to be advancing rapidly,[138] although there remains an intention to facilitate a process to reaching a decision and (eventually) a definite solution.[139]

United Nations *Declaration on the Rights of Indigenous Peoples* (UNDRIP)

The United Nations *Declaration of the Rights of Indigenous Peoples* (UNDRIP) expressly provides that 'States shall take effective measures to recognize and protect the exercise' of the following 'rights' (Art. 31.2), noting that this 'right' extends to 'the right to practice and revitalize their cultural traditions and customs' and the provision of 'redress through effective mechanisms' (Art. 11):

> Indigenous peoples have the right to maintain, control, protect and develop their cultural heritage, traditional knowledge and traditional cultural expressions, as well as the manifestations of their sciences, technologies and cultures, including human and [GRs], seeds, medicines, knowledge of the properties of fauna and flora, oral traditions, literatures, designs, sports and traditional games and visual and performing arts. They also have the right to maintain, control, protect and develop their intellectual property over such cultural heritage, [TK], and traditional cultural expressions (Art. 31.1).

The UNDRIP reflects a broader engagement of indigenous peoples in the work of the United Nations that has achieved: 'a new awareness of indigenous peoples' concerns and human rights', 'recognition of indigenous peoples' invaluable contribution to humanity's cultural diversity and heritage, not least through their [TK]', and 'an awareness of the need to address the issues of indigenous peoples through policies, legislation and budgets'.[140] Importantly, however, UNDRIP is not considered to establish new rights, but instead, 'elaborates upon [existing] fundamental rights in the specific cultural, historical, social and economic circumstances of indigenous peoples'.[141] Within these broader debates about UNDRIP and its implementation, TK is subsumed in the broader analyses of social, economic, and cultural rights even though TK remains under active discussions.[142] Thus, UNDRIP is considered from the perspective of 'indigenous peoples' right to development in accordance with their own aspirations and needs',[143] and the array of human rights and other issues that go to these social, economic, and cultural rights, of which TK is *only* one of many.

This perspective is reflected in the IGC, where indigenous people and local community representation has been encouraged through opportunities to participate in the IGC (including a voluntary fund to promote and enable these opportunities),[144] with the result that these groups have taken up the IGC as another forum to address their broader concerns about self-determination, human rights, customary law and land rights (see Chapter 10 by Tobin).[145] The following quote illustrates the founding of those broader perspectives, including the effect of UNDRIP, on the IGC's deliberations:

> The representative of Tupaj Amaru, reporting on the challenges faced by Indigenous peoples with regard to fair and equitable recognition of their cultural heritage by the international community, recalled the mandate given to the IGC. He believed there had been a lack of tangible progress since its establishment over 10 years ago owing to the lack of political will of Member States and the different economic and geopolitical interests on the part of the developed countries from the North. Despite years of general discussions, informal consultations, and the creation of [inter-sessional working groups], the Member States, insensitive to plundering Indigenous peoples' spiritual and material wealth, undermined the legal essence of and limited the scope of the draft instruments under negotiation, on the pretext of 'simplifying' the texts. The process suffered from procedural failures in form and substance. Representatives of Indigenous peoples were relegated to being observers, and could not participate in the deliberations on an equal footing with Member States. As such, proposals by Indigenous peoples could not be taken into consideration, without the explicit agreement or support of one or more Member States. He stated that this was not what had happened in the negotiating process that had led to ILO Convention 169 or the UNDRIP, and it was, therefore, in his view contrary to the spirit of the *Vienna Convention on the Law of Treaties*. He explained the urgent need for an international legal framework with a coherent universal definition, acceptable to all, and in line with other international instruments, providing legal protection in particular to intangible TK and the sacred and secret knowledge of Indigenous peoples and communities. At a time when bio-prospecting and bio-piracy had reached dramatic proportions, it was disappointing to observe the systematic opposition by developed countries to the harmonization and adoption of standards for the protection of TK. Despite the fact that more than 90 per cent of the patent applicants originated from the developed countries, double standards in policies were evident by the fact that IP law was a Western concept, which, by its very nature and scope, did not recognize the owners of TK as legal subjects, nor was it designed to protect the cultural heritage of ancestral civilizations. According to the report of the CBD concerning access to benefits stemming

> from the use and misuse of the indigenous cultural heritage, the following 10 huge multinationals were plundering TK, GRs and TCEs: Pfizer (USA), GlaxoSmithKline (UK), Novartis (Switzerland), Sanofi-Aventis (France), Johnson & Johnson (USA), AstraZeneca (UK), Merck & Co (USA), Roche (Switzerland), Abbott (USA) and Amgen (USA). As a result, holders' natural resources and knowledge had been robbed, without prior consent, of their living knowledge and traditions, and were deprived of the right to fair sharing of the benefits derived from the use and misuse of their knowledge.[146]

The issue of indigenous representation, however, remains vexed. Despite extending procedures to indigenous groups[147] and funding their participation,[148] they are not able to submit proposals, amendments and motions, and may only speak at the invitation of the chair.[149] The effect is that indigenous peoples have been able to express their opinions without actually bringing them into the formal negotiations and decisions. Unhappy with the position that indigenous groupings did not have the same equal status as WIPO Member States, they withdrew – declaring they would not return until they had 'full and equitable participation at all levels of the IGC and until the instruments recognize and are consistent with the existing international frameworks for the rights of Indigenous Peoples within the scope of the IGC'.[150] After a short standoff the indigenous groups have returned with the IGC committing to take note of their interventions and perhaps address them at some time in the future[151] and made no changes to 'full and equitable participation' although some minor practical, procedural and budgetary measures have been made.[152] The concern remains that indigenous peoples are being 'relegated to the periphery of discussions while their substantive rights were being threatened'.[153]

The way that UNDRIP frames intellectual property as 'to recognize and protect the exercise' (Art. 31.2) the 'right to maintain, control, protect and develop their intellectual property over such cultural heritage, traditional knowledge, and traditional cultural expressions' (Art. 31.1) incorporates both the formal recognition of GRs, TK and TCEs as intellectual property and the protections against intellectual property claims by others misappropriating GRs, TK and TCEs. There already exist a number of proposed measures addressing misappropriation, including databanks, registries, catalogues, and the like to confound the novelty and inventive step claims for patent applications about GRs, TK and TCEs,[154] and disclosure requirements about GRs, TK and TCEs in patent applications, such as the proposed TRIPS, Art. 29*bis*.[155]

United Nations Educational, Scientific and Cultural Organization

There are UNESCO agreements, such as the *Universal Declaration on Cultural Diversity* (2001), *Convention for the Safeguarding of the Intangible Cultural Heritage* (2003), and *Convention on the Protection and Promotion of the Diversity of Cultural Expressions* (2005) that also affect the likely form and substance of measures to protect TK and TCEs (see also Chapter 8 by Ruiz Muller on TK and

Chapter 9 by Perlman on TCEs). The agreements build on earlier agreements regulating and protecting tangible cultural and natural heritage, such as the *Convention for the Protection of Cultural Properties in the Event of Armed Conflict* (1954) and its two protocols, the *Convention on the Means of Prohibiting and Preventing the Illicit Import, Export and Transfer of Ownership of Cultural Property* (1970), the *Convention on the Protection of the World Cultural and Natural Heritage* (1972) and the *Convention on the Protection of Underwater Cultural Heritage* (2001). The sum of these UNESCO agreements is a scheme for the protection of heritage and support for cultural diversity, although they do not specifically address the nexus between intellectual property and cultural expressions.

UNESCO and WIPO have addressed intellectual property and cultural expressions through a long history of mutually addressing folklore as intellectual property and intangible cultural heritage. WIPO and UNESCO in 1985 together developed the *Model Provisions for National Laws on the Protection of Expressions of Folklore Against Illicit Exploitation and Other Prejudicial Actions*[156] and UNESCO in 1989 adopted the *Recommendation on the Safeguarding of Traditional Culture and Folklore*.[157] Both these measures, however, relied on national laws and guidance for their effective implementation, and this proved a weakness that continues to be agitated through further developments including the IGC's current work at WIPO. Most significantly, the UNESCO and WIPO's model provisions have not yet been adopted by any state.

Of particular relevance to the IGC is UNESCO's *Convention for the Safeguarding of the Intangible Cultural Heritage* that was UNESCO's response to the concern that earlier approaches were not effective. This convention defines 'intangible cultural heritage' as 'the practices, representations, expressions, knowledge, skills – as well as the instruments, objects, artefacts and cultural spaces associated therewith – that communities, groups and, in some cases, individuals recognize as part of their cultural heritage' (Art. 2.1) that is 'manifested' in:

(a) oral traditions and expressions, including language as a vehicle of the intangible cultural heritage;
(b) performing arts;
(c) social practices, rituals and festive events;
(d) knowledge and practices concerning nature and the universe;
(e) traditional craftsmanship (Art. 2.2(d)).

The convention's objective is, among other things, to ensure respect, safeguard and raise awareness about intangible cultural heritage (Art. 1). The convention required that each State Party shall 'take the necessary measures to ensure the safeguarding of the intangible cultural heritage present in its territory' and 'identify and define the various elements of the intangible cultural heritage present in its territory, with the participation of communities, groups and relevant non-governmental organizations' (Art. 11). This is assisted through the Intergovernmental Committee for the Safeguarding of the Intangible Cultural Heritage that promotes the objectives, encourages and monitors the implementation, and lists and reports on programmes, projects and activities for the safeguarding of intangible cultural heritage under the convention (Art. 7). The main focus of UNESCO has been on developing the Register of Best Safeguarding Practices and, in recent time, this has been expanded to preparing materials on guidance on best practices and making recommendations on measures for the safeguarding of the intangible cultural heritage.[158] The intention, however, is that 'UNESCO's cooperation with WIPO [be] strengthened to ensure ongoing exchange and learning between the two organizations and their Member States concerning traditional knowledge and cultural expressions'.[159] And more recently UNESCO and WIPO have collaborated to hold workshops, prepare materials and participate in seminars about GRs, TK and TCEs.[160]

In the IGC the focus has been on developing an instrument for the protection of GRs, TK and TCEs[161] while recognising that UNESCO's agreements do already address some of the relevant issues that WIPO may not need to address.[162] The UNESCO agreements are also proving useful for defining terms that might be relevant for any WIPO instruments addressing GRs, TK and TCEs.[163]

Discussion and conclusions

The IGC was an outcome of the Diplomatic Conference for the Adoption of the Patent Law Treaty in 2000, formally bringing together various discussions within WIPO about GRs, TK and TCEs.[164] These include the following significant milestones:

(a) The 1992 United Nations' *Convention on Biological Diversity* (CBD) expressly recognising access and equitable benefit sharing from using GRs in the context of biodiversity conservation (Art. 15) and an obligation to 'respect, preserve and maintain knowledge, innovations and practices of indigenous and local communities embodying traditional lifestyles relevant for the conservation and sustainable use of biological diversity' (Art. 8(j)).

(b) The 1994 United Nations General Assembly proclamation of the *International Decade of the World's Indigenous Peoples*[165] and the responding activities by WIPO[166] including establishing the Global Intellectual Property Issues Division (GIPID; now known as the Traditional Knowledge Division) within WIPO,[167] and the Policy Advisory Commission (PAC) as an external and independent advisory body to 'identify and comprehensively review issues that may need to be dealt with by WIPO, or that may have bearing on its operations, and formulate recommendations for policy development and enhancement of WIPO's operations'.[168] The GIPID conducted a fact-finding mission and prepared a report on the breadth and scope of the GRs, TK and TCEs and intellectual property[169] and the PAC elaborated and adopted the *World Intellectual Property Declaration* that expressly recognised the importance of traditional knowledge.[170]

The IGC was to address 'access to [GRs] and benefit sharing', 'protection of [TK], whether or not associated with those resources' and 'the protection of expressions of folklore' as they cut across the existing WIPO forums for addressing intellectual property issues and because 'the three themes are closely interrelated, and none can be addressed effectively without considering aspects of the others'.[171] Perhaps significantly, the IGC was addressing an issue that was of special importance to developing country members,[172] and WIPO (probably) needed to engage these members after its conclusion of the pro-developed nation *Patent Law Treaty*[173] and the then recent commitment of the international community to the United Nations *Millennium Declaration*, and in particular, the commitments to development and poverty eradication.[174]

From these origins the IGC has essentially become a forum for trying to find some coherence in international norms between intellectual property and ABS arrangements, with the apparent driving imperative being to negotiate an agreement to impose ABS objectives, principles and obligations onto intellectual property schemes. And there also remained the ongoing conflict between dealing with TK and TCEs as cultural heritage or intellectual property,[175] most evident in the failure of the UNESCO and WIPO *Model Provisions for National Laws on the Protection of Expressions of Folklore Against Illicit Exploitation and Other Prejudicial Actions* (1985)[176] and then the adoption by UNESCO of the *Convention for the Safeguarding of the Intangible Cultural Heritage* without addressing intellectual property. Despite the intensive negotiations in the IGC since 2000, the outcome of an agreed text and a Diplomatic Conference remain elusive. And importantly, the IGC's deliberations have been carried out in parallel with similar discussions and debates in other forums, and in some instances, overtaken by developments in these other forums, such as the conclusion of the *Nagoya Protocol* addressing TK associated with GRs. Interestingly these various forums have remained distinctly separate with limited acknowledgement in the submissions that similar, and perhaps often the same, proposals are being presented in other forums. The effect is that each forum presents as a distinct and coherent negotiation. Importantly, however, the outcomes in these different forums have distinctly different implications and impacts. So, for example, were TRIPS to be amended with the proposed Art. 29*bis* requiring the disclosure of GRs and TK in patent applications[177] that commitment would be backed by the WTO's dispute settlement arrangements with possible trade retaliatory action (TRIPS, Art. 64), while failure to implement the CBD and *Nagoya Protocol* national laws recognising TK associated with GRs at best attracts arbitration or submission to the International Court of Justice (CBD, Art. 27).

The analyses in this chapter considered the TK issues in the forums of the CBD, the Plant Treaty, TRIPS, UNDRIP and UNESCO. The analyses showed that the CBD, the Plant Treaty and TRIPS have grappled with the TK issues in a way that directly impacts the kinds of considerations actively being considered by the IGC, and the substance of any instruments articulating rights and obligations about TK. The CBD processes provided a comprehensive negotiation of the GRs and associated TK issues with a series of evolving agreements, including the *Akwé:*

Kon Guidelines setting out guidance about how to take into account TK, innovations and practices as part of the impact-assessment processes[178] and the *Tkarihwaié:ri Code* setting out aims to promote respect, preservation and the maintenance of TK, innovations and practices relevant for the conservation and the sustainable use of biodiversity,[179] and culminating in the *Nagoya Protocol*.

The major advance of the *Nagoya Protocol* was to identify the subject matter of the TK associated with the GRs as being the 'indigenous and local communities' customary laws, community protocols and procedures', and establish an enforcement mechanism for TK associated with GRs, including 'situations of non-compliance' and 'co-operat[ion] in cases of alleged violation'.[180] The *Nagoya Protocol* also establishes a Clearing House mechanism and checkpoints[181] that provide a means of both tracking and enforcing any obligations about TK interests associated with GRs. The future work of the Ad Hoc Open-ended Inter-sessional Working Group to Address the Implementation of Article 8(j) and Related Provisions is also likely to provide valuable insights for the WIPO IGC through the development of best practice guidelines,[182] and a likely *sui generis* system for the protection, preservation and promotion of TK, innovations and practices.[183]

The challenge faced by the IGC, however, is likely to be more complex. Under the CBD and *Nagoya Protocol* arrangements the GRs are accessed according to an agreement (a contract) with the prior informed consent and mutually agreed terms of the resource holder (CBD, Arts 15(4) and (5)).[184] This process enables the holders of the GRs, and including the indigenous and local communities where national laws make this provision, to set the terms and conditions of access that will deal with the TK (and TCEs) associated with the GRs as it is defined and described by the resource holder. In short, the resource holder takes responsibility for describing and defining the TK (and TCEs) associated with the GR according to what they identify is TK (and TCEs) and how they want to deal with them. For the IGC, however, it needs to deal with TK and TCEs more broadly and find a suitable definition of TK and TCEs that is sufficient to clarify the subject matter of protectable TK and TCEs, and there are many and variable subject matters.[185] So far, settling a definition has been highly contested.[186]

The developments under the Plant Treaty are likely to be insightful as the CGIAR Centres grapple with implementing their commitments to Farmers' Rights.[187] With the major focus of the Plant Treaty presently being *ex situ* germplasm collections, and having a limited scope to dealing with facilitated access for 'the purpose of utilization and conservation for research, breeding and training for food and agriculture, provided that such purpose does not include chemical, pharmaceutical and/or other non-food/feed industrial uses' (Art. 12(3)(a)), the TK issues are likely to be confined. When the Plant Treaty starts dealing more broadly with *in situ* germplasm there will likely be some very insightful developments as those responsible for the *in situ* sites will likely be the TK holders. Meanwhile the developments under TRIPS are more muted and almost certainly awaiting further guidance from WIPO about how TRIPS might be amended or implemented in ways that address TK, primarily through applying the existing standards taking account of TK and adequate TK disclosures in patent applications.[188]

In contrast to the developments at the CBD, Plant Treaty and TRIPS, however, UNDRIP poses a different kind of challenge asserting the rights of indigenous peoples (being their existing recognised rights in other international agreements and instruments rather than new rights) that may not be compatible with intellectual property as conceived by WIPO. The following quote captures the conception of TK as something beyond intellectual property:

> The [IGC] has played a leading role in the push for the recognition of indigenous traditional knowledge and its protection from misuse and misappropriation. However, the pre-eminent role of WIPO has meant that the international debate has occurred primarily within the parameters of intellectual property law. International intellectual property law provides protection for creators of certain works, whether in the areas of literature, music, dance or art. Although this does at times suffice, for the most part intellectual property law fails to protect indigenous rights and interests because western constructs of intellectual property focus on individual knowledge and creativity rather than on communal trans-generational knowledge. Attempting to alter intellectual property law so that it accommodates traditional knowledge, knowledge that is completely different in essence, is reminiscent of the proverb, 'You can't fit a round peg in a square hole'. No matter how one tries, it just does not fit. It is for this reason that a completely new and customized approach is needed.[189]

How this will be resolved by the IGC remains unclear, even though the elements of broader concerns about self-determination, human rights, customary law and land rights are keenly pursued as other contributions to this collection demonstrate. This is 'potentially confronting, and certainly provokes new ways of thinking about old subjects'.[190] And most importantly, these are issues that have predominantly not been canvassed in other forums because the UNDRIP has only recently entered into the discussions and those other forums essentially pre-date the UNDRIP crystallising these issues.

Notes

* This project has been generously supported by the Australian Research Council (DP120101434).

1 WO/GA/26/10 (2000) World Intellectual Property Organization General Assembly, *Twenty-Sixth WIPO General Assembly: Report* (WIPO, Geneva), [71]; WO/GA/26/6 (2000) World Intellectual Property Organization General Assembly, *Matters Concerning Intellectual Property and Genetic Resources, Traditional Knowledge and Folklore* (WIPO, Geneva), [4]-[19].

2 WO/GA/26/10, above n 1, [28]-[70]; WO/GA/26/9 (2000) World Intellectual Property Organization General Assembly, *Traditional Knowledge and the Need to Give it Adequate Intellectual Property Protection* (WIPO, Geneva), [3]-[4] and Annex I; WO/GA/26/6, above n 1, [1]-[2].

3 WIPO/GRTKF/IC/1/13 (2001) Intergovernmental Committee on Intellectual Property and Genetic Resources, Traditional Knowledge and Folklore, *Report: First Session* (WIPO, Geneva), [1]; WO/GA/26/10, above n 1, [27]-[71].

4 WIPO/GRTKF/IC/13/11 (2008) Intergovernmental Committee on Intellectual Property and Genetic Resources, Traditional Knowledge and Folklore, *Report: Thirteenth Session* (WIPO, Geneva), [89]; WIPO/GRTKF/IC/13/6 (2008) Intergovernmental Committee on Intellectual Property and Genetic Resources, Traditional Knowledge and Folklore, *Summary of Options for the International Dimension of the Committee's Work* (WIPO, Geneva), [5]-[13].

5 WO/GA/38/20 (2009) World Intellectual Property Organization General Assembly, *Thirty-Eighth WIPO General Assembly: Report* (WIPO, Geneva), [217]; WO/GA/30/8 (2003) World Intellectual Property Organization General Assembly, *Thirtieth WIPO General Assembly: Report* (WIPO, Geneva), [93].

6 WO/GA/30/8, above n 5, [93].

7 For example the issues associated with GRs: WIPO/GRTKF/IC/13/8(b) (2008) Intergovernmental Committee on Intellectual Property and Genetic Resources, Traditional Knowledge and Folklore, *Genetic Resources: Factual Update of International Developments* (WIPO, Geneva); WIPO/GRTKF/IC/12/8(b) (2008) Intergovernmental Committee on Intellectual Property and Genetic Resources, Traditional Knowledge and Folklore, *Genetic Resources: Factual Update of International Developments* (WIPO, Geneva); WIPO/GRTKF/IC/11/8(b) (2007) Intergovernmental Committee on Intellectual Property and Genetic Resources, Traditional Knowledge and Folklore, *Genetic Resources: Factual Update of International Developments* (WIPO, Geneva); WIPO/GRTKF/IC/10/7 Prov. 2 (2006) Intergovernmental Committee on Intellectual Property and Genetic Resources, Traditional Knowledge and Folklore, *Revised Draft Report* (WIPO, Geneva), [255].

8 WIPO/GRTKF/IC/14/6 (2009) Intergovernmental Committee on Intellectual Property and Genetic Resources, Traditional Knowledge and Folklore, *Summary of Options for the International Dimension of the Committee's Work* (WIPO, Geneva), [4]-[13]; WIPO/GRTKF/IC/13/6, above n 4, [4]-[13]; WIPO/GRTKF/IC/12/6 (2007) Intergovernmental Committee on Intellectual Property and Genetic Resources, Traditional Knowledge and Folklore, *Summary of Options for the International Dimension of the Committee's Work* (WIPO, Geneva), [4]-[13]; WIPO/GRTKF/IC/11/6 (2007) Intergovernmental Committee on Intellectual Property and Genetic Resources, Traditional Knowledge and Folklore, *Summary of Options for the International Dimension of the Committee's Work* (WIPO, Geneva), [4]-[13]; WIPO/GRTKF/IC/10/6 (2006) Intergovernmental Committee on Intellectual Property and Genetic Resources, Traditional Knowledge and Folklore, *Options for Giving Effect to the International Dimension of the Committee's Work* (WIPO, Geneva), [5]-[13]; WIPO/GRTKF/IC/9/6 (2006) Intergovernmental Committee on Intellectual Property and Genetic Resources, Traditional Knowledge and Folklore, *Practical Means of Giving Effect to the International Dimension of the Committee's Work* (WIPO, Geneva), [7]-[14]; WIPO/GRTKF/IC/8/6 (2005) Intergovernmental Committee on Intellectual Property and Genetic Resources, Traditional Knowledge and Folklore, *Practical Means of Giving Effect to the International Dimension of the Committee's Work* (WIPO, Geneva), [29]; WIPO/GRTKF/IC/6/6 (2003) Intergovernmental Committee on Intellectual Property and Genetic Resources, Traditional Knowledge and Folklore, *Traditional Knowledge, Traditional Cultural Expressions and Genetic Resources: The International Dimension* (WIPO, Geneva), [34].

9 WO/GA/38/20, above n 5, [217].

10 UNEP/CBD/COP/1/17 (1995) Conference of the Parties to the Convention on Biological Diversity, *Report of the First Meeting of the Conference of the Parties to the Convention on Biological Diversity* (CBD, Montreal), p 64; UNEP/CBD/COP/2/19 (1995) Conference of the Parties to the Convention on Biological Diversity, *Report of the Second Meeting of the Conference of the Parties to the Convention on Biological Diversity* (CBD, Montreal), p 76.

11 UNEP/CBD/COP/3/38 (1997) Conference of the Parties to the Convention on Biological Diversity, *Report of the Third Meeting of the Conference of the Parties to the Convention on Biological Diversity* (CBD, Montreal), [79].
12 UNEP/CBD/COP/3/38, above n 11, pp 92–93.
13 UNEP/CBD/COP/4/27 (1998) Conference of the Parties to the Convention on Biological Diversity, *Report of the Fourth Meeting of the Conference of the Parties to the Convention on Biological Diversity* (CBD, Montreal), p 111.
14 UNEP/CBD/COP/5/23 (2000) Conference of the Parties to the Convention on Biological Diversity, *Report of the Fifth Meeting of the Conference of the Parties to the Convention on Biological Diversity* (CBD, Montreal), pp 140 and 143–146.
15 UNEP/CBD/COP/5/23, above n 14, pp 143–144.
16 UNEP/CBD/COP/5/23, above n 14, p 144.
17 UNEP/CBD/COP/5/23, above n 14, pp 144–145.
18 UNEP/CBD/COP/5/23, above n 14, p 146.
19 UNEP/CBD/COP/5/23, above n 14, p 145.
20 UNEP/CBD/COP/5/23, above n 14, p 145.
21 UNEP/CBD/COP/5/23, above n 14, p 146.
22 UNEP/CBD/COP/6/20 (2002) Conference of the Parties to the Convention on Biological Diversity, *Report of the Sixth Meeting of the Conference of the Parties to the Convention on Biological Diversity* (CBD, Montreal), pp 152–154.
23 UNEP/CBD/COP/6/20, above n 22, p 154.
24 UNEP/CBD/COP/6/20, above n 22, pp 155–156.
25 UNEP/CBD/COP/6/20, above n 22, pp 156–159.
26 UNEP/CBD/COP/6/20, above n 22, pp 174–175.
27 UNEP/CBD/COP/6/20, above n 22, pp 253–269.
28 UNEP/CBD/COP/6/20, above n 22, p 254.
29 UNEP/CBD/COP/6/20, above n 22, p 258.
30 UNEP/CBD/COP/7/21 (2004) Conference of the Parties to the Convention on Biological Diversity, *Report of the Seventh Meeting of the Conference of the Parties to the Convention on Biological Diversity* (CBD, Montreal), pp 255–260.
31 UNEP/CBD/COP/7/21, above n 30, pp 261–276.
32 UNEP/CBD/COP/7/21, above n 30, pp 276–279.
33 UNEP/CBD/COP/7/21, above n 30, pp 280–283.
34 UNEP/CBD/COP/7/21, above n 30, pp 300 and 301.
35 UNEP/CBD/COP/8/31 (2006) Conference of the Parties to the Convention on Biological Diversity, *Report of the Eighth Meeting of the Conference of the Parties to the Convention on Biological Diversity* (CBD, Montreal), p 139.
36 UNEP/CBD/COP/8/31, above n 35, p 140.
37 UNEP/CBD/COP/8/31, above n 35, p 142.
38 UNEP/CBD/COP/8/31, above n 35, p 130.
39 UNEP/CBD/COP/9/29 (2008) Conference of the Parties to the Convention on Biological Diversity, *Report of the Ninth Meeting of the Conference of the Parties to the Convention on Biological Diversity* (CBD, Montreal), pp 121–122.
40 UNEP/CBD/COP/9/29, above n 39, p 123.
41 UNEP/CBD/COP/9/29, above n 39, pp 124–125.
42 UNEP/CBD/COP/9/29, above n 39, p 125.
43 UNEP/CBD/COP/9/29, above n 39, pp 127–128.
44 UNEP/CBD/COP/9/29, above n 39, pp 128–137.
45 UNEP/CBD/COP/10/27 (2010) Conference of the Parties to the Convention on Biological Diversity, *Report of the Tenth Meeting of the Conference of the Parties to the Convention on Biological Diversity* (CBD, Montreal), pp 85–109.
46 UNEP/CBD/COP/10/27, above n 45, p 87.
47 UNEP/CBD/COP/10/27, above n 45, pp 309–310.
48 UNEP/CBD/COP/10/27, above n 45, pp 311–319.

49 UNEP/CBD/COP/10/27, above n 45, p 320; UNEP/CBD/COP/5/23, above n 14, pp 143–144.
50 UNEP/CBD/COP/10/27, above n 45, p 320; UNEP/CBD/COP/5/23, above n 14, p 144.
51 UNEP/CBD/COP/10/27, above n 45, p 320; UNEP/CBD/COP/5/23, above n 14, p 146.
52 UNEP/CBD/COP/10/27, above n 45, p 320; UNEP/CBD/COP/5/23, above n 14, p 145.
53 UNEP/CBD/COP/10/27, above n 45, pp 320–321; UNEP/CBD/COP/5/23, above n 14, p 146.
54 UNEP/CBD/COP/12/29 (2014) Conference of the Parties to the Convention on Biological Diversity, *Report of the Twelfth Meeting of the Conference of the Parties to the Convention on Biological Diversity* (CBD, Montreal), pp 77–79; UNEP/CBD/COP/11/35 (2012) Conference of the Parties to the Convention on Biological Diversity, *Report of the Eleventh Meeting of the Conference of the Parties to the Convention on Biological Diversity* (CBD, Montreal), pp 149–150.
55 UNEP/CBD/COP/12/29, above n 54, pp 77–79; UNEP/CBD/COP/11/35, above n 54, pp 150–153.
56 UNEP/CBD/COP/12/29, above n 54, pp 89–90; UNEP/CBD/COP/11/35, above n 54, pp 153–154.
57 UNEP/CBD/COP/12/29, above n 54, pp 87–88; UNEP/CBD/COP/11/35, above n 54, pp 154–156.
58 UNEP/CBD/COP/12/29, above n 54, pp 90–91; UNEP/CBD/COP/11/35, above n 54, pp 156–157.
59 UNEP/CBD/COP/10/27, above n 45, p 95.
60 UNEP/CBD/COP/10/27, above n 45, pp 95, 97 and 100.
61 UNEP/CBD/COP/10/27, above n 45, p 96.
62 UNEP/CBD/COP/10/27, above n 45, pp 96–97.
63 UNEP/CBD/COP/10/27, above n 45, pp 98–99.
64 UNEP/CBD/COP/10/27, above n 45, p 100.
65 UNEP/CBD/COP/10/27, above n 45, pp 100–101.
66 UNEP/CBD/COP/10/27, above n 45, pp 101–102.
67 UNEP/CBD/COP/10/27, above n 45, p 97.
68 UNEP/CBD/COP/10/27, above n 45, pp 96–97.
69 UNEP/CBD/COP/10/27, above n 45, pp 98–99.
70 UNEP/CBD/COP/10/27, above n 45, p 92.
71 UNEP/CBD/COP/10/27, above n 45, p 93.
72 UNEP/CBD/COP/10/27, above n 45, pp 94–95.
73 UNEP/CBD/COP/10/27, above n 45, p 95.
74 UNEP/CBD/COP/10/27, above n 45, p 96.
75 UNEP/CBD/COP/10/27, above n 45, p 100.
76 UNEP/CBD/COP/10/27, above n 45, pp 100–101.
77 UNEP/CBD/COP/10/27, above n 45, p 101.
78 UNEP/CBD/COP/10/27, above n 45, pp 89–90.
79 UNEP/CBD/COP/10/27, above n 45, p 91.
80 UNEP/CBD/COP/10/27, above n 45, p 92.
81 UNEP/CBD/COP/10/27, above n 45, p 92.
82 UNEP/CBD/COP/10/27, above n 45, p 93.
83 UNEP/CBD/COP/10/27, above n 45, p 94.
84 UNEP/CBD/COP/10/27, above n 45, p 95.
85 UNEP/CBD/COP/10/27, above n 45, p 95.
86 UNEP/CBD/COP/10/27, above n 45, p 98.
87 UNEP/CBD/COP/10/27, above n 45, p 98.
88 UNEP/CBD/COP/10/27, above n 45, p 92.

89 UNEP/CBD/COP/7/21, above n 30, pp 261–276.
90 UNEP/CBD/COP/7/21, above n 30, p 264.
91 UNEP/CBD/COP/10/27, above n 45, pp 313–314.
92 UNEP/CBD/COP/10/27, above n 45, p 314.
93 UNEP/CBD/COP/12/29, above n 54, pp 89–90; UNEP/CBD/COP/11/35, pp 153–154.
94 UNEP/CBD/COP/12/29, above n 54, pp 87–88; UNEP/CBD/COP/11/35, pp 154–156.
95 UNEP/CBD/COP/12/29, above n 54, pp 90–91; UNEP/CBD/COP/11/35, pp 156–157.
96 WIPO/GRTKF/IC/27/10 (2014) Intergovernmental Committee on Intellectual Property and Genetic Resources, Traditional Knowledge and Folklore, *Report* (WIPO, Geneva), p 18.
97 WIPO/GRTKF/IC/27/10, above n 96, p 24.
98 WIPO/GRTKF/IC/27/10, above n 96, pp 34, 48 and 49.
99 WIPO/GRTKF/IC/27/10, above n 96, p 79.
100 WIPO/GRTKF/IC/27/10, above n 96, p 24.
101 see WIPO/GRTKF/IC/27/10, above n 96, pp 13, 19, 21, 23, 26, 27 and so on.
102 WIPO/GRTKF/IC/27/10, above n 96, p 33.
103 See Charles Lawson, 'Implementing Farmers' Rights: Finding Meaning and Purpose for the International Treaty on Plant Genetic Resources for Food and Agriculture Commitments?' (2015) 37(7) *European Intellectual Property Review* pp 442–454.
104 IT/GB-5/13/Report (2013) Fifth Session of the Governing Body of the International Treaty on Plant Genetic Resources for Food and Agriculture, *Report of the Fifth Session of the Governing Body* (FAO, Rome), [25]-[26] and Appendix A.2; IT/GB-4/11/Report (2011) Fourth Session of the Governing Body of the International Treaty on Plant Genetic Resources for Food and Agriculture, *Report of the Fourth Session of the Governing Body* (FAO, Rome), [21]-[23] and Appendix A.3; IT/GB-3/09/Report (2009) Third Session of the Governing Body of the International Treaty on Plant Genetic Resources for Food and Agriculture, *Report of the Governing Body of the International Treaty on Plant Genetic Resources for Food and Agriculture* (FAO, Rome), [30] and Appendix A.3; IT/GB-2/07/Report (2007) Second Session of the Governing Body of the International Treaty on Plant Genetic Resources for Food and Agriculture, *Report of the Governing Body of the International Treaty on Plant Genetic Resources for Food and Agriculture* (FAO, Rome), [45] and Appendix D (Annexes 1–3) (Funding Strategy); IT/GB-1/06/Report (2006) First Session of the Governing Body of the International Treaty on Plant Genetic Resources for Food and Agriculture, *Report of the Governing Body of the International Treaty on Plant Genetic Resources for Food and Agriculture* (FAO, Rome), [12] and Appendix F (Funding Strategy) and Appendix G (Standard Material Transfer Agreement).
105 IT/GB-5/13/Report, above n 104, [25]-[26] and Appendix A.2; IT/GB-4/11/Report, above n 104, [21]-[23] and Appendix A.3; and so on.
106 IT/GB-5/13/Report, above n 104, [39] and Appendix A.8; IT/GB-4/11/Report, above n 104, [29] and Appendix A.6; IT/GB-3/09/Report, above n 104, [49] and Appendix A.6.
107 For examples IT/GB-5/13/18 (2013) Fifth Session of the Governing Body of the International Treaty on Plant Genetic Resources for Food and Agriculture, *Report of the Compliance Committee* (FAO, Rome); IT/GB-4/11/7; and so on.
108 IT/GB-5/13/Report, above n 104, [3] and Appendix D, Preamble [(9)] and [1(g)] (Muscat Declaration).
109 IT/GB-5/13/Report, above n 104, [39] and Appendix A.8; IT/GB-5/13/10 (2013) Fifth Session of the Governing Body of the International Treaty on Plant Genetic Resources for Food and Agriculture, *Implementation of Article 9, Farmers' Rights* (FAO, Rome).
110 Consultative Group on International Agricultural Research (2011) *CGIAR Principles on the Management of Intellectual Assets (approved by the Fund Council on 7 March 2012 and Consortium Board on 1 March 2012)* (CGIAR, Montpellier), Art. 3.2.

111 Consultative Group on International Agricultural Research (2013) *Implementation Guidelines for the CGIAR IA Principles on the Management of Intellectual Assets (approved by the Consortium Board on 14 June 2013)* (CGIAR, Montpellier), pp 1–3.
112 CGIAR 2013, above n 111, p 2.
113 CGIAR 2013, above n 111, p 2; Juanita Chaves Posada (2013) *Mechanisms by Which Centers of the CGIAR Consortium Can Support the Development of Appropriate Policies and Procedures for the Recognition and Promotion of Farmers' Rights* (Global Forum on Agricultural Research, Rome).
114 CGIAR Consortium Legal Counsel (2013) *CGIAR Intellectual Assets (IA) Report for 2012 – Annual Report from the CGIAR Consortium to the Fund Council on the implementation of the CGIAR Principles on the Management of Intellectual Assets* (CGIAR Consortium, Montpellier), p 9.
115 Justin Malbon, Charles Lawson and Mark Davison (2014) *The WTO Agreement on Trade-Related Aspects of Intellectual Property Rights: A Commentary* (Edward Elgar, 2014), pp 390–395 and 398–400.
116 See IP/C/M/36/Add.1 (2002) Council for Trade-Related Aspects of Intellectual Property Rights, *Minutes of Meeting Held in the Centre William Rappard on 25–27 June 2002 – Addendum* (WTO, Geneva), [194]-[206]; IP/C/M/35 (2002) Council for Trade-Related Aspects of Intellectual Property Rights, *Minutes of Meeting Held in the Centre William Rappard on 5–7 March 2002* (WTO, Geneva), [206]-[227].
117 WT/MIN(01)/DEC/1 (2001) Ministerial Conference (WTO), *Ministerial Declaration* (WTO, Geneva), [12]; WT/MIN(01)/17 (2001) Ministerial Conference (WTO), *Implementation-related Issues and Concerns* (WTO, Geneva), [13] (footnote); JOB (01)/152/Rev.1 (2001) WTO Secretariat, *Compilation of Outstanding Implementation Issues Raised by Members: Revision* (WTO, Geneva), [10] (Tirets 88, 91 and 93–95).
118 WT/MIN(01)/DEC/1, above n 117, [19]; WT/MIN(05)/DEC (2005) Ministerial Conference (WTO), *Ministerial Declaration: Doha Work Programme* (WTO, Geneva), [39] and [44].
119 IP/C/M/79 (2015) Council for Trade-Related Aspects of Intellectual Property Rights, *Minutes of Meeting Held in the Centre William Rappard on 11 June 2015* (WTO, Geneva), [5.1]-[5.5]; IP/C/M/78 (2015) Council for Trade-Related Aspects of Intellectual Property Rights, *Minutes of Meeting Held in the Centre William Rappard on 4 March 2015* (WTO, Geneva), [5.1]-[5.4]; IP/C/M/77 (2014) Council for Trade-Related Aspects of Intellectual Property Rights, *Minutes of Meeting Held in the Centre William Rappard on 26 November 2014* (WTO, Geneva), [5.1]-[5.5]; IP/C/M/76 (2014) Council for Trade-Related Aspects of Intellectual Property Rights, *Minutes of Meeting Held in the Centre William Rappard on 27 June 2014* (WTO, Geneva), [6.1]-[6.6]; IP/C/M/75 (2014) Council for Trade-Related Aspects of Intellectual Property Rights, *Minutes of Meeting Held in the Centre William Rappard on 7 March 2014* (WTO, Geneva), [5.1]-[5.6]; and so on.
120 IP/C/M/79, above n 119, [5.1]-[5.5]; IP/C/M/78, above n 119, [5.1]-[5.4]; IP/C/M/77, above n 119, [5.1]-[5.5]; and so on.
121 IP/C/W/370/Rev.1 (2006) Council for Trade-Related Aspects of Intellectual Property Rights, *The Protection of Traditional Knowledge and Folklore: Summary of Issues Raised and Points Made* (WTO, Geneva).
122 IP/C/W/370/Rev.1, above n 121, [8].
123 IP/C/W/370/Rev.1, above n 121, [9].
124 IP/C/W/370/Rev.1, above n 121, [10] and [11].
125 IP/C/W/370/Rev.1, above n 121, [13]-[14].
126 IP/C/W/370/Rev.1, above n 121, [5].
127 IP/C/W/370/Rev.1, above n 121, [18]-[19].
128 IP/C/W/370/Rev.1, above n 121, [20]-[24].

129 IP/C/W/370/Rev.1, above n 121, [25]-[26].
130 IP/C/W/370/Rev.1, above n 121, [27] and [28]; IP/C/W/491 (2007) Council for Trade-Related Aspects of Intellectual Property Rights, *The Relationship Between the TRIPS Agreement, the Convention on Biological Diversity and the Protection of Traditional Knowledge: Amending the TRIPS Agreement to Introduce an Obligation to Disclose the Origin of Genetic Resources and Traditional Knowledge in Patent Applications – Answers From Norway to Questions Posed by Switzerland in the TRIPS Council* (WTO, Geneva); IP/C/W/475 (2006) Council for Trade-Related Aspects of Intellectual Property Rights, *Response to Questions Raised on the Draft Amendment to TRIPS – Article 29bis* (WTO, Geneva); IP/C/W/474 (2006) Council for Trade-Related Aspects of Intellectual Property Rights, *Doha Work Programme – The Outstanding Implementation Issue on the Relationship Between the TRIPS Agreement and the Convention on Biological Diversity* (WTO, Geneva) (and Add.1-Add.9); IP/C/W/473 (2006) Council for Trade-Related Aspects of Intellectual Property Rights, *The Relationship Between the TRIPS Agreement, the Convention on Biological Diversity and the Protection of Traditional Knowledge – Amending the TRIPS Agreement to Introduce an Obligation to Disclose the Origin of Genetic Resources and Traditional Knowledge in Patent Applications* (WTO, Geneva); and so on.
131 IP/C/W/370/Rev.1, above n 121, [28] and [31]-[32].
132 IP/C/W/370/Rev.1, above n 121, [28].
133 IP/C/W/370/Rev.1, above n 121, [27] and [28]; IP/C/W/491, above n 130; IP/C/W/475, above n 130; IP/C/W/474, above n 130 (and Add.1-Add.9); IP/C/W/473, above n 130; and so on.
134 IP/C/W/370/Rev.1, above n 121, [28] and [33]-[40].
135 IP/C/W475; IP/C/W/474, above n 130 (and Add.1-Add.9).
136 IP/C/W/474, above n 130, [2].
137 TN/C/W/52 (2008) Trade Negotiations Committee, *Draft Modalities for TRIPS Related Issues* (WTO, Geneva), [4]-[6]; IP/C/W/474, above n 130, [1]-[5]; IP/C/W/473, above n 130; IP/C/W/433 (2004) Council for Trade-Related Aspects of Intellectual Property Rights, *Further Observations by Switzerland on its Proposals Regarding the Declaration of the Source of Genetic Resources and Traditional Knowledge in Patent Applications* (WTO, Geneva); IP/C/W/383 (2002) Council for Trade-Related Aspects of Intellectual Property Rights, *Communication from the European Communities and their Member States* (WTO, Geneva).
138 For example IP/C/M/79, above n 119, [5.1]-[5.5]; IP/C/M/78, above n 119, [5.1]-[5.4]; IP/C/M/77, above n 119, [5.1]-[5.5]; and so on.
139 IP/C/M/79, above n 119, [5.3].
140 ST/ESA/328 (2009) Secretariat of the Permanent Forum on Indigenous Issues, *State of the World's Indigenous Peoples*, p 1.
141 A/69/267 (2014) General Assembly, *Report of the Special Rapporteur on the Rights of Indigenous Peoples* (United Nations, New York), [12]; A/68/317 (2014) General Assembly, *Report to the General Assembly* (United Nations, New York), [70]; A/64/338 (2009) General Assembly, *Report to the General Assembly* (United Nations, New York), [47].
142 A/RES/69/159 (2015) General Assembly, *Resolution Adopted by the General Assembly on 18 December 2014* (United Nations, New York), p 2; A/RES/68/149 (2014) General Assembly, *Resolution Adopted by the General Assembly on 18 December 2013* (United Nations, New York), pp 2 and 3; A/RES/67/153 (2013) General Assembly, *Resolution Adopted by the General Assembly on 20 December 2012* (United Nations, New York), p 2; and so on.
143 A/69/267, above n 141, [9].
144 WIPO/GRTKF/IC/28/11 Prov.2 (2014) Intergovernmental Committee on Intellectual Property and Genetic Resources, Traditional Knowledge and Folklore, *Draft Report* (WIPO, Geneva), [21]-[43]; WIPO/GRTKF/IC/27/10, above n 96, [16]-[32]; WIPO/

GRTKF/IC/26/8 (2014) Intergovernmental Committee on Intellectual Property and Genetic Resources, Traditional Knowledge and Folklore, *Report* (WIPO, Geneva), [68]-[70]; and so on.

145 Vivas-Eugui, David, *Bridging the Gap on Intellectual Property and Genetic Resources in WIPO's Intergovernmental Committee (IGC)*, ICTSD's Programme on Innovation, Technology and Intellectual Property, Issue Paper No 34 (2012), pp 24–25.

146 WIPO/GRTKF/IC/18/11 (2011) Intergovernmental Committee on Intellectual Property and Genetic Resources, Traditional Knowledge and Folklore, *Report* (WIPO, Geneva), [381].

147 see WIPO/GRTKF/IC/1/13, [18].

148 See WO/GA/32/13 (2005) World Intellectual Property Organization General Assembly, *Thirty-Second WIPO General Assembly: Report* (WIPO, Geneva), [168].

149 see WIPO/GRTKF/IC/1/13, [17].

150 International Indigenous Forum (2012) *Final Statement of the International Indigenous Forum*, 21 February 2012. www.docip.org/fileadmin/user_upload/OMPI_STATEMENT_FEB21_ENG.pdf

151 WIPO/GRTKF/IC/20/10 (2012) Intergovernmental Committee on Intellectual Property and Genetic Resources, Traditional Knowledge and Folklore, *Declaration of the Source of Genetic Resources and Traditional Knowledge in Patent Applications: Proposals by Switzerland* (WIPO, Geneva), [800].

152 WIPO/GRTKF/IC/21/7 Prov.2 (2012) Intergovernmental Committee on Intellectual Property and Genetic Resources, Traditional Knowledge and Folklore, *Draft Report* (WIPO, Geneva), [618].

153 WIPO/GRTKF/IC/25/8 Prov.2 (2013) Intergovernmental Committee on Intellectual Property and Genetic Resources, Traditional Knowledge and Folklore, *Draft Report* (WIPO, Geneva), [431].

154 See WIPO/GRTKF/IC/6/8 (2004) Intergovernmental Committee on Intellectual Property and Genetic Resources, Traditional Knowledge and Folklore, *Defensive Protection Measures Relating to Intellectual Property, Genetic Resources and Traditional Knowledge: An Update* (WIPO, Geneva); WIPO/GRTKF/IC/5/6 (2003) Intergovernmental Committee on Intellectual Property and Genetic Resources, Traditional Knowledge and Folklore, *Practical Mechanisms for the Defensive Protection of Traditional Knowledge and Genetic Resources within the Patent System* (WIPO, Geneva).

155 See IP/C/W/474, above n 130 (and Add.1-Add.9); Jacques De Werra (2009) 'Fighting against Biopiracy: Does the Obligation to Disclose in Patent Applications Truly Help?', *Vanderbilt Journal of Transnational Law*, 42(1), pp 143–179.

156 United Nations Educational, Scientific and Cultural Organization and World Intellectual Property Organization (1985) *Model Provisions for National Laws on the Protection of Expressions of Folklore Against Illicit Exploitation and Other Prejudicial Actions* (UNESCO and WIPO, Geneva), [5]-[24].

157 General Conference of UNESCO (1989) *Recommendation on the Safeguarding of Traditional Culture and Folklore*, Twenty-fifth Session. http://portal.unesco.org/en/ev.php-URL_ID=13141&URL_DO=DO_TOPIC&URL_SECTION=201.html (accessed 21 December 2016)

158 ITH/14/9.COM/6 (2014) Intergovernmental Committee for the Safeguarding of the Intangible Cultural Heritage, *Report by the Secretariat on its Activities* (UNESCO, Paris), [25].

159 ITH/14/9.COM/6, above n 158, [25] (footnote).

160 ITH/15/10.COM/7.b (2015) Intergovernmental Committee for the Safeguarding of the Intangible Cultural Heritage, *Report by the Secretariat on its Activities* (UNESCO, Paris), [66].

161 See, for example, WIPO/GRTKF/IC/27/10, above n 96, [249].

162 See, for example, WIPO/GRTKF/IC/27/10, above n 96, [65], [68] and [249].

163 See, for example, WIPO/GRTKF/IC/27/10, above n 96, [240] and [256].
164 WO/GA/26/6, above n 1, [9]-[12]; Graham Dutfield, *Intellectual Property, Biogenetic Resources and Traditional Knowledge* (Routledge, 2010), pp 134–135.
165 A/RES/48/163 (1994) General Assembly, *International Decade of the World's Indigenous People* (United Nations, New York).
166 See E/2004/82 (2004) Economic and Social Council, *Report of the Secretary-General on the Preliminary Review by the Coordinator of the International Decade of the World's Indigenous People on the Activities of the United Nations System in Relation to the Decade* (United Nations, New York), [58].
167 See WO/BC/18/2 – A/32/2 (1998) WIPO Budget Committee, *Draft Program and Budget 1998–1999* (WIPO, Geneva), Main Program 11.
168 WO/GA/26/4 (2000) WIPO General Assembly, *Policy Advisory Commission (PAC) and Industry Advisory Commission (IAC): Memorandum of the Director General* (WIPO, Geneva), [1]-[2].
169 World Intellectual Property Organization (2001) *Intellectual Property Needs and Expectations of Traditional Knowledge Holders: WIPO Report on Fact-finding Missions on Intellectual Property and Traditional Knowledge (1998–1999)* (WIPO, Geneva).
170 WO/GA/26/4, above n 168, Annex IV, [1(v)].
171 WO/GA/26/6, above n 1, [15].
172 WO/GA/26/10, above n 1, [28]-[70]; World Intellectual Property Organization (2012) *Intellectual Property and Genetic Resources, Traditional Knowledge and Traditional Cultural Expressions – An Overview* (WIPO, Geneva), p 5.
173 PT/DC/47 (2000) World Intellectual Property Organization, *Patent Law Treaty, Regulations Under the Patent Law Treaty and Agreed Statements by the Diplomatic Conference* (WIPO, Geneva).
174 A/RES/55/2 (2000) General Assembly, *United Nations Millennium Declaration* (United Nations, New York), [11]-[20].
175 See, for example, Antoniette Di Blase (2015) 'Traditional Knowledge: Cultural Heritage or Intellectual Property Right?' in Valentina Vadi and Bruno de Witte, *Culture and International Economic Law* (Routledge, United Kingdom), pp 145–159.
176 UNESCO and WIPO 1985, above n 156.
177 See IP/C/W/474, above n 130.
178 UNEP/CBD/COP/7/21, above n 30, pp 261–276.
179 UNEP/CBD/COP/10/27, above n 45, pp 313–314.
180 UNEP/CBD/COP/10/27, above n 45, pp 95 and 98.
181 UNEP/CBD/COP/10/27, above n 45, pp 93 and 96.
182 UNEP/CBD/COP/12/29, above n 54, pp 87–88; UNEP/CBD/COP/11/35, pp 154–156.
183 UNEP/CBD/COP/12/29, above n 54, pp 90–91; UNEP/CBD/COP/11/35, pp 156–157.
184 UNEP/CBD/COP/10/27, above n 45, p 93.
185 See, for example, WIPO/GRTKF/IC/5/INF/4 (2003) Intergovernmental Committee on Intellectual Property and Genetic Resources, Traditional Knowledge and Folklore, *Comparative Summary of Existing National* Sui Generis *Measures and Laws for the Protection of Traditional Knowledge* (WIPO, Geneva).
186 See, for example, WIPO/GRTKF/IC/28/INF/7 (2014) Intergovernmental Committee on Intellectual Property and Genetic Resources, Traditional Knowledge and Folklore, *Glossary of Key Terms Related to Intellectual Property and Genetic Resources, Traditional Knowledge and Traditional Cultural Expressions* (WIPO, Geneva), p 40.
187 CGIAR 2013, above n 114, p 2.
188 IP/C/W/370 Rev.1.
189 E/C.19/2007/10 (2007) Permanent Forum on Indigenous Issues, *Report on Indigenous Traditional Knowledge* (United Nations, New York), [20].
190 Anthony Taubman (2012), 'Indigenous Innovation: New Dialogues, New Pathways' in Peter Drahos and Susy Frankel (eds), *Indigenous People's Innovation: Intellectual Property Pathways to Development* (Canberra, ANU Press), p xv.

Part 2

Substantive issues

5 Of disclosure 'straws' and IP system 'camels'

Patents, innovation, and the disclosure of origin requirement

Margo A. Bagley[1]

Introduction

Turning trash into treasure is an appealing prospect and a developing reality. Biomass conversion involves converting low value cellulosic trash, such as wood chips, straw, hay, and fruit and seed hulls, into high value products such as ethanol and lactic acid, and is a promising technique for fuel production.[1] Not surprisingly, patent applications have been and are being filed in patent offices around the globe on biomass conversion methods and components.

One such document, published in 2014 as Patent Cooperation Treaty (PCT) application number WO2013050582, discloses novel heat-loving bacterial species from the genus *Thermoanaerobacter*, useful in producing ethanol from cellulose.[2] While the application notes that a sample of the claimed invention was appropriately deposited in a Budapest Treaty compliant facility, the application gives no indication of the geographical origin of the original bacteria from which the claimed strains were developed.[3] This is perhaps not surprising, as neither the PCT, nor any other multilateral intellectual property treaty, mandates that countries impose a genetic resource disclosure of origin requirement in patent applications; it is solely a matter of national and/or regional law. As such, there is no easy way to ascertain whether the inventors are in compliance with any applicable access and benefit-sharing laws in the country from which the bacteria originated.

The disclosure of such information may soon become routine, however, depending on the ultimate success of negotiations taking place in the World Intellectual Property Organization's (WIPO) Intergovernmental Committee (IGC) on Intellectual Property and Genetic Resources (GR), Traditional Knowledge (TK) and Folklore.

1 Hardy Cross Dillard Professor of Law and Joseph C. Carter, Jr. Research Professor of Law, University of Virginia School of Law; Asa Griggs Candler Professor of Law, Emory University School of Law. This chapter benefited greatly from comments received during presentations at the WIPIP Workshop, U.S. Patent & Trademark Office, Arlington, VA; the European Policy on Intellectual Property (EPIP) Conference, European Parliament, Brussels, Belgium; and the International Intellectual Property Scholars Roundtable at DePaul Law School, Chicago, IL. Thanks also to Joseph Babitz and the University of Virginia Law librarians for excellent and timely research assistance. The author currently serves as Lead Facilitator and Friend of the Chair in the WIPO IGC.

The current mandate of the IGC is to engage in text-based negotiations leading to one or more international legal instruments.[4] Recent negotiations have yielded three draft texts: a GR text that would include provisions, such as a requirement that inventors seeking patent protection disclose the origin of genetic resources and associated TK (GRAATK) used in developing a claimed invention, and TK and traditional cultural expressions (TCE) texts that would include, among other things, a suite of moral and economic rights for certain categories of TK and TCEs.[5] The focus of this chapter is the proposed disclosure of origin (DOO) provision in the draft GR text.

Some demandeurs seeking a legally binding GR agreement argue that such a requirement will facilitate compliance with access and benefit-sharing (ABS) agreements in relation to GRAATK and reduce the incidence of patents that should not have been granted because the subject matter lacks novelty or is obvious.[6] Such erroneously granted patents can create barriers preventing access to foreign markets for products from demandeur countries.[7]

Many non-demandeurs have resisted a binding agreement, or want any agreement on GRs to employ only defensive measures, such as traditional knowledge databases, to avoid the grant of erroneous patents and *not* to facilitate ABS. Such countries claim that a mandatory GR DOO would be unworkable, as it would unacceptably reduce legal certainty and place too much of a burden on patent applicants, resulting in diminished innovation.[8] Both sides are reacting to and operating within a pre-existing landscape of national laws that, in some cases, stymie transparency goals and in others exacerbate legal uncertainty.

Modern patent systems are inherently uncertain regimes, and a new DOO requirement will be an additional obligation for affected patent applicants, and could be imposed by member countries in ways that may increase uncertainty in relation to patent rights. However, DOO requirements can be designed to be reasonable, negligible in their impact on legal certainty and no more burdensome than other obligations that patent applicants are currently required to complete. Moreover, there are many existing requirements and policy levers in developed country patent systems which regularly impose burdens and create uncertainty of equal or greater magnitude to that of a DOO requirement, yet such systems survive and thrive with no clear indication of a diminished capacity to absorb additional reasonable requirements.

This chapter takes a closer look at developed country concerns regarding a mandatory GRAATK DOO requirement in the context of existing burdens and uncertainties in the patent system. The next section provides additional background on the motivations of demandeurs and the validity of concerns articulated by opponents of a mandatory DOO requirement that such a requirement will impose untenable, innovation-deterring burdens and legal uncertainty on the global patent system. It goes on to highlight several examples of current burdens and areas of uncertainty in the patent system and distinguishes between the kinds of burdens and uncertainties that are necessary to a well-functioning patent system and those that are more problematic. The following section describes an alternative approach that non-demandeur countries should consider adopting in IGC negotiations in light of the increasing inclusion of DOO provisions in national patent laws. The final part concludes that far from being the straw that may break the camel's back, a properly constructed, formal, treaty-based DOO requirement need not harm, and may even enhance, innovative activity in user and provider countries.

Burdens and uncertainties: the disclosure of origin requirement for genetic resources and associated traditional knowledge (GRAATK)

A useful starting point for understanding the concerns of DOO proponents and opponents is the Convention on Biological Diversity (CBD or 'the Convention'). Ratified in 1993, the Convention was designed to facilitate the conservation of biological diversity, the sustainable use of its components, and the fair and equitable sharing of the benefits generated by the use of genetic resources.[9] The CBD is an environmental treaty in the sense that its creation was motivated by a desire to stem the uncontrolled depletion of flora and fauna in the global South. It establishes that genetic resources are the property of sovereigns who should provide access to them under principles of prior informed consent (PIC) on mutually agreed terms (MAT), with appropriate benefit sharing.[10] The 2010 Nagoya Protocol on Access and Benefit Sharing to the Convention on Biological Diversity[11] was necessary because, while the CBD obligated Parties to ensure the fair and equitable sharing of benefits arising from the utilization of genetic resources, it provided almost no detail on how ABS should be accomplished in practice.[12] Consequently, provider countries have either not yet implemented ABS provisions or implemented widely varying legislation, often resulting in confusion and uncertainty for users faced with rules for ABS/PIC that differ, sometimes significantly, by country.

The Protocol, which came into effect on October 12, 2014, was designed to reduce uncertainty and provide increased transparency for both users and providers of genetic resources and associated traditional knowledge. It is a binding agreement and is '*the* instrument for implementation of the access and benefit sharing provisions of the [CBD].'[13] The Nagoya Protocol, *inter alia*, obligates each member to take appropriate steps to 'provide that genetic resources utilized within its jurisdiction' have been accessed in accordance with the domestic ABS/PIC/MAT requirements of another Party.[14] Furthermore, members must cooperate, as far as possible and as appropriate, in cases where another Party's domestic ABS legislation has been violated.[15] As of this writing, seventy-eight countries are party to the Protocol, and forty-three countries have submitted ABS legislative or policy instruments to the new Protocol ABS Clearinghouse, but many of those documents (other than the EU member states' submissions of the EU implementing directive) are pre-Nagoya Protocol laws, indicating that much work remains to be done in member states to pass implementing legislation for the Protocol.[16]

One of the driving forces behind both the CBD and the Nagoya Protocol has been a desire to prevent the misappropriation or misuse of GRAATK, often broadly characterized as 'biopiracy'. 'Biopiracy' is a rather pejorative term that is often used as a shorthand label for some form of egregious misappropriation.[17] By one definition, it is 'the patenting of (often spurious) inventions based on biological resources and/or traditional knowledge that are extracted without adequate authorization and benefit sharing from other (usually developing) countries, indigenous or local communities'.[18] This definition creates a clear connection to the word 'piracy' which is defined as 'the unauthorized use of another's production, invention, or conception especially in infringement of a copyright'.[19] As explained by one commentator: 'What developing tropical nations are saying is that if the West cries foul over piracy of intellectual property, [such as] computer software, then biopiracy in Western labs of jungle extracts should also be considered a high economic crime.'[20]

For many years, a number of CBD member countries rich in genetic resources (so-called 'provider countries') have been pressing for a GRAATK DOO requirement for patent applications in multiple multinational fora. Such efforts have generally been consistent with, and designed to give effect to, the CBD's ABS goals.[21] As a 2004 Queen Mary Institute report explains, there are a variety of reasons why developing countries and indigenous peoples are seeking DOO requirements and other obligations/protections in WIPO and elsewhere, namely:

1 The fact that patent claims in various countries may incorporate biological and genetic material including life forms within their scope.
2 The conviction – widely held among developing countries and NGOs – that biodiversity and associated traditional knowledge have tremendous economic potential.
3 The belief, also shared by developing countries and NGOs, that this feature of the patent system enables corporations to steal, misappropriate or unfairly free-ride on genetic resources and associated traditional knowledge.
4 The ability of modern intellectual property law to protect the innovations produced by industries based mainly in the developed world and its inability to protect adequately those in which the developing countries are relatively well endowed.
5 The perception that as a consequence of reasons 1 to 4, the unequal distributions and concentrations of patent ownership and the unequal share of benefits obtained from industrial use of biogenetic resources are closely related.[22]

Under the Demsetzian view that property rights arise to allow parties to internalize externalities,[23] it is not surprising that individuals and groups that provide access to, and conservation of, genetic resources and associated traditional knowledge might also wish to share in internalizing some of the benefits to which they have contributed. Yet not all groups are financially motivated. Some are primarily motivated by concerns regarding control and overexploitation of the genetic resources in their region.[24] The implementation of a GRAATK disclosure of origin (DOO) requirement for patent applications may aid possibly aggrieved parties in determining if an ABS law was broken by the inventor and if a claim for relief is justified.

At WIPO, DOO concerns were first raised in the Standing Committee on Patents, but were later moved to the, then newly created, IGC for deliberation in 2000.[25] The IGC's first meeting was in 2001, and while there has been much talk in successive meetings, real progress largely began with the start of text-based negotiations in 2009.[26] Parties took a hiatus from formal IGC discussions in 2015; however, during the WIPO General Assemblies October 5–14, 2015, agreement was reached to resume IGC negotiations in the 2016–2017 biennium toward development of one or more international legal GR, TK, and TCE texts. Many WIPO member states are hopeful that consensus will continue to grow, in particular around the draft GR text, especially as the language is refined and improved during the drafting process.[27]

In relation to the draft GR text and its DOO requirement, several developed countries (which tend to have many companies and other entities which use genetic resources in inventive activity – 'users'), have consistently resisted efforts by demandeurs to reach agreement on an IGC instrument. These countries have cited concerns that such a treaty would introduce legal uncertainty into the intellectual property system and impose innovation-deterring burdens on inventors. Consider, for example, the following statements made during the Twenty-sixth IGC Session[28]:

> A number of proposals in all three texts would add burdens, create uncertainty and add new patentability requirements, based on compliance with the administrative rules, which [are] unrelated to patentability itself.[29]
>
> DOO requirements have already resulted in 'serious problems', such as 'processing delays and uncertainty in obtaining and exercising patent rights'.[30]
>
> One of the key concerns related to the potential burden on the IP system and possible consequences, which could limit access to GRs and TK associated with GRs, cause uncertainty in the IP system and, thereby, impede innovation and achievement of economic benefits.[31]
>
> Any instrument would need to ensure that patent systems and related IPRs would not be threatened by *any* element of legal uncertainty. . . . A disclosure requirement which would create legal uncertainty in the patent system would not be in the interest of Member States and patent users.[32]

A close examination of these comments reveals several concerns: (1) that a DOO requirement will impose a new burden on patent applicants; (2) that such a burden could affect an applicant's ability to obtain a patent but would be unrelated to the substantive patentability of the claimed invention; (3) the requirement would impose a burden on patent offices that would be difficult to administer; (4) non-compliance with the requirement could result in patent invalidity or other sanctions which would create uncertainty regarding the value of patents and perhaps reduce interest in obtaining patents, thus reducing the kinds of innovation for which patents provide incentives; and (5) a DOO requirement might inhibit applicants from accessing genetic resources and traditional knowledge for fear of incurring significant economic liability.[33]

These are legitimate issues, and the DOO requirements in some provider country laws do provide a basis for concern. For example, the disclosure of the source or geographical origin of genetic resources (and traditional knowledge) used in creating an invention is required by the Indian Patent Act and failing to disclose or wrongly stating the source can result in revocation of an issued patent.[34] Also, a 2001 Brazilian law requires that the origin of genetic material used in creating an invention must be disclosed in the patent application with patent revocation as a possible penalty for non-compliance.[35]

On the other hand, some other existing patent system DOO requirements do not appear to be creating excessive uncertainty or placing unreasonable burdens on applicants or patent offices. China, the country that currently receives more patent applications each year than any other country, provides a pertinent example. Article 26(5) of the Chinese Patent Act (3rd Revision) requires patent applicants to disclose the origin of genetic resources used in creating an invention. According to a recent analysis of China's genetic resource DOO requirement, between October 1, 2009 and June 30, 2013, genetic resource source forms were filed in 7,149 patent applications, most after the examiner requested submission of the form.[36] While the vast majority were filed by Chinese domestic applicants, thirty-eight were from abroad, and filed by applicants from South Korea, Japan, France, Germany, Finland, India, Canada, Switzerland, Indonesia and the United States.[37] Thus, researchers and multinationals, including from the U.S., are already having to deal with such a requirement if they are seeking patent protection in DOO countries such as China.[38] Importantly, the authors of the study conclude that the new genetic resource disclosure requirements are not placing an 'undue burden' on patent applicants.[39]

Another successful DOO example comes from Switzerland. Since 2008, the Swiss Patent Act has included a genetic resource and traditional knowledge DOO requirement for patent applications, including Patent Cooperation Treaty (PCT) applications entering the national phase in Switzerland.[40] The law requires applicants to disclose the source[41] of genetic resources and traditional knowledge in patent applications, if the claimed invention is directly based on the resources/knowledge. Failure to comply with the DOO requirement interrupts processing of the patent application until the requirement is met or the applicant discloses that the source is unknown. If, after the patent issues, the patentee is found to have lied either about not knowing the source, or intentionally disclosing an incorrect source, the patent would not be revoked, but the applicant would face criminal penalties in the form of fines, up to 100,000 Swiss Francs, and publication of the ruling.[42] The requirement is perceived as functioning well by the Swiss Industrial Property Office, although their sample size is small: only eleven disclosures as of 2014, two objections, which were corrected, and no criminal sanctions.[43]

A critical difference between the DOO regimes in India and Brazil on the one hand and China and Switzerland on the other, is that the former are substantive and the latter are formal. That is, non-compliance with the DOO requirement can result in revocation of a patent in India and Brazil. However, in Switzerland and China, an application will not be processed without the required information being disclosed but, once a patent issues, the fact that an applicant provided erroneous information regarding DOO will not invalidate the patent. Of course, if ABS laws were violated in creating a patented invention, then those national laws may contain provisions that would affect a patent grant, but invalidation would not result solely from a violation of the formal DOO requirement.

These examples show that DOO patent requirements are already a fact of life for patent applicants in many countries. In fact, at least twenty CBD member countries already have genetic resource DOO requirements for patent applications.[44] And more countries can be expected to adopt DOO requirements as part of implementing the Nagoya Protocol. These DOO requirements can be problematic for a variety of reasons, including because they are not harmonized on critical issues such as the formal or substantive nature of the requirement, whether only domestic GRAATK need be disclosed, and whether the disclosed information is made available to the interested public.[45]

The above snapshot of DOO provisions does suggest that a treaty mandating a DOO requirement in national patent laws will impose a new obligation on patent applicants in countries that did not previously require such disclosure. Nevertheless, assessing whether such developments would truly be cause for concern should not be done in a vacuum, but in the context of the structures and dynamics of the current global patent system. As the following selected examples of existing burdens and uncertainties in the patent system show, there is reason to believe this system is not so perfectly calibrated that it could not reasonably absorb the benefits and drawbacks of carefully designed, reasonable DOO requirements.

Thriving in the fog: current patent system burdens and uncertainties

To a certain degree, current debates in the IGC regarding a GRAATK DOO requirement are conflating burdens with uncertainties; but they are not the same thing. One can design a formal DOO requirement that is a necessary and manageable obligation, but does not create, on balance, detrimental uncertainty. What is more, all burdens and all uncertainties are not created equal. The following examples (most existing, one proposed) show differing kinds of burdens and uncertainties that patent applicants may or do face currently. Some, which bear a close resemblance to what a well-designed GRAATK DOO requirement would entail, are more justifiable and less harmful than others.

Patent application disclosure burdens

The United States is not a party to the CBD or the Nagoya Protocol, and its longstanding aversion to a genetic resource DOO requirement is well-known. In a communication to the TRIPs Council in 2010, the United States expressed the following objections to a genetic resource DOO requirement for patent applications:

> [N]ew patent disclosure requirements would be unworkable given the absence of knowledge of country of origin for numerous publicly available

> resources. . . . such new disclosure requirements create legal uncertainty and increased burdens on the patent system . . . new patent disclosure requirements will not achieve the desired objectives and . . . incorporation of such requirements in the patent system would stifle innovation and undermine the patent system.[46]

Considering the opposition of the United States to a mandatory DOO requirement, it may seem somewhat surprising that three provisions, two current, one proposed, contain disclosure requirements with features akin to a GRAATK DOO requirement.

Requirements for Information: 37 C.F.R. 1.105, known as Rule 105, is called 'Requirement for Information' (ROI), and allows an examiner to require an applicant to disclose a variety of different types of information on *any* issue if it will help the examiner to examine the patent application. Moreover, failure to respond to a requirement for information results in the application being declared abandoned by the office.[47] Importantly, according to the Court of Appeals for the Federal Circuit, the threshold for the requirement is a low one: information required need only be 'reasonably necessary to treat a matter in an application. The matter may or may not be related to patentability'.[48]

Thus an examiner could currently require an applicant to disclose the geographical origin of genetic resources used in the claimed invention, as such information could help in the assessment of the novelty and non-obviousness of the claims by identifying areas to search for prior art. If an applicant does not know the answer to the information requested, he or she can simply say so and the examiner will accept that as a complete response.[49] Of course, lying to the examiner would be a violation of the duty of candor and potentially render any resulting patent unenforceable due to inequitable conduct.[50] Interestingly, the United States Patent and Trademark Office (USPTO) Manual of Patent Examining Procedure calls Rule 105 'a substantial burden' on patent applicants.[51] But society is comfortable with such a burden because of the goals it is designed to serve in improving examination and making it less likely that a patent will issue on an undeserving invention.

The United States is not the only country that allows examiners to require the submission of information relevant to examination and such requirements, along with most other aspects of substantive examination, create legal uncertainty for patent applicants. The twenty-year term clock begins ticking at the filing date, and an application is normally published after eighteen months. Due to application backlogs and inefficiencies, application publication often occurs before substantive examination has taken place. Thus an applicant has given up the right to pursue trade secret protection without any certainty that she or he will be granted a patent on the claimed invention. Such uncertainty, while not optimal, is necessary for societies that value examining an invention for patentability before issuing a patent. A formal GRAATK DOO requirement would generate no greater uncertainty (and perhaps less) than an ROI, as prosecution would continue on the application once the disclosure information was provided. What is more, this minimal uncertainty is necessary to place the burden of facilitating transparency where it is most easily carried – on the patent applicant who is in the best position to know the provenance of the GRAATK.

The Bayh Dole Act Disclosure Requirement: The patent system serves a variety of purposes, including as a legislative policy tool to effectuate non-patent societal objectives. A clear example of this is the U.S. Bayh-Dole Act, in which Congress explicitly states '[i]t is the policy and objective of the Congress to use the patent system to promote the utilization of inventions arising from federally supported research or development.'[52] When Congress passed the Bayh-Dole amendments to the Patent Act in 1980, it gave universities and their inventors the ability to gain presumptive title to inventions produced with federal funds, as long as they complied with specific requirements.[53] By allowing patent title to initially vest outside of the federal government for such inventions, the Bayh-Dole Act paved the way for more interaction between universities and companies, which could now obtain exclusive licences to commercialize academic research that previously might have remained unused.[54]

The impetus for Bayh-Dole was the belief that a wealth of basic, useful research developed in U.S. universities was languishing in academic ivory towers as it took, on average, fifteen to twenty years for basic research disclosed in publications to result in marketed products.[55] This delay was attributed to reluctance by private industry to invest in commercializing federally funded research because they could not obtain exclusive rights to it.[56] The Bayh-Dole Act addressed this reluctance by allowing universities and/or their inventors to take title to inventions developed with federal funds and grant exclusive licences to entities willing to commercialize such technology.[57] Goals for Bayh-Dole included expediting the transfer of publicly supported R&D results to the marketplace by promoting collaborations between universities and industry, and creating new jobs for regional economies through the commercialization of inventions made in the United States.[58] The Act brought to universities a potential influx of new capital from licensing revenue derived from transferred technology. But in order to capitalize on the opportunity, universities had to comply with the myriad rules mandated by Bayh-Dole, including seeking patents on inventions.[59]

One of these rules is a disclosure requirement, found in 35 U.S.C. §202(c)(6). It requires federal funding agreements to include:

> [a]n obligation on the part of the contractor, in the event a United States patent application is filed by or on its behalf or by any assignee of the contractor, to include within the specification of such application and any patent issuing thereon, a statement specifying that the invention was made with Government support and that the Government has certain rights in the invention.[60]

Thus patents claiming inventions produced with federal funding must reflect that funding source *on their face*. This is not a negligible requirement; determining inventorship for claims in a patent is no easy matter, as researchers can move in and out of labs, universities, contractors, and companies, often taking aspects of research with them and complicating the Bayh-Dole disclosure requirement.[61]

Nevertheless, the disclosure requirement of Section 202(c)(6) appears to have worked well for several decades and could be a model for a genetic resource DOO requirement either as a simple formal requirement during the application process, or as a statement that would appear on an issued patent. As proposed by Oldham *et al.*:

> In our view it is reasonable to expect that patent applicants should state where they obtained the genetic resources and associated traditional knowledge that is material to an invention. . . . much could be achieved by following the model for disclosure of federal funding in patent applications developed in the United States under the 1980 Bayh-Dole Act. The Bayh-Dole system requires recipients of federal funding to include a 'Statement Regarding Federally Sponsored Research and Development' setting out the source of federal funding and contract number at the beginning of a patent application. This model for disclosure has caused no reported difficulties in the patent system and could be readily adapted to improve certainty for countries that access and benefit-sharing laws are being recognized and respected. We propose that this model could be adapted as a Statement or Declaration on Access and Benefit-Sharing in patent applications involving genetic resources and associated traditional knowledge that are material to the claimed invention.[62]

If this statement approach were adopted, the contents of such a statement would need to be tailored to address the variety of situations that the particular national law defines as giving rise to a disclosure obligation, while not unduly delaying processing of the patent application or requiring significant extra work for patent examiners. Regardless, in the same way that the patent system can be used to facilitate the aims of the Bayh-Dole Act, the patent system can appropriately be used by governments to facilitate ABS/PIC compliance using a GRAATK DOO requirement. To the extent that access to GRAATK is limited in various countries due to concerns about misappropriation or biopiracy, a multilateral, formal DOO requirement may encourage broader access and collaboration and effectively enhance cross-border innovative activity.

The 'Burdensome' USPTO Attributable Ownership Rules: On January 24, 2014 the USPTO announced proposed rules for patent applicants and owners that, if adopted, would likely have created a greater disclosure burden for patent applicants than a new genetic resource DOO requirement. The new rule, to be codified at 37 C.F.R. 1.271, would have required the disclosure of any and all owners of a patent or patent application. The proposed definition of an 'attributable owner' reached far beyond simple assignees, and included parent entities,[63] parties with standing to enforce the patent (including exclusive licensees), and beneficial owners whose identities might be otherwise hidden. Moreover, the identification of attributable owners would have been required at a number of points during *and* after prosecution: on filing of an application (or shortly thereafter), when there is a change in the attributable owner during the pendency of an application, at the time of issue fee and maintenance fee payments, and when a patent is involved in supplemental examination, *ex parte* reexamination, or a trial proceeding before the Patent Trial and Appeal Board (PTAB).[64]

These proposed 'Attributable Owner' rules would create significant burdens for patent applicants and owners alike as the information they require can be complex and difficult to determine. Standing to sue is one such issue. As one commentator explains:

> The proposed rules would require the disclosure of entities having standing to enforce a patent, or who are necessary parties to litigation. These can be complex legal questions that can depend on the terms of confidential license agreements. Even non-assigned patent applications can raise complex issues relating to marital property laws and divorce decrees.[65]

Furthermore, the rules would be retroactive in nature and impose an ongoing requirement to disclose all 'attributable owners' of a pending application or granted patent. Importantly, failure to comply would be penalized by abandonment of the application.

Interestingly, just as the genetic resource DOO requirement is alleged by some to be inappropriate as 'unrelated to patentability criteria' such as novelty and non-obviousness, the proposed ownership disclosure rules (and the Bayh-Dole disclosure requirement) are unrelated to patentability criteria. They were expressly designed to give effect to several White House initiatives, namely, to 'increase transparency of the patent system, ensure the highest-quality patents, reduce abusive patent litigation and level the playing field for innovators' in part by making the 'real party in interest the new default.'[66] As one commentator noted, 'the fact that the USPTO defines "attributable owner" with reference to parties who may have the right to enforce a patent shows that any need for such information pertains more to patent litigation than patent examination'.[67]

After receiving numerous comments critical of the proposed rules, the USPTO announced on February 5, 2015, that it would not advance the rules package 'at this time'.[68] This leaves open the possibility that the rules could be revisited for adoption at a later date. Nevertheless, even if this particular disclosure obligation ultimately is never implemented, the fact remains that the patent system is evolving, it is not static. Just as we continue to expand benefits to applicants (e.g. lengthening the term of design patents from fourteen to fifteen years from issuance), we will continue imposing new burdens on applicants when society perceives they are necessary to accomplish important objectives. And, as compared to the attributable owner rules, the Chinese and Swiss examples show that a formal, pre-grant, GRAATK DOO requirement need not be an overly burdensome obligation for patent applicants to fulfil.[69]

Post-grant patent uncertainties

Uncertainties regarding patent rights can be *ex ante* – before the patent issues, or *ex post*, after the patent issues. The *ex ante* uncertainties such as might be generated by a Rule 1.105 Requirement for Information, or a formal GRAATK DOO requirement, are of a kind that are necessary to a well-functioning (though unfortunately inefficient) patent system. A far greater concern for inventors are *ex post* uncertainties.

The grant of a patent does not end the quest for certainty regarding patent rights – far from it. Re-examinations, oppositions, nullity actions, and patent infringement/validity litigation all provide opportunities for patent rights to be strengthened, weakened, or lost completely, as well as for the meaning, scope, and coverage of patent claims to be broadened or narrowed in ways that neither patentees nor third parties can effectively anticipate.[70] Consider the following statistics:

- 69 per cent of challenged European Patent Office (EPO) patents are revoked or amended[71]
- One or more claims were invalidated in 86 per cent of U.S. patent cases where validity was challenged in the Federal District Court (between 2007–2011)[72]
- 40 per cent of patents challenged in the Japanese Patent Office (JPO) were cancelled or modified.[73]

Also in the U.S., the ability to challenge patents in Inter Partes Review (IPR) proceedings created by the 2011 America Invents Act has resulted in increasing numbers of such petitions being filed and patents being successfully challenged in the USPTO. As one commentator notes, 'overall, as of the IPR three-year anniversary, if your patent claims are targeted in an IPR petition, the odds of each claim surviving is now almost a flip of the coin: 51 per cent'.[74] Moreover, in the U.S., several recent Supreme Court decisions on patent subject matter eligibility have effectively invalidated claims in thousands of patents on isolated DNA, software, business methods, and diagnostic methods.[75] And the eligibility of such subject matter for patent protection varies in other countries as well.[76]

It may come as a surprise to some that such a high percentage of challenged patents are invalidated after issuance, especially considering the many factors in a patentee's favour such as a presumption of patent validity in several jurisdictions,[77] the ability to modify claims, often throughout the entire term of a patent,[78] the ability to choose the forum for infringement suits, and more.[79] Professor Lemley attributes the fact that claims in nearly half of U.S. patents litigated to judgment are invalidated, to the 'fractioning' of patent law, traceable to the nineteenth century switch from central to peripheral claiming. As he explains:

> In our fractional patent system, we break the patent infringement lawsuit into an enormous number of different inquiries, and . . . the patentee has to win each of those arguments to win its case. Patentees lose most of their cases because they may need to win ten different issues [novelty, non-obviousness, written description, enablement, subject matter, etc.] to prevail, while defendants generally need only win one.[80]

Another source of significant uncertainty regarding patent rights is their probabilistic nature. Essentially, patent rights and boundaries are uncertain as long as the patent remains valid and enforceable. According to Professors Lemley and Shapiro, there are at least two types of uncertainty that make patents probabilistic: uncertainty regarding the claimed invention's commercial significance, and uncertainty regarding the scope and validity of the right granted.[81]

And a related, yet distinct, further source of uncertainty is the malleability of patent rights and the fact that the scope and strength of the rights themselves can be altered by patentees and third parties interacting with the patent. According to Professor Rantenen, '[p]atent rights are not fixed . . . rather they can be pushed and pulled . . . their very existence and shape can be changed through an array of legal mechanisms'.[82]

Consequently, there is significant uncertainty regarding whether a granted patent will be upheld as valid over time, and what scope it will have, at any given point.[83] Such post-grant uncertainty arguably is the most debilitating kind of uncertainty for patent owners as it arises when one would expect rights to be fixed.

Yet some amount of post-grant uncertainty is necessary for a well-functioning patent system. Patent examination is an imperfect practice performed by examiners of varying skill and incomplete information who are dealing in an *ex parte* setting with applicants who may choose to hide or misrepresent material information regarding the patentability of an invention. Moreover, increasing numbers of patent applications continue to be filed worldwide each year, indicating the global patent system's ability to absorb and adjust for such uncertainty while still providing meaningful benefits to applicants.[84] A formal GRAATK DOO requirement would not create this most detrimental form of extant uncertainty, as remedies for violation of the DOO would not involve patent revocation, but some other type of penalty for non-compliance.

Helping, not hurting the camel

As noted above, opponents of a DOO requirement in patent applications are justifiably concerned about legal uncertainty being a detriment to patent systems and innovation. Yet inclusion of such a requirement in a legally binding international agreement could actually enhance legal certainty in at least two ways.

First, a DOO requirement is already a fact of life for U.S., EU and other researchers seeking patent protection in many countries even though the U.S., for example, does not have a genetic resource disclosure of origin requirement and is not a party to the CBD or the Nagoya Protocol. With the entry into force of the Nagoya Protocol, additional countries are likely to include a DOO obligation in their implementing legislation to facilitate determinations of compliance with ABS laws. It is not a coincidence that a country's patent or intellectual property office is a logical Nagoya Protocol checkpoint, and a genetic resource DOO requirement is a logical transparency tool to aid in effectuating the goals of the Protocol. As of this writing, three of the eleven countries notifying checkpoints to the CBD ABS Clearinghouse so far have specified intellectual property offices as checkpoints.[85] Thus, while designating an intellectual property or plant variety protection office as a checkpoint is not a Protocol requirement, defining the contours of a patent application genetic resource DOO requirement in an international treaty could actually enhance certainty for all researchers working with genetic material.[86] As explained by Professors Sarnoff and Correa:

> Rather than cause uncertainty, new international treaty provisions addressing disclosure of origin requirements may help to make more coherent existing and future national laws regarding misappropriation, including their recognition and enforcement in other countries. At least such an instrument may make existing uncertainties more transparent and predictable with regard to national access and benefit-sharing and to intellectual property laws that are applicable to transboundary resource and information flows.[87]

DOO requirements should not be an onerous burden on applicants, and should not require them to provide information that they do not know. Instead, they should optimally be formal mechanisms to provide necessary transparency that can facilitate several policy goals. While countries may impose a due diligence requirement on applicants to make reasonable efforts to ascertain origin or source, others, including China, will accept as true an applicant's assertion that the source and/or origin of the genetic material is unknown. As discussed above, even the United States has a requirement that allows an examiner to request any information from an applicant that might aid in examining the application (including disclosure of origin), even if it does not relate to patentability, and to accept as a complete answer that the applicant does not know the requested information.

Moreover, administering a formal DOO requirement should not impose a significant burden on patent offices, as there would be no obligation on an office to verify the accuracy or veracity of the substance of the information disclosed. The office would simply need to provide a way (e.g. a form) for applicants to provide the information, and perhaps a mechanism (which could be facilitated by WIPO) for the provided information to be made publicly accessible and, optimally, searchable at a designated time.

Second, the current WIPO IGC draft GR text contains many bracketed provisions, on which parties are not in agreement. However, a growing number of IGC participants appear to be somewhat open to a minimum standard or 'floor' in the agreements which, for the GR text, could involve a mandatory DOO requirement as a formal transparency mechanism that leaves several of the details of scope to be addressed under national law. Consequently, instead of opposing a DOO requirement in the IGC, non-demandeurs might be better off pursuing such a formal requirement containing both a floor and a *ceiling* on the scope of DOO protections and penalties in national laws. This could result in a ratcheting back of some of the most onerous national provisions while providing both transparency and legal certainty. This could be particularly important for facilitating patent transactions, as the origin of some resources may not be known by all parties that could be charged with DOO violations, such as downstream patent/patent application owners in a chain of title.

Resistance to a genetic resource DOO requirement may be a strawman, covering for a more malevolent force, such as lingering imperialism, discrimination, racism or 'other-ism' that tacitly disrespects sovereign claims to ownership of genetic resources for commercial gain[88] and devalues certain types of creative activity on the basis of origin. Then again, it may be a result of the more legitimate view that GRAATK tends to be used upstream in the development process, and considerable value is added to arrive at the patentable invention and/or downstream commercialized product that the original knowledge holders and source of the GRAATK had nothing to do with. While that argument has merit, it seems more relevant to the *amount* of compensation, not the need *for* compensation. In copyright law, a second comer may add a significant creative contribution to the modification of an original work in creating a new, derivative work. Nevertheless, the improver has no rights to the derivative work without the permission of the original creator. And even in patent law, while one may obtain a patent on an improvement of an invention, if practice of the improvement requires making or using the underlying, dominant patented invention, such making or using is infringement. Similarly, using a reasonable genetic resource DOO requirement to combat GRAATK misappropriation seems a justifiable and appropriate response by sovereigns that can be fully consistent with existing treaty obligations and compatible with existing patent system practices.

Conclusion

Uncertainty pervades developed country patent systems. Divergent approaches to issues such as subject matter eligibility and post-grant challenges create or exacerbate legal uncertainty for patentees and third parties within and across jurisdictions. Nevertheless, players in these systems survive and thrive with this state of affairs and apparently reap significant benefits that, by and large, appear to outweigh the negative impacts. There seems no compelling reason to expect the addition of a reasonable formal DOO requirement in these systems, when one is already (or will soon be) present in so many other patent systems, to either suddenly or slowly create an untenable level of uncertainty or an unbearable burden for patentees.

Although even a carefully crafted, DOO requirement technically is a burden for patent applicants, it is unlikely to be the metaphorical 'straw that breaks the camel's back' in terms of harming either innovative output or patenting levels in developed countries. Moreover, the 'burden' and pre-grant uncertainty a formal GRAATK DOO requirement would engender are both of a type that patent applicants are quite used to dealing with and which are necessary to effectuate important transparency-related objectives.

DOO opponents should consider redirecting their efforts towards developing a binding instrument that reaches a middle ground of providing both floors *and* ceilings for disclosure through the mechanism of a formal requirement. Even with the inevitable uncertainties that can be expected to accompany distinctions in national implementation, such a treaty has the potential to deter true thieves from engaging in piratical practices and encourage innovators to equitably and profitably access genetic and knowledge resources around the globe to develop new advances for the betterment of all societies.

Notes

1 *See, e.g.,* Biomass Energy Centre: Conversion Technologies available at www.biomassenergycentre.org.uk/portal/page?_pageid=75,15179&_dad=portal&_schema=PORTAL (last visited Jan. 24, 2016).
2 *See* www.google.com/patents/WO2013050582A2?cl=en (last visited Jan. 24, 2016).
3 *Id. See also* Edward Hammond (2014), *Patent Claims on Genetic Resources of Secret Origin*, p. 20, Third World Network.
4 Assemblies of Member States of WIPO, Fifty-Fifth Session, Oct. 5–14, 2015, Agenda Item 17: Matters Concerning the Intergovernmental Committee on Intellectual Property and Genetic Resources, Traditional Knowledge and Folklore, *available at* www.wipo.int/export/sites/www/tk/en/igc/pdf/igc_mandate_1617.pdf (last visited Jan. 3, 2016).
5 *See* Intergovernmental Committee on Intellectual Property and Genetic Resources, Traditional Knowledge and Folklore, Joint Recommendation on Genetic Resources and Associated Traditional Knowledge, WIPO/GRTKF/IC/28/7 (May 9, 2014); Intergovernmental Committee on Intellectual Property and Genetic Resources, Traditional Knowledge and Folklore, The Protection of Traditional Knowledge: Draft Articles WIPO/GRTKF/IC/28/5 (June 2, 2014); Intergovernmental Committee on Intellectual Property and Genetic Resources, Traditional Knowledge and Folklore, The Protection of Traditional Culture Expressions: Draft Articles WIPO/GRTKF/IC/28/6 (June 2, 2014).
6 Intergovernmental Committee on Intellectual Property and Genetic Resources, Traditional Knowledge and Folklore, *Report*, WIPO/GRTKF/IC/27/10 (July 9, 2014) (hereinafter '27th IGC Report') at para. 44.
7 *See, e.g.,* discussion of the Enola bean patent controversy in Margo A. Bagley (2003), 'Patently Unconstitutional: Geographical Limitations on Prior Art in a Small World', 87 *Minnesota Law Review* 679, 701–703.
8 *See* discussion *infra*. *See also* Council for Trade-Related Aspects of Intellectual Property Rights, Communication from the United States: Views of the United States on the Relationship Between the Convention on Biological Diversity and the TRIPS Agreement, IP/C/W/257 (June 13, 2001); Council for Trade-Related Aspects of Intellectual Property Rights, Communication from the United States: Article 27.3(B), Relationship Between the TRIPS Agreement and the CBD, and the Protection of Traditional Knowledge and Folklore, para. 4, IP/C/W/469 (Mar. 13, 2006) (noting in paragraph 5 'legal uncertainty and other negative consequences'). *See also* Council for Trade-Related Aspects of Intellectual Property Rights, Minutes of Meeting, para. 34, IP/C/M/44 (July 19, 2004) (European Communities arguing PIC would be hard to verify); Council for Trade-Related Aspects of Intellectual Property Rights, Minutes of Meeting, para. 155, IP/C/M/29 (Mar. 6, 2001) (Japan arguing that DOO would violate TRIPS).
9 Convention on Biological Diversity, art. 1, June 5, 1992, 1760 U.N.T.S. 143, 144.
10 Convention on Biological Diversity, art. 15, June 5, 1992, 1760 U.N.T.S. 143, 152; *see also* Cynthia M. Ho, *Biopiracy and Beyond: A Consideration of Socio-Cultural Conflicts with Global Patent Policies*, 39 U. MICH. J. L. REFORM, 433, 473 (2006).
11 Nagoya Protocol on Access to Genetic Resources and the Fair and Equitable Sharing of Benefits Arising from their Utilization (ABS) to the Convention on Biological Diversity, Oct. 29, 2010, C.N.782.2010.TREATIES-1 (Depositary Notification) [hereinafter 'Nagoya Protocol'].

12 *See* European Commission (2012), Proposal for a Regulation of the European Parliament and of the Council on Access to Genetic Resources and the Fair and Equitable Sharing of Benefits Arising from their Utilization in the Union, *available at* www.cbd.int/doc/measures/abs/post-protocol/msr-abs-eu-en.pdf (last visited Jan. 16, 2017).

13 Nagoya Protocol, *supra* note 12 at Art. 4(4). *See also* Thomas Geiber, Sonia Peña Moreno, Mattias Åhrén, Jimena Nieto Carrasco, Evanson Chege Kamau, Jorge Cabrera Medaglia, Maria Julia Oliva & Frederic Perron-Welch in cooperation with Natasha Ali and China Williams, *An Explanatory Guide to the Nagoya Protocol on Access and Benefit-sharing*, IUCN Environmental Policy and Law Paper No. 83, 2012.

14 *See* Nagoya Protocol Art. 15. However, the mandatory 'shall' language in such Protocol provisions is weakened by the insertion of vague, broad phrases such as 'effective and appropriate legislative, administrative or policy measures', 'as far as possible', and 'as appropriate'. In fact, the Protocol has been called 'a masterpiece in creative ambiguity' on a variety of topics. Maria Julia Oliva (2011), 'Sharing the Benefits of Biodiversity: A New International Protocol and Its Implications for Research and Development', 77 *Planta Medica* 1221, 1223.

15 *See* Nagoya Protocol Art. 15. Denmark's recent NP implementing legislation provides one approach to complying with this obligation, as it prohibits the utilization of GRs when the use is based on GRs acquired in violation of GR access regulations in the country where the GRs were accessed. Violations are punishable by fines, or up to two years in prison if willful or grossly negligent, and foreign states and persons appear to have standing to bring relevant actions in Danish courts. *See* 'Implementing the Nagoya Protocol in Denmark', Denmark Ministry of the Environment, *available at* www.cbd.int/abs/side-events/cop-11/denmark-en.pdf (last visited Dec. 21, 2016).

16 *See* CBD, *Measures on Access and Benefit-Sharing for Implementing the Nagoya Protocol, available at* www.cbd.int/abs/progress/default.shtml (last visited Jan. 3, 2016).

17 *See Peru Accuses Chinese Companies of Biopiracy Involving Maca Root*, MOELLER IP ADVISORS (Aug. 4, 2014), www.moellerip.com/peru-accuses-chinese-companiesof-biopiracy-involving-maca-root/. *See also* Sanofi, *Sanofi Position on Biodiversity and Biopiracy*, (posted May 2013), *available at* http://en.sanofi.com/csr/download_center/download_center.aspx. (last visited Dec. 21, 2016).

18 Daniel F. Robinson (2010), *Confronting Biopiracy: Challenges, Cases and International Debates*, p. 21 (Routledge/Earthscan).

19 Piracy, *Merriam Webster Dictionary*, *available at* www.merriam-webster.com/dictionary/piracy (last visited Jan. 3, 2016).

20 Anthony Faiola, 'Amazon Cash Crop', *Washington Post*, July 9, 1999.

21 Joshua D, Sarnoff & Carlos M. Correa, *Analysis of Options for Implementing Disclosure of Origin Requirements in Intellectual Property Applications* (UNCTAD ed., 2005), *available at* http://digitalcommons.wcl.american.edu/pijip_trad_knowledge/3 (last visited Jan. 16, 2017); G. Kristin Rosendal, *The Convention on Biological Diversity and Developing Countries*, 159–161 (1st ed. 2000).

22 Queen Mary Intellectual Property Research Institute, *Report on Disclosure of Origin in Patent Applications*, 3, 5, 53–54 (2004).

23 *See* Harold Demsetz, 'Toward a Theory of Property Rights', 57 *American Economic Review* 347, 348 (1967).

24 Naomi Roht-Arriaza, 'Of Seeds and Shamans: The Appropriation of the Scientific and Technical Knowledge of Indigenous and Local Communities', 17 *Michigan Journal of International Law* 919, 955 (1996) ('The goal is not simply to receive money in exchange for access to knowledge and resources, but to control whether, and how, such knowledge is commercialized, while also leaving it available for noncommercial uses').

25 Ahmed Abdel-Latif, 'Genetic Resources, Patents and Benefit Sharing: State of Play and Challenges Facing Multilateral Discussions', in *Intellectual Property in the Pharmaceutical Industry*, 59–88 (Jacque de Werra ed., 2012).

26 International Centre for Trade and Sustainable Development, *WIPO Traditional Knowledge Committee Pushes Toward Text-Based Talks* 13 *Bridges* (Dec. 16, 2009) *available at* www.ictsd.org/bridges-news/bridges/news/wipo-traditional-knowledge-committee-pushes-toward-text-based-talks (last visited Jan. 16, 2017).

27 It should be noted that at least 108 countries, more than two-thirds of the WTO's members, expressly endorsed a proposed DOO amendment to the TRIPS Agreement. *See* WTO Council for Trade-Related Aspects of Intellectual Property Rights, 'Minutes of Meeting Held in the Centre William Rappard on 26–27 October 2010' (17 February 2011) IP/C/M/64, para. 22, *available at* www.wto.org/english/tratop_e/trips_e/intel6_e.htm (search IP/C/M/64 in meeting minutes) (last visited Jan. 18, 2017).

28 *See* Intergovernmental Committee on Intellectual Property and Genetic Resources, Traditional Knowledge and Folklore, *Report*, WIPO/GRTKF/IC/26/8 (Mar. 24, 2014) (hereinafter '26th IGC Report').

29 Ibid. at para. 50.

30 Ibid. at para. 47.

31 Ibid. at para. 24.

32 Ibid. at para. 18 (emphasis added).

33 As Oldham *et al.* explain:

> Recent estimates suggest that global receipts from intellectual property across all areas of technology stood at USD180 billion in 2009. Intellectual property involving genetic resources and traditional knowledge will represent an unknown percentage of global receipts. The implication of demands for disclosure of origin is that developing countries would receive a share of receipts when a patent involved a genetic resource or traditional knowledge originating from their country. . . The lack of clarity on the actual economic values of biodiversity and traditional knowledge related intellectual property and related products . . . creates a situation of expectations about the high economic value of biodiversity and traditional knowledge among 'provider' countries that is pitted against the fear of exposure to the economic costs of disclosure among 'user' countries.

Paul Oldham, Stephen Hall, and Oscar Forero (Nov. 13, 2013), *Biological Diversity in the Patent System*, PLoS ONE 8(11), e78737. doi:10.1371/journal.pone.0078737, at p.11.

34 Patent (Amendments) Act, 2005 at Sec. 25 (India).

35 Brazilian Provisional Act, No. 2,186–16, Title IX, Art. 31, August 23, 2001. The 2015 Brazilian Biodiversity Act mentions IP rights but does not appear to contain a specific patent application DOO requirement.

36 Qingkui Zhang and Dongcheng Pang (May 30, 2014), 'Chinese Patent Law and Protection for Genetic Resources, IP in the Life Sciences Industries', 66 *Intellectual Asset Management Magazine*, 113, 116 *available at* www.iam-media.com/Magazine/Issue/66/Management-report/Chinese-patent-law-and-protection-for-genetic-resources (last visited Jan. 16, 2017). For context, the authors report that 825,000 total patent applications (utility, design, and utility model) were filed in the Chinese patent office in 2013 alone.

37 Ibid.

38 The Decision of the Standing Committee of the National People's Congress on Amending the Patent Law of the People's Republic of China (promulgated by the Standing Comm. Eleventh National People's Cong., December 27, 2008, effective October 1, 2009) (China) [hereinafter 'Chinese Patent Law'] at art. 5 and art. 26.

39 Zhang & Pang, *supra* note 36 at 113.

40 SCHWEIZERISCHES ZIVILGESETZBUCH [ZGB], CODE CIVIL [CC], June 25, 1954, SR 232.14 *as amended* Jan. 1, 2012 (Switz.), Loi fédérale du 25 juin 1954 sur les brevets d'invention (état le 1er janvier 2012), Articles *49a,* 8la and 138 subpara. B.

41 The term 'source' is broadly defined; in comments on the Act, the Swiss government noted that:

> The generic term source should be understood in its broadest sense, covering any type of provenance or origin: it includes national governments and their authorities, including the 'country of origin of genetic resources' and the 'country providing genetic resources' according to Article 2 of the CBD; the 'geographic origin' according to recital 27 of the EU Biotech Directive; the Multilateral System of the FAQ International Treaty; natural and legal persons; indigenous and local communities; *ex situ* collections such as gene banks and botanical gardens; scientific publications and books; and databases.

Intergovernmental Committee on Intellectual Property, Genetic Resources, Traditional Knowledge and Folklore, Sixteenth Session, Policies, Measures and Experiences Regarding Intellectual Property and Genetic Resources: Submission by Switzerland, WIPO/GRTKF/IC/16/INF/14, Annex, p. 2 (Feb. 19, 2010).

42 Ibid. at p. 4.

43 See Dr. Martin Girsberger, *Presentation of Dr. Martin Girsberger, Roundtable 4: National Experiences with Disclosure Requirements Related to Genetic Resources and Associated Traditional Knowledge* (Mar. 2015), *available at* www.wipo.int/meetings/en/doc_details.jsp?doc_id=300304 (last visited Jan. 3, 2016).

44 For example, Belgium, Brazil, Chile, China, Egypt and Switzerland all require DOO as a condition of patentability. Loi modifiant la loi du 28 mars 1984 sur les brevets d'invention, en particulier la brevetabilité des inventions biotechnologiques [Law Modifying the Law on Patentability of Biotechnological Inventions] of April 28, 2005, MONITEUR BELGE [Official Gazette of Belgium], *available at* www.wipo.int/wipolex/fr/text.jsp?file_id=180969; Brazil: Brazilian Provisional Act, No. 2,186–16, Title II, Art. 7, para. I, August 23, 2001, Lei Provisional No. 2.186–16, de Agosto de 2001, CONSTITUIÇÃO FEDERAL [C.F.] [CONSTITUTION] art. 225 (Braz.); Law No. 20569, Febrero 6, 2012, REPERORIO DE LEGISLACIÓN Y JURISPRUDENCIA CHILENAS [REP. LEG. JURISP.] (Chile); Law No. 82 of 2005 (Law on the Protection of Intellectual Property Rights) *Al-Jarida Al-Rasmiyya,* June 3, 2002 (Egypt); SCHWEIZERISCHES ZIVILGESETZBUCH [ZGB], CODE CIVIL [CC], June 25, 1954, SR 232.14 *as amended* Jan. 1, 2012 (Switz.); Chinese Patent Law, *supra* note. Other countries with DOO requirements include Denmark, Norway, Bolivia, Colombia, Ecuador, Peru, Costa Rica, India, Cuba, Djibouti, Namibia, Kyrgyzstan, the Philippines and South Africa.

45 Of course, it is possible to make DOO requirements more useful. As Dr. Oldham notes 'Enhancing disclosure is important because taxonomic information about existing species and biodiversity is, and will always be, imperfect.' He suggests, inter alia, enhancing disclosure through, among other things, requiring identification of the source country in the application text and whether any indigenous and local communities were involved in the identification and/or invention process. Paul Oldham, *Tracking & Monitoring Genetic Resources in the Patent System,* ESRC CENTRE FOR ECONOMIC AND SOCIAL ASPECTS OF GENOMICS (CESAGEN) & UNU, *available at* www.abs-initiative.info/fileadmin/user_upload/Activities/2011/Addis_IP_09-2011/Documents/2_-_Tracking___Monitoring__GRs_in_the_Patent_System___Paul_Oldham__ESRC_Cesagen___UNU_.pdf (last visited Jan. 16, 2017).

46 Council for Trade-Related Aspects of Intellectual Property Rights, Communication by the United States, Article 27.3(b), Relationship Between the TRIPS Agreement and the CBD, and the Protection of Traditional Knowledge and Folklore, IP/C/W/469, para. 37 (Mar. 13, 2006). *See also* Jonathan Carr, 'Agreements That Divide: TRIPS vs. CBD and Proposals for Mandatory Disclosure of Source and Origin of Genetic Resources in

Patent Applications', 18 *Journal of Transnational Law and Policy* 131, 144 (2008) (discussing U.S. opposition to proposed TRIPS disclosure of origin amendments).

47 *See* Star Fruits S.N.C. v. United States, 393 F.3d 1277 (Fed. Cir. 2005).

48 Ibid.

49 The Rule states 'Any reply to a requirement for information pursuant to this section that states either that the information required to be submitted is unknown to or is not readily available to the party or parties from which it was requested may be accepted as a complete reply.'

50 *See* 37 C.F.R. 1.56 (2014).

51 USPTO, Manual of Patent Examining Procedure 704.11 [R-08.2012].

52 35 U.S.C. §200 (2014). As noted by Congressman Brown: 'Policy tools like patents, the Bayh-Dole Act, the Stevenson-Wydler Act, and incentives for commercialization, are important links in the bench to bedside chain.' *NIH: Moving Research from the Bench to the Bedside: Hearings Before the Subcomm. On Health of the Committee on Energy and Commerce*, 108 Cong. (2003) (statement of Sherrod Brown, D-OH).

53 *See* 35 U.S.C. §§200–211 (2014).

54 Useful summaries of the history of the Bayh-Dole Act and the varying forces leading up to its enactment can be found in: D. Mowery, R. Nelson, B. Sampat & A. Ziedonis (2004) *Ivory Tower and Industrial Innovation: University-industry Technology Transfer Before and After the Bayh-Dole Act* 9–34 (Stanford Business Books); Rebecca S. Eisenberg, 'Public Research and Private Development: Patents and Technology Transfer in Government-Sponsored Research', 82 *Virginia Law Review* 1663 (1996). Some of the material in this section is adapted from Margo A. Bagley, 'Academic Discourse and Proprietary Rights: Putting Patents in Their Proper Place', 47 *Boston College Law Review* 217 (2006).

55 Ibid. at 9–34.

56 While the funding agency could make the decision to allow licensing, such decisions were rare and were made on a case by case basis resulting in significant uncertainty regarding the likelihood of a favourable result. It was well understood, of course, that commercialization was well beyond the mission, resources, and expertise of university researchers and should be handled by the private sector.

57 *See also* Stevenson-Wydler Technology Innovation Act of 1980, Pub. L. No. 96–480, 94 Stat. 2311–2320 (codified as amended at 15 U.S.C. 3701–3714 (1994)) (addressing technology transfer in government laboratories).

58 35 U.S.C. §200 (2014).

59 *See* Council on Governmental Relations, *The Bayh-Dole Act: A Guide to the Law and Implementing Regulations* at 4, *available at* www.umventures.org/sites/umventures.com/files/COGR_Bayh_Dole.pdf (last visited Jan. 16, 2017).

60 35 U.S.C. §202(c)(6) (2014). Section 310 of the USPTO Manual of Patent Examining Procedure (MPEP) further specifies the wording of the disclosure: 'This invention was made with government support under (identify the contract) awarded by (identify the Federal agency). The government has certain rights in the invention', and 37 C.F.R. 1.77(b)(3) specifies the statement's location in the application/patent. *See also* Government Documents Atlantic Information Services, *A Guide to Managing Federal Grants for Colleges and Universities* Section 1822, *available at* www.managingfederalgrants.com/1800_intellect/1800_intellect_sample.html#Sec1822 (last visited Jan. 1, 2016).

61 *See, e.g.*, Ethicon v. U.S. Surgical Corp., 135 F.3d 1456 (Fed. Cir. 1998); Fina Oil & Chem. Co. v. Ewen, 123 F.3d 1466 (Fed. Cir. 1997).

62 Paul Oldham, Stephen Hall & Oscar Forero, *Biological Diversity in the Patent System*, pp. 11–12, PLoS ONE 8(11): e78737. doi:10.1371/journal.pone.0078737 (2013).

63 Defined in 16 C.F.R. § 801.1(a)(3).

64 Department of Commerce United States Patent and Trademark Office, Changes to Require Identification of Attributable Owner 79 Fed. Reg. 4105 (Jan. 24, 2014) (to be codified at 37 C.F.R. pt. 1).

65 Simon J. Elliott, 'The Complexities of the USPTO Proposed Attributable Ownership Rules' in *PharmaPatents*, www.pharmapatentsblog.com/2014/04/11/the-complexities-of-the-uspto-proposed-attributable-ownership-rules/ (online April 11, 2014) (last visited Jan. 16, 2017).

66 Department of Commerce United States Patent and Trademark Office, Interim Guidance on Patent and Subject Matter Eligibility, 79 Fed. Reg. 16 (proposed January 24, 2014) (to be codified at 37 C.F.R. pt. 1).

67 Courtenay C. Brinkerhoff, 'The USPTO Does Not Need the Onerous Proposed Attributable Ownership Rules', in *PharmaPatents* (online, April 4, 2014) www.pharmapatentsblog.com/2014/04/04/the-uspto-does-not-need-the-onerous-proposed-attributable-owner-rules/ (last visited Jan. 16, 2017).

68 United States Patent & Trademark Office, 'Attributable Ownership', United States Patent and Trademark Office (February 5, 2015), *available at* www.uspto.gov/patent/initiatives/attributable-ownership (last visited Dec. 27, 2015).

69 The Swiss delegation to IGC 31 submitted a document which helpfully provides additional information on the operation of the Swiss DOO requirement. See *The Declaration of the Source of Genetic Resources and Traditional Knowledge in the Swiss Patent Act and Related Swiss Regulations on Genetic Resources – Submission by Switzerland in Response to Document WIPO/GRTKF/IC/30/9*, Document WIPO/GRTKF/IC/31/8 (Sept. 12, 2016).

70 See Margo Bagley, Ruth L. Okediji & Jay A. Erstling, *International Patent Law and Policy*, Chapters 9–10 (West 2013) (discussing post grant proceedings and infringement, claim construction, purposive construction, doctrine of equivalents, etc.).

71 *See* EPO, *Annual Report 2013 – Searches, Examinations, Oppositions*, 2014 European Patent Office, www.epo.org/about-us/annual-reports-statistics/annual-report/2013/statistics-trends/searches.html#tab=3; Steven Seidenberg, 'Patent Lawyers Ponder the Changed Post-Grant Process', www.abajournal.com/magazine/article/patent_lawyers_ponder_the_changed_post-grant_process (posted Feb. 1, 2013); Steven Seidenberg, *US Perspectives: New USPTO Post-Grant Review A Small Step For Patent Harmonisation*, IPWatch.com, www.ip-watch.org/2012/10/18/new-uspto-post-grant-review-a-small-step-for-patent-harmonisation/ (posted Oct. 18, 2012) (last visited Jan. 16, 2017).

72 Robert Smyth, *White Paper Report: United States Patent Invalidity Study 2012* (presented at the 2012 IPO Annual Meeting), *available at* www.morganlewis.com/~/media/files/publication/presentation/speech/smyth_uspatentinvalidity_sept12.ashx (last visited Jan. 16, 2017).

73 Japan Patent Office, Annual Report 2012, 175, www.jpo.go.jp/shiryou_e/toushin_e/kenkyukai_e/pdf/annual_report2012/all.pdf (last visited Jan. 16, 2017).

74 Matt Cutler, '3 Years of IPR: A Look at the Stats', Law 360 (Oct. 9, 2015), *available at* www.law360.com/articles/699867/3-years-of-ipr-a-look-at-the-stats (last visited Dec. 27, 2015).

75 *See* Ass'n for Molecular Pathology v. Myriad Genetics, 133 S. Ct. 2107 (2013); Alice Corp v. CLS Bank Int'l, 134 S. Ct. 734 (2013); and Mayo Collaborative Services v. Prometheus Labs, 132 S. Ct. 1289 (2012). In the same way, earlier decisions by that Court and the Court of Appeals for the Federal Circuit cast doubt on the validity of thousands of other patents on expressed sequence tags and 'obvious' inventions and considerably reduced the certainty of injunctive relief for patent infringement that had previously been routinely granted. See KSR Int'l Co. v. Teleflex Inc., 550 U.S. 398 (2007); eBay Inc. v. MercExchange, 547 U.S. 338 (2006); In re Fisher, 421 F.3d 1365 (Fed. Cir. 2005). For a further discussion of subject matter eligibility challenges see Margo A. Bagley (2014), 'Patent Barbarians at the Gate: The Who, What, When, Where, Why & How of U.S. Patent Subject Matter Eligibility Challenges', in *Patent Law in Global Perspective*, 149 (Ruth L. Okediji & Margo A. Bagley eds., Oxford University Press).

76 *See, e.g.*, D'Arcy v. Myriad Genetics [2015] HCA 35 (Australia), European Patent Office Case Number G 0003/08, October 16, 2009; Case C-34/10, Brüstle v. Greenpeace, 2011 E.C.R. I-09821 [2011].

77 *See* Microsoft v. i4i Ltd., 131 S. Ct. 2238, (2011); ALCOA v. Amerola Products Corp., 552 F.2d 1020, 1021 (3d Cir. 1977); American Hoist & Derrick Co. v. Sowa & Sons, Inc., 725 F.2d 1350 (CA Fed. 1984); Joblove v. Barr Labs., Inc. (In re Tamoxifen Citrate Antitrust Litig.), 466 F.3d 187 (2d Cir. N.Y. 2006); Grefco, Inc. v. Kewanee Indus., 499 F. Supp. 844, 845 (D. Del. 1980).

78 *See, e.g.*, Yoshinari Oyama, *Standard of Proof for Patent Invalidation in the U.S. and Japan*, 13 Chi.-Kent J. Int'l & Comp. L. 25 (2012): Malwina Mejer & Bruno van Pottelsberghe de la Potterie, 'Economic Incongruities in the European Patent System', 34 *European Journal of Law and Economics* 215, 217 (2012); Michael Carrier, 'Post-Grant Opposition: A Proposal and a Comparison to the America Invents Act', 45 *UC Davis Law Review* 103, 112–120 (2011); Alexander Harguth, *Patents in Germany and Europe: Procurement, Enforcement and Defense,* p. 107 (Walters Kluwer, 2011); Tun-Jen Chiang, 'Fixing Patent Boundaries', 108 *Michigan Law Review* 523, 531 (2010); Mark A. Lemley & Kimberly A. Moore, 'Ending Abuse of Patent Continuations', 84 *Boston University Law Review* 63 (2004); and Japan Patent Attorneys Association, *Unique JPO Practices*, JPAA.com, www.jpaa.or.jp/english/patent/unique_jpo_practices.html#subcontents3 (last visited Jan. 16, 2017).

79 Mark A. Lemley (2013), 'The Fractioning of Patent Law', in *Intellectual Property and the Common Law*, pp. 504–505 (Shyamkrishna Balganesh, ed., Cambridge University Press).

80 Ibid. at 506.

81 Mark A. Lemley & Carl Shapiro, 'Probabilistic Patents', 19 *Journal of Economic Perspectives* 75 (2005).

82 Jason Rantenen, 'Patent Malleability', *Michigan State Law Review* (forthcoming 2016).

83 As Professor Crouch insightfully notes: We only learn the true scope of a claim once it is interpreted in court – and often only after an appeal to the Federal Circuit. . . . Claim construction is an issue in the vast majority of patent infringement lawsuits and, over the years, the Federal Circuit has altered a district court's claim construction in hundreds of different cases. In many of those, the district court's opinion was a reasonable interpretation of the claim scope. Dennis Crouch, *Oral Arguments: Can the Supreme Court Force Clarity in Claim Scope?* PATENTLYO, http://patentlyo.com/patent/2014/04/arguments-supreme-clarity.html (discussing Nautilus, Inc. v. Biosig Instruments, Inc., 134 S. Ct. 896 (2014)).

84 In 2012 patent applications filed in the top five largest patent offices; China, United States of America, Japan, Republic of Korea and the European Patent Office, rose by 24%, 7.8%, 0.1%, 5.6% and 4.0%, respectively. WIPO, *IP Facts and Figures*, Economics & Statistics Series, 2013, *available at* www.wipo.int/ipstats/en/statistics/patents/ (last visited Jan. 16, 2017). Furthermore, patent applications overall rose by 4.5% from 2013 to 2014, yet throughout the previous four years the application growth rate had been between 7% and 10%. WIPO, *World IP Indicators 2015*, *available at* www.wipo.int/ipstats/en/wipi/#sections (last visited Jan. 16, 2017). This is not to suggest that this uncertainty is not problematic, it is. And particular decisions and trends can exert a cumulative dampening effect on applications at any given time. For example, a reduction in U.S. patent application filings by domestic applicants in 2014 may be attributable, at least in part, to a series of district court and USPTO decisions invalidating patents on subject matter eligibility grounds based on the above-mentioned Supreme Court precedents. From 2013 to 2014 the filing of U.S. patent applications declined by 0.9%, the first decline in domestic applications since the time period from 2008 to 2009. USPTO, *U.S. Patent Statistics Chart*, www.uspto.gov/web/offices/ac/ido/oeip/taf/us_stat.htm (last visited Jan. 16, 2017).

85 *See* CBD, *The Access and Benefit-Sharing Clearing-House*, https://absch.cbd.int/.
86 Joshua D. Sarnoff and Carlos M. Correa, *Analysis of Options for Implementing Disclosure of Origin Requirements in Intellectual Property Applications* (UNCTAD ed., 2005), *available at* http://digitalcommons.wcl.american.edu/pijip_trad_knowledge/3/ (last visited Dec. 21, 2016).
87 Ibid.
88 Several examples of companies and universities utilizing genetic resources acquired from developing countries both pre-and-post CBD, without clear PIC/ABS/MAT are chronicled in Edward Hammond, 'Biopiracy Watch: A Compilation of Some Recent Cases', *Third World Network*, Vol. 1 (2013), *available at* www.twn.my/title2/books/Biopiracy%20Watch.htm (last visited Jan. 1, 2016).

6 The role of databases, contracts and codes of conduct

*Daniel F. Robinson and Claudio Chiarolla**

Introduction

This chapter focuses on the potential role of databases, contracts and codes of conduct in preventing the erroneous grant of patents on, and the possible misappropriation of, genetic resources (GRs) and associated traditional knowledge (TK). It outlines some recent proposals and discussions in the IGC about the use of these tools. It then examines in further detail the potential benefits and limitations of using each of these tools as mechanisms largely for providing defensive protection to GRs and TK.

Recent proposals and discussions in the IGC

In the past few years, considerable consolidation of ideas surrounding the protection of traditional knowledge and on genetic resources have been achieved in the IGC in the form of consolidated texts that have the potential to become new international legal instruments.[1] The most recent IGC meeting to consider the protection of IP and genetic resources (prior to the adoption of the new mandate)[2] was the 26th session of the IGC, which took place from February 3 to 7, 2014.[3] The outcome document, namely, the *Consolidated Document Relating to Intellectual Property and Genetic Resources Rev. 2*, developed during the session, was transmitted to the 2014 General Assembly. Similarly, at the close of the 27th and 28th sessions, respectively, the Committee also transmitted the outcome documents concerning *The Protection of Traditional Knowledge: Draft Articles, Rev. 2* and *The Protection of Traditional Cultural Expressions (TCEs): Draft Articles Rev. 2* to the 2014 General Assembly.[4]

Despite these important advancements on consolidating and duly reflecting the proposed normative options in the Draft Articles on TK and TCEs, as well as in the Consolidated Document on IP and GRs, these three texts are heavily bracketed as to the scope and methods of protections, among other things, with deep divisions evident between key countries' coalitions.

* The views expressed in this chapter are the author's own and do not necessarily reflect the views of WIPO or any of its Member States, or of the authors' affiliated organizations.

This chapter focuses specifically on GRs and associated TK and the proposals that databases, contracts and codes of conduct can be used to prevent the erroneous grant of patents over inventions that are derived from, or utilize, GRs and also associated TK. The use of databases and codes of conduct have been argued as potential solutions to this problem by several countries. However, in the WIPO IGC normative process, their use has been advocated most strongly by the countries of 'Group B' and, in particular, in submissions by the Delegations of Canada, Japan, the Republic of Korea and the United States of America (US)[5] as an *alternate solution* to the proposed introduction of new patent disclosure of origin requirements related to GR and TK (see Chapter 5 by Bagley, this volume, for a detailed discussion of disclosure or origin). The use of contracts has also been emphasized by countries such as the US and Japan in both the IGC deliberations and the TRIPS Council.[6] Many biodiverse developing countries such as India and Brazil, however, have often argued that there are inherent 'power imbalances' in the negotiation of contracts and that they may not be enforceable internationally, emphasizing the need for a new patent disclosure requirement (ICTSD, 18 May 2011). This was heavily debated at the 18th and 19th sessions of the IGC in 2011 when the WIPO Secretariat prepared a document of 'Options for Future Work on Intellectual Property and Genetic Resources' (WIPO Secretariat, 2011) reflecting the diverging views of experts from around the world in the Inter-sessional Working Groups. The WIPO Secretariat organized the discussions into three clusters, namely:

a) Options on defensive protection of genetic resources (including databases, information systems and guidelines)
b) Options on disclosure requirements (including mandatory disclosure, further examination of issues relating to disclosure, guidelines and alternative mechanisms)
c) Options on IP issues in mutually agreed terms (MAT) for fair and equitable benefit-sharing (focusing on contracts, including online databases of IP clauses in MAT on access and benefit sharing [ABS], guidelines for contractual practices, and licensing practices on GRs).

The subsequent negotiations were exemplary of the split between countries over a disclosure of origin requirement, particularly on 'mandatory disclosure' proposals, which in the *demandeurs*' views should be coupled with remedies such as the rejection of patent applications that have failed to disclose, or have wrongfully disclosed, the required information regarding the origin or legal provenance of GRs and associated TK. For example, South Africa, speaking on behalf of the African Group and supported by Brazil, made clear that they were 'not prepared to discuss cluster A and C if it did not come with a strong disclosure proposal' with India also supporting this, arguing: 'mandatory disclosure is a must – everything else can follow' (ICTSD, 18 May 2011). During the talks at the 18th session of the IGC and in many subsequent sessions, Group B often sought to broaden out the discussion away from disclosure to focus on rules, procedures and mechanisms for the prevention of erroneous grant of patents including the role of databases, codes of conduct and the use of clearly designed/model contracts. This impasse is reflected in the bracketed options included in the 'Consolidated Document on IP and GRs', which does contain a possible framework for a patent disclosure requirement, but also an option for 'no disclosure requirement' at all, as well as draft provisions on 'defensive measures' including text on databases and codes of conduct.

A similar contrast between alternatives regarding the assessment of the need for mandatory patent disclosure requirements *vis-à-vis* the sufficiency of other defensive measures also emerged during the negotiation of the 'Draft Articles for Protection of Traditional Knowledge'. These divisions contributed to the – at least temporary – breakdown on negotiations towards the convening of a potential diplomatic conference (see Chapter 3 by Wendland) and stalled the programme of work of the IGC between the WIPO General Assemblies of September 2014 and October 2015.

The following sections examine the potential role of databases, contracts and codes of conduct, with reference to IGC country positions, Intersessional Working Group expert opinions and case studies from around the world. While these mechanisms and tools could have been clustered and discussed under the rubric of 'non-disclosure options', mainly for reasons that reflect the predominant way in which relevant proposals have been bargained and framed in the context of the IGC, we prefer to refrain from doing so, since it is also important to highlight that their use is not in principle incompatible with, and can, in fact, be mutually supportive of patent discourse requirements related to GR and TK.[7]

Databases

At the time of writing, the most recent submission on the use of databases for defensive protection was made at the 31st session of the IGC (September 19 to 23, 2016) by the Delegations of Canada, Japan, the Republic of Korea and the US. This is in fact a resubmission by the group, and closely reflects earlier submissions, including a first proposal circulated by Japan in the IGC as far back as April 2006 (Japan 2006, WIPO/GRTKF/IC/9/13). The submission suggests that the 'erroneous granting of patents can be effectively addressed by improving databases for storing genetic resources and non-secret traditional knowledge associated with genetic resources that are used for prior art or reference material searches' (Canada *et al.* 2016, WIPO/GRTKF/IC/31/6).

At the 30th session of the IGC (May 30 to June 3, 2016), developing country delegates introduced new bracketed text in the form of Article 7.2, with regards to databases being 'complementary to mandatory disclosure' as part of 'defensive measures' for prevention of the erroneous grant of patents (sometimes referred to as misappropriation in the negotiations). Article 7.2 in brackets states:

> As a complement to the disclosure obligation provided for in Article 3, and in the implementation of this instrument, the Contracting State may consider the

> use of databases on traditional knowledge and genetic resources in accordance with its needs, priorities, and safeguards as may be required under national laws and special circumstances.

The idea of databases as 'complementary to mandatory disclosure' has been rejected by most Group B countries who have preferred language such as 'due diligence', with this debate reflecting the diverging opinions of member states regarding the purpose of these proposed provisions.

A brief digression is worthwhile to explain key conceptual issues that underpin the possible functions of databases and the problems that can be prevented by avoiding the erroneous grant of patents *and/or* the alleged misappropriation of GRs and TK. It is important to note that the policy goal being pursued by the above proposal is limited to help with addressing the core issue of the erroneous grant of patents, without any further pretentions that databases shall promote compliance by patent applicants with other legal requirements exogenous to the patent system – i.e. requirements non-material to the patentability of inventions, such as ABS requirements that may nonetheless affect the legal status of GRs and TK.

The former goal is notably narrower than the dual policy goal that is generally pursued by countries which are also *demandeurs* of compulsory patent disclosure requirements.[8] In the latter's view, databases and registries of GRs and TK can be used as operational mechanisms for both the defensive and positive legal protection of GR and TK.[9] Hence, the argument goes, the policy goal of preventing the misappropriation of GRs and TK, in a comprehensive way, shall be achieved, along with the avoidance of erroneous patent granting as such.

Admittedly, these two policy goals may sometimes overlap (see below Figure 6.1). The possible prevention of erroneous patents, by making available to patent examiners information (about existing TK or other GR-related information) that would defeat the novelty of the claimed invention, may also help in preventing their misappropriation. This would be realized in such cases where access to relevant GRs and associated TK by a user (i.e. the patent applicant) has occurred in violation of applicable PIC and MAT requirements of the provider country. However, this overlap is not always present in each and every given case. There may be instances where prior art exists that would destroy novelty of a GR or TK-related invention, even if no alleged violation of PIC and MAT requirements is involved. *Vice versa* a GR or TK may have been acquired in violation of the applicable PIC and MAT requirements although from a strict patentability standpoint the claimed invention does meet the patentability requirement of novelty, inventive step, industrial application and sufficiency of disclosure.

Against this backdrop, the proposal by Canada *et al.* advocates a 'one-click' database search system, whereby national databases are maintained and controlled by each participating member State and are made accessible through a WIPO portal site by examiners from around the world. The proposal suggests that member States would develop their databases according to their own countries' sensitivities and considerations, and it would therefore be their responsibility to take into account issues such as customary law, conditions for providing information and

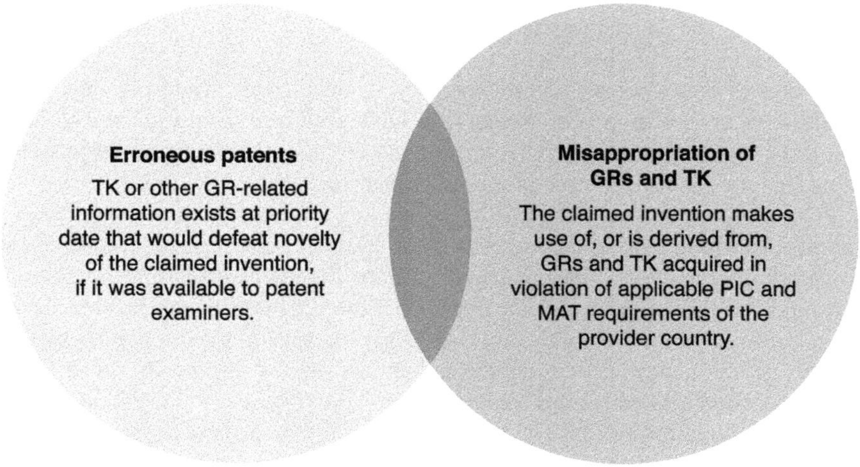

Figure 6.1 Overlap between prior art issues affecting novelty of the claimed invention and the possible violation of PIC and MAT requirements concerning the legal status of background GRs and TK used in the R&D process.

coordination among multiple parties with common knowledge. Anticipating some broader issues concerning the management of such a database system, the proposal also suggests that third party access (i.e. by people other than patent examiners) could be authorized only from registered IP (internet protocol) addresses (or also, presumably, by password-protected user accounts) (Canada *et al.* 2016, WIPO/GRTKF/IC/29/6). Licensing access to national databases – and the specific terms under which such access may be provided to patent offices and other stakeholders, including non-disclosure provisions – would seemingly be an important management aspect of the proposed system.

Some other important issues that need to be considered with respect to the development of a global database search system and its interface with national-level databases include, in particular: the detailed responsibilities and resource requirements for compiling and maintaining national-level databases;[10] the cost of setting up and operating the database search system; its structure and content; the form in which that content would be expressed; its interoperability with other database systems nationally (e.g. national biobank repositories, which may also contain GR-related information), regionally (e.g. the PINTO database)[11] and internationally (e.g. the ABS clearing house of the CBD);[12] a determination of the purpose(s) of the database (e.g. purely defensive or also positive protection of registered TK);[13] the type of protection afforded to the information included therein, and the management of the rights pertaining to the database.

As regards the harmonized contents of a proposed 'one-click' database search system accessible through a WIPO portal site, the following categories of information could be included:

- name of the database and contact details;
- member State or organization(s) responsible for administering the database and its status (whether formal or informal);[14]
- scope and type of information included in the database, such as, but not limited to, TK *stricto sensu* and/or other GR-related information;[15]
- the categories of people authorized to access the content of the database (for example, IP offices and/or other stakeholders) and the conditions governing such access for these different categories, if they are non-confidential.

Recently, at the second Seminar on Intellectual Property and Genetic Resources, Traditional Knowledge and Traditional Cultural Expressions, which was organized by the WIPO Secretariat from June 23 to 25, 2015, a specific roundtable included presentations on collections, registers and databases.[16] There were several papers by experts and officials from patent offices, including papers from Australia, the EU, Japan and South Africa. Presentations by officials from the JPO (Mr Yoshinari Oyama) and EPO (Mr Enrico Luzzatto) highlighted that the existing use of the Indian Traditional Knowledge Digital Library (TKDL) and Traditional Chinese Medicine Database, among others, were helping to reduce erroneous grant of patents and post-grant oppositions. Mr Steven Bailie from IP Australia noted the possibility of having a database that linked to maps of *governance* of knowledge relating to genetic resources, rather than the knowledge itself, with some development already under way with the Atlas of Living Australia. While the papers at this seminar are not an exhaustive sample of recent developments on databases, they highlight some of the current approaches and proposals of policymakers from different regions.

Following the papers, Mr Preston Hardison, a policy analyst from Tulalip Tribes, United States, commented on the potential for databases to lead to unwanted use and misappropriation, and issues related to who owns the database, and the way they are protected. He pointed out that in the US if a database is publicly funded, the Freedom of Information Act would allow anybody to ask for access to the information contained in the database (also noted by Hardison at IGC 29; and cited by Saez, 2015). He also expressed concern about petty users of the registered TK without permissions (e.g. herbalists developing commercial herbal medicines from traditional uses). On this note, the Group B proposal shows consideration of these sorts of issues, suggesting an IP (Internet Protocol) Address Authentication System (IPAAS) so that only registered IP addresses of patent examiners or other authorized parties can access the databases.[17]

At IGC 29, Mr Preston Hardison from Tulalip Tribes also asked if the US had consulted with indigenous peoples about the collection of information to populate such databases, if collection would be acceptable to knowledge holders, or if they had studied the potential benefits or harms of databases. The US replied only that 'consultations are ongoing'. Other authors have also raised similar concerns, such as Praveen Raj (2015), who notes that a 'Traditional Knowledge Docketing System' might be a preferred system to indicate the location where knowledge of a specific genetic resource or medicine might be found instead of registers that document the knowledge (as in the case of the TKDL – see Chakravarty, 2010).

This is a similar idea to that in the Australian presentation noted above, which might improve clarity regarding which peoples to seek prior informed consent from for access to GRs and associated traditional knowledge.

There are a number of concerns and potential limitations associated with a reliance on databases. First, as anticipated above in this section, the identification of a genetic resource and associated traditional knowledge by a patent examiner does not necessarily provide any information as to whether prior informed consent and benefit-sharing has been agreed to through mutually agreed terms (as per the United Nations Convention on Biological Diversity (CBD) and its Nagoya Protocol) (discussed in Chapter 15 by Dhar, this volume). One possible solution here is for the examiners to also search the ABS Clearing House of the CBD when they identify GRs and TK that are not accompanied by evidence which shows that they have been lawfully obtained. Notably, the possibility to directly link or include ABS-related information on the legal status of GRs in patent databases is an option which may deserve further consideration. Second, as noted by delegates from Namibia, Indonesia, Nigeria and India at IGC 29, databases are typically likely to be incomplete as some knowledge is kept secret or some GRs are rare or relatively unknown. This means that searches by examiners may not identify all cases where GRs and associated TK are used.

The nature of biotechnological and pharmaceutical research may also be limiting these preventative measures. For instance, the biochemical derivatives might be listed in the patent without noting the GR that they are extracted from – thus the applicant could avoid identification by an examiner searching GRs databases. With advances in synthetic biology and the ability to create synthetic natural mimics or synthetic analogues, inventors might omit mention of the GRs and associated TK that were the original lead for a particular drug or useful product (see Newman and Cragg, 2012; Parry, 2004). Further, after a patent is granted, what recourse would traditional knowledge holders or GRs providers have if a misappropriation had occurred in the first place? It seems likely they would have to make a legal challenge at some cost to the providers/knowledge holders.[18]

Contracts

The use of contracts for access to genetic resources is embedded in the CBD and Nagoya Protocol's insistence on achieving 'Mutually Agreed Terms' (MAT), and is found in other international instruments, such as the Standard Material Transfer Agreement (SMTA) of FAO International Treaty on Plant Genetic Resources for Food and Agriculture (see Chiarolla, 2008). As such, contracts are already an important tool used for ABS and to license the use of GRs.

In the context of the IGC deliberations, 'Mutually Agreed Terms for Fair and Equitable Benefit-Sharing' have been discussed as a set of options that were grouped under the so-called 'Cluster C'.[19] At the 19th session of the IGC, held from July 18 to 22, 2011, the Committee requested the Secretariat to finalize, and regularly update, the activities referred to in 'Cluster C' and to provide information to the

Committee at each session.[20] At present, four reports on the implementation of 'Cluster C' activities have been prepared by the Secretariat for IGCs 20, 23, 26 and 29.[21]

In particular, the WIPO Secretariat has provided a list of model agreements and actual agreements that can be searched by the clauses that they contain.[22] These indicate that already there are dozens of existing ABS agreements and related contracts for transfer of GRs (e.g. Material Transfer Agreements) that have been adopted between relevant providers and users (see also Robinson, 2015 and CBD Secretariat, 2015). These agreements are often complex, require significant negotiation, may involve transfer of associated knowledge and/or IP, and will have sectoral differences (e.g. for health, agriculture, forestry, fisheries; see Tvedt, 2014). As Bhatti *et al.* (2009, p5) note: '[a]ttempts to determine and apply contract law to ABS have long been recognized as quite complex by experts working in natural resource industries and related commercial sectors, such as forestry, agriculture, and fisheries. Consequently, the government officials and legal experts who have actually applied relevant commercial law to ABS contracts in non-theoretical situations (i.e., "in real life") generally disagree with the "ABS expert" view that the contractual and legal concerns of ABS are relatively simple and few.'

Although generally seen as a useful, if not essential, tool for regulating access to genetic resources and associated knowledge,[23] there are some concerns about their limitations. For example, there have been agreements whereby the terms are not really enforceable once the GRs have left the country, once the term of the contract ends, and/or the contract specifications cannot really be monitored. Robinson (2015, Chapter 8) examines an agreement between the Government of Vanuatu and the French National Museum of Natural History for a biodiversity expedition to the Island of Santo, whereby this sort of contractual ambiguity has occurred. In this case the memorandum of agreement was made in 2006 and only applies for three years. Having now lapsed, a number of government officials have been left wondering if duplicate samples will eventually be returned to Vanuatu and they have not received subsequent publications written about the identification of new species in Santo.

Another concern relates to the potential for there to be an imbalance in power relations between researchers and provider communities of GRs and associated TK with regards to negotiation of the contract, monitoring of its terms, and seeking dispute resolution if there is a perceived breach of terms. In some cases it has been suggested that agreements can end up being made to the exclusion of other relevant peoples that might be TK holders or GR providers (e.g. see Tobin, 2001, and Greene, 2004, on agreements made in Peru; and Hayden, 2003 and 2007 on bioprospecting in Mexico). In a survey of experts by WIPO, some indigenous representatives raised these and similar concerns, and Mr Pierre Du Plessis, an expert from Namibia, noted disclosure requirements would still be needed to track whether contracts were being adhered to.[24]

Therefore, while there is considerable agreement that contracts play a role in regulating access to GRs and TK, and benefit-sharing, some differences of opinion remain about the extent to which contracts – in and of themselves – can effectively limit misappropriation and allow for monitoring of GRs uses once they have been collected and used for R&D. However, among the countries that reject the idea of introducing compulsory patent disclosure requirements, the United States of America has consistently disagreed with the view that the patent system should be used as a mechanism to enforce ABS contracts.[25] In particular, at IGC 23, their delegation believed instead that 'a contractual requirement could be useful in promoting transparency if it required a report to be made to a specified authority where a patent application was filed relating to the subject matter of the contract'.[26] At IGC 29 the US suggested that the patent system should not be used to enforce another international agreement (PIC and MAT under the CBD), and have previously argued that contracts are the appropriate way to ensure agreements for benefit-sharing are made. Along the same line of argument, the Republic of Korea also emphasized that 'PIC and benefit-sharing requirements should be implemented through a contractual system rather than within the patent system'.[27]

Finally, it is also worth highlighting that, at the request of the IGC, draft guidelines for contractual practices were developed between 2002 and 2010, and that, in 2013, they were made available as *Draft Intellectual Property Guidelines for Access to Genetic Resources and Equitable Sharing of the Benefits arising from their Utilization.*[28] The WIPO Secretariat is currently reviewing the existing draft of the guidelines with a view to updating and refining them, and making them available in a more user-friendly format.[29] The updated guidelines would continue to be a neutral and non-prescriptive source of practical information for those entering into ABS agreements.

Codes of conduct

Codes of conduct have also been suggested as having a role in encouraging ethical conduct relating to research or work that utilizes GRs, TK or also traditional cultural expressions (TCEs). The WIPO Secretariat provides an extensive list of the codes and statements of ethics of professional societies from around the world, including many in relation to TCEs, cultural artefacts, archives and museum collections (see Chapter 9 by Perlman on TCEs).[30] Similarly, the CBD Secretariat has a list of codes of conduct and guidelines that relate to different sectors that utilize GRs (and in some cases associated TK).[31] Many of these have been in existence for decades. For example, in 1988 the First International Congress of Ethnobiology met and produced the Declaration of Belém, which outlined explicitly the responsibilities of scientists and environmentalists in addressing the needs of local communities and acknowledged the central role of indigenous peoples in all aspects of global planning. It was reputedly the first time that an international scientific organization recognized a basic obligation that 'procedures be developed to compensate native peoples for the utilization of their knowledge and their biological resources' (Statement 4).[32]

While these codes of conduct play an important role in guiding the behaviour of researchers and associated staff in their work with GRs, TK and/or TCEs, in many cases they may not be strictly enforceable. Rather, the codes typically establish a benchmark of expected ethical norms. Consequences for researches that breach ethical codes might in some cases mean sanctions, suspension or expulsion from professional societies. Within some countries their university systems or governments have established ethical protocols for research – such as the Institutional Review Boards (IRB) in the US, or the Human Research Ethics Committees set up by the National Statement on Ethical Conduct in Human Research in Australia.[33] Compliance with this is generally a requirement of employment within the university systems of the US and Australia, with non-compliance resulting in penalties or even expulsion from a university in serious circumstances.

Interestingly, under the so-called 'due diligence approach' that was adopted by the EU Regulation No. 511/2014 (of 16 April 2014 on *compliance measures for users from the Nagoya Protocol on Access to Genetic Resources and the Fair and Equitable Sharing of Benefits Arising from their Utilization in the Union*), best practices, such as those included in codes of conduct, are a key tool to implement compliance-related ABS obligations by EU Member States. In particular, the above Regulation emphasizes that: 'Best practices developed by users should play an important role in identifying due diligence measures that are particularly suitable for achieving compliance with the system of implementation of the Nagoya Protocol at an affordable cost and with a high level of legal certainty. [. . .] Competent authorities of the Member States should consider that the implementation of a *recognised* best practice by a user reduces that user's risk of non-compliance and justifies a reduction in compliance checks' (Preamble, para. 24). The Regulation further provides that: 'Associations of users or other interested parties may submit an application to the Commission to have a combination of procedures, tools or mechanisms, developed and overseen by them, recognised as a best practice. . .' (Article 8 of Regulation (EU) No. 511/2014).

Finally, 'where the Commission receives substantiated information regarding repeated or significant cases of non-compliance [. . .] by a user implementing a best practice', the Commission may examine possible deficiencies in a best practice and cases of non-compliance with the relevant obligations. The association of users or the other interested parties subject to examination shall cooperate with the Commission and assist it in its actions. The examination shall include any remedial actions to be taken by the association of users and may result in a decision to withdraw recognition of the best practice. (See Article 11 of the Commission Implementing Regulation (EU) 2015/1866 of 13 October 2015 laying down detailed rules for the implementation of Regulation (EU) No. 511/2014 as regards the register of collections, monitoring user compliance and best practices.) However, only 'Member States shall lay down the rules on penalties applicable to infringements of [relevant obligations by users of GRs]. . .'. Such penalties shall be effective, proportionate and dissuasive (Article 11 of Regulation (EU) No. 511/2014).

Conclusions

Many delegates to the WIPO IGC have indicated support for the role of databases, contracts and codes of conduct for regulating activities (research and development, ethnobotanical fieldwork, archaeological activities, museum curation, and other related work) that might result in the utilization of GRs, TK and TCEs.

Regarding the prevention of misappropriation of GRs, and associated TK where relevant, many delegates and experts at IGC meetings have expressed concern that these might not be sufficient to prevent erroneous patents and have been arguing for legal text towards a patent disclosure of origin requirement. Despite these differences of opinion, databases, contracts and codes of conduct are generally seen as valuable support tools for patent examination, for the regulation of the behaviour of those working with GRs, TK and TCEs, and they can be complementary to a potential patent disclosure requirement. Current negotiations at the WIPO IGC in 2016–17 urge the IGC to 'expedite its work, with a focus on narrowing existing gaps' such as this perceived gap between the role of databases and contracts and the need for a disclosure requirement. Arguably, if the IGC achieves international legal instrument(s) in the near future, the use of a combination of the above-mentioned tools is likely to be on the menu.

Notes

1 The recently renewed IGC Mandate for the biennium 2016/2017, as agreed by the Assemblies of Member States of WIPO at their 55th session, which took place from October 5 to 14, 2015, states that: 'The Committee will, during the next budgetary biennium 2016/2017, continue to expedite its work, with a focus on narrowing existing gaps, with open and full engagement, including text-based negotiations, with the objective of reaching an agreement on an *international legal instrument(s)*, without prejudging the nature of outcome(s), relating to intellectual property which will ensure the balanced and effective protection of GRs, TK and traditional cultural expressions (TCEs).' Emphasis added. See para. (a) at: www.wipo.int/export/sites/www/tk/en/igc/pdf/igc_mandate_1617.pdf (accessed 10/01/2017).

2 Ibid.

3 www.wipo.int/meetings/en/doc_details.jsp?doc_id=266143 (accessed 10/01/2017).

4 All three consolidated documents were submitted subject to agreed adjustments or modifications arising on cross cutting issues at IGC 28. These documents and the meeting details can be found here: www.wipo.int/tk/en/igc/snapshot.html (accessed 10/01/2017).

5 These four countries have jointly tabled a 'Joint Recommendation on the Use of Databases for the Defensive Protection of Genetic Resources and Traditional Knowledge associated with Genetic Resources' at various IGC as contained in documents WIPO/GRTKF/IC/23/7, WIPO/GRTKF/IC/24/7, WIPO/GRTKF/IC/26/6, WIPO/GRTKF/IC/27/7, WIPO/GRTKF/IC/28/8. This joint recommendation draws from earlier submissions made by the Delegation of Japan in documents WIPO/GRTKF/IC/9/13 on 'The patent system and genetic resources', and WIPO/GRTKF/IC/11/11 and WIPO/GRTKF/IC/20/INF/11 concerning 'Additional explanation from Japan regarding the document WIPO/GRTKF/IC/9/13 on the patent system and genetic resources'.

6 Japan, The Patent System and Genetic Resources, TRIPS Council Submission, 17 October 2007, IP/C/W/504 available here: www.wto.org/english/tratop_e/trips_e/art27_3b_e.htm (accessed 25/11/2015); United States, Article 27.3(b), Relationship Between the TRIPS Agreement and the CBD, and the Protection of Traditional Knowledge and Folklore, 13 March 2006, IP/C/W/469.

7 It is worth noting, for instance, that the first time that the proposed use of databases and registries of TK and biological/genetic resources was put on the agenda of the IGC, it was due to the merit of a 'position paper' submitted to the 2nd session of the IGC by India, on behalf of the Asian Group and China (India *et al.*, 2001, WIPO/GRTKF/

IC/2/10). India, on behalf of the Asian Group, subsequently submitted a 'technical proposal' on the same subject matter to the 4th session of the IGC in December 2002 (India *et al.*, 2002, WIPO/GRTKF/IC/4/14). Interestingly, India, China and other Like-Minded Megadiverse Asia Pacific countries are also strong proponents of compulsory patent disclosure requirements in WIPO as well as in the WTO (see, for instance, WT/GC/W/590).

8 See above, for instance, India *et al.*, 2002, WIPO/GRTKF/IC/4/14, supra n. 7.

9 Ibid.

10 For instance, it could be practically difficult, exponentially expensive, and ethically complex to make respectfully available TK-related content that has never been codified in written form and that is scattered in multiple remote locations and only transmitted orally in local languages. See, for more details, *WIPO Traditional Knowledge Documentation ToolKit* (November 1, 2012), www.wipo.int/export/sites/www/tk/en/resources/pdf/tk_toolkit_draft.pdf (accessed 10/01/2017).

11 http://pinto.euroseeds.eu/(accessed 10/01/2017). The PINTO database of the European Seed Associations provides information on how plant varieties in the EU relate to patents and patent applications. It comprises varieties which have or had EU market authorization, as long as they are linked to a non-expired, non-rejected patent or patent application.

12 https://absch.cbd.int/(accessed 10/01/2017). As regards the potential relevance of developments in other international fora for GRs databases, it is recalled that the *Global Information System* (GLIS) of the FAO International Treaty on Plant Genetic Resources for Food and Agriculture is well positioned to become the largest existing integrated network of GRs databases worldwide. Following a recommendation of the FAO *Task Force on Permanent Global Unique Identifiers (PGUIs) for PGRFA* to adopt Digital Object Identifiers (DOIs) as the PGUIs strategy for the GLIS, the recently adopted work programme on the GLIS 2016–2022 focuses *inter alia* on interoperability of databases. The stated goal is to promote the adoption of DOIs as the global common standard for PGUIs. The CBD, which has participated in the work of the *Task Force*, has reported to have also adopted DOIs as the PGUIs for the ABS Clearing House under the Nagoya Protocol.

13 See WIPO/TK/MCT/11/INF/2, pp. 16–17 and *WIPO Traditional Knowledge Documentation ToolKit* (November 1, 2012) pp. 19–20.

14 For example, these responsibilities may be formally assigned to research institutions or relevant authorities representing indigenous peoples and local communities (from whom TK has originated and to whom it belongs), or they may be simply exercised *de facto* by institutions and authorities that hold relevant information.

15 In a broad sense, GR-related databases may include information on: GR samples in physical repositories; morphological, genetic and/or biochemical characterization of GRs; associated TK and other known uses of GRs; derivatives and by-products; relevant scientific articles; and other metadata categories including contextual data such as the legal status of the material, the providing country, the geographic location of the sampling site, but also environmental data and images obtained from biogeochemical parameters analysis, hydrological and climatic parameters measurement related to the materials.

16 See, in particular, Roundtable 4 on 'Collections, Registers and Databases relating to Genetic Resources, Traditional Knowledge and Traditional Cultural Expressions: Issues and Practical Experiences', www.wipo.int/meetings/en/details.jsp?meeting_id=36502 (accessed 10/01/2017).

17 WIPO/GRTKF/IC/27/7, as per note 5.

18 This highlights the potential importance of pre-grant submissions of evidence. For example, under S.27 of the Australian Patents Act, a pre-examination submission of evidence can be made regarding the novelty, obviousness or usefulness of a patent by a third party, within a specified time frame.

19 WIPO/GRTKF/IC/19/7, 'Options for Future Work on Intellectual Property and Genetic Resources', Annex, pp. 37–39.
20 See WIPO/GRTKF/IC/19/12, 'Report of the Nineteenth Session of the Committee'.
21 See, respectively, WIPO/GRTKF/IC/20/INF/14, WIPO/GRTKF/IC/23/INF/7 REV., WIPO/GRTKF/IC/26/INF/7 and WIPO/GRTKF/IC/29/INF/8.
22 www.wipo.int/tk/en/databases/contracts/ (accessed 18/01/2016).
23 Several experts from different regions of the world supported options to include draft guidelines for contractual practices for future work relating to the relationship between IP and GRs in 2011. See WIPO/GRTKF/IC/19/7 May 20, 2011 'Options for Future Work on Intellectual Property and Genetic Resources'.
24 WIPO/GRTKF/IC/19/7 May 20, 2011 'Options for Future Work on Intellectual Property and Genetic Resources'.
25 WIPO/GRTKF/IC/23/8 PROV. 2, p. 19.
26 Ibid., p. 20.
27 Ibid., p. 50.
28 www.cbd.int/financial/mainstream/wipo-guidelines.pdf (accessed 10/01/2017).
29 WIPO/GRTKF/IC/29/INF/8.
30 www.wipo.int/tk/en/databases/creative_heritage/codes/ (accessed 10/01/2017).
31 www.cbd.int/abs/instruments/ (accessed 10/01/2017).
32 The International Society for Ethnobiology: www.ethnobiology.net/what-we-do/core-programs/global-coalition/declaration-of-belem/ (accessed 19/01/2016).
33 See www.nhmrc.gov.au/guidelines-publications/e72 (accessed 10/01/2017) for details. The Australian guidelines cite additional relevant codes, such as the AIATSIS (2012) Guidelines for Ethical Research in Australian Indigenous Studies.

References

Bhatti, S., Carrizosa, S., McGuire, P., & Young, T. (eds) (2009) *Contracting for ABS: The Legal and Scientific Implications of Bioprospecting Contracts*, IUCN, Gland. Available here: https://portals.iucn.org/library/efiles/edocs/EPLP-067-4.pdf (accessed 18/01/2016).

Canada, Japan, the Republic of Korea and the United States of America (delegations from), (August 23, 2016) 'Joint Recommendation on the use of Databases for the Defensive Protection of Genetic Resources and Traditional Knowledge Associated with Genetic Resources', WIPO/GRTKF/IC/31/6, www.wipo.int/meetings/en/details.jsp?meeting_id=40386 (accessed 10/01/2016).

CBD Secretariat (2015) Model ABS Agreements and Contractual Clauses, www.cbd.int/abs/resources/contracts.shtml (accessed 19/1/2016).

Chakravarty, R. (2010) 'Preserving Traditional Knowledge: Initiatives in India', *IFLA Journal*, 36(4), 294–299.

Chiarolla, C. (2008) 'Plant Patenting, Benefit Sharing and the Law Applicable to the FAO Standard Material Transfer Agreement', *The Journal of World Intellectual Property*, 11(1), 1–28.

Greene, S. (2004) 'Indigenous People Incorporated? Culture as Politics, Culture as Property in Pharmaceutical Bioprospecting', *Current Anthropology*, 45(2), 211–237.

Hayden, C. (2003) *When Nature Goes Public: The Making and Unmaking of Bioprospecting in Mexico*. Princeton University Press.

Hayden, C. (2007) 'Taking as Giving: Bioscience, Exchange, and the Politics of Benefit-sharing', *Social Studies of Science*, 37(5), 729–758.

ICTSD (2011) 'Govts Divided over Disclosure at WIPO Talks on Genetic Resources, Traditional Knowledge', *Bridges* Vol. 15, No. 18: www.ictsd.org/bridges-news/bridges/

news/govts-divided-over-disclosure-at-wipo-talks-on-genetic-resources (accessed 25/11/2015).

Newman, D. J., & Cragg, G. M. (2012) Natural products as sources of new drugs over the 30 years from 1981 to 2010, *Journal of Natural Products*, 75(3), 311–335.

Parry, B. (2004) *Trading the Genome: Investigating the Commodification of Bio-information*. Columbia University Press.

Praveen Raj, R. S. (2015) 'Inside views: No Need of IPRs for Protecting Traditional Knowledge', *IP-Watch*, www.ip-watch.org/2015/09/03/no-need-of-ipr-route-for-protecting-traditional-knowledge/ (accessed 19/1/2016).

Robinson, D. F. (2015) *Biodiversity, Access and Benefit-Sharing: Global Case Studies*. Routledge, Oxon.

Saez, C. 'WIPO: Databases To Protect GRs, TK, Useful But Some Controversy', *IP-Watch*, www.ip-watch.org/2015/06/29/wipo-databases-to-protect-grs-tk-useful-but-some-controversy/ (accessed 25/11/2015).

Tobin, B. (2001) 'Redefining perspectives in the search for protection of traditional knowledge: A case study from Peru', *Review of European Community & International Environmental Law*, 10(1), 47–64.

Tvedt, M. W. (2014) 'Beyond Nagoya. Towards a Legally Functional System of Access and Benefit-Sharing', in Oberthur, S. & Rosendal, G. K. (eds) *Global Governance of Genetic Resources: Access and Benefit Sharing after the Nagoya Protocol*, Routledge, Oxon, pp. 158–177.

WIPO Secretariat (2011) 'Options for Future Work on Intellectual Property and Genetic Resources', 20 May 2011, WIPO/GRTKF/IC/19/7.

7 The legal protection of widely shared and dispersed traditional knowledge

Manuel Ruiz Muller

Introduction

The protection of traditional knowledge (TK) of indigenous peoples remains a contentious issue in the international agenda. However, this is not because of disagreement on the basic premise that TK *needs* legal protection, but rather on the type of legal instrument, mechanisms and tools which may help achieve this goal. The recently approved Draft Articles for TK protection developed by the Intergovernmental Committee on Intellectual Property and Genetic Resources, Traditional Knowledge and Folklore (IGC) of the World Intellectual Property Organization (WIPO), reflect both the consensus and divergences between countries regarding substantial and formal dimensions of this process.[1]

Growing evidence seems to demonstrate that shared and widely dispersed TK is the prevailing rule rather than an exceptional situation. Though certain TK is still secretly guarded by specific leaders and figures in communities, such as the shaman, the elder, the healer, or specific men and women, it is not surprising that a portion of this special knowledge is also kept by other leaders and figures in neighbouring communities, often dispersed across jurisdictions. Traditional knowledge which is not secret or confidential, but still part of the cultural heritage and livelihood of communities, is also often shared and dispersed among non-indigenous individuals, sometimes codified in research papers and books. For most scholars this latter TK would typically fall under the public domain (see Chapter 8 by Okediji on the public domain, this volume).[2]

Some international instruments such as the Nagoya Protocol on Access to Genetic Resources and the Fair and Equitable Sharing of Benefits Arising from their Utilization (2010), consider the possibility for the development and future negotiation of specific multilateral approaches and options for cases of shared genetic resources and TK under articles 10 and 11.[3] The WIPO Draft Articles for TK protection also refer to trans-boundary situations or shared TK.[4] Paradoxically, the debate on shared TK echoes some elements of the discussions regarding shared and widely disseminated genetic resources.

Both the Nagoya Protocol and the WIPO Draft Articles present as an exceptional situation what in reality is a general rule and principle: this is, that biodiversity, TK and genetic resources are widely shared and dispersed across and among eco-regions such as the Amazon, Andes or Mesoamerica, and between indigenous peoples and communities.

This chapter analyses the case of widely shared and dispersed TK and some of the key issues and elements which may require attention if an effective TK protection regime is sought – whether through the international regime proposed by WIPO and the IGC, the Nagoya Protocol and its openings in articles 10 and 11, or through regional or national initiatives. The first section addresses some of the existing trends and initiatives to protect TK worldwide. The second section will analyse some of the challenges confronted by TK protection frameworks in the context of shared and dispersed TK. The following section offers some possible alternatives as to how to legally protect shared and dispersed TK. Some final reflections on this issue are provided as a corollary to the chapter.

The prevailing trends to protect traditional knowledge

The term 'protection' was initially used in the early 1990s very loosely, meaning different things to different people. Only recently has a more systematic approach been developed to understand what exactly 'protection' means (McManis and Terán, 2011). In a classic IP context, protection in general refers to granting exclusionary rights to inventors and creators through IP tools – patents, breeders' rights and copyright. Protection may also mean compensation, social recognition through moral rights, sharing benefits, and maintaining, preserving and controlling access to and uses of TK through unfair competition principles.[5] Defensive protection is yet another form of safeguarding rights pertaining to TK and genetic resources.

Although TK protection initiatives and measures vary considerably in their form and substance, there are some common features which stand out. To begin with, most policy and legal instruments, such as the Peruvian law for TK protection, Costa Rica Law 7788, and Panama Law 21, recognize prior informed consent (PIC) as a critical condition which needs to be met as a pre-requisite for accessing and using TK for any purpose. This involves some form of bilateral interaction with or negotiation between a user and indigenous peoples to define access conditions to their TK.

Almost invariably, TK-related policies and instruments include registers as a tool to support protection measures, whether defensively or to positively help in assigning rights to indigenous peoples.[6] This is the case of existing laws in Peru, Panama, Costa Rica, the Traditional Knowledge Digital Library (TKDL) initiative in India, among others. This is not to say that registers are free from controversy, especially in regards to the fact that they systematize TK under certain pre-established criteria and provide an informational platform which is often alien to indigenous peoples and communities – in content and process.

When discussing and developing these policies and laws, it has been acknowledged that classic IP tools – mainly patents, breeders' rights, copyright – are intrinsically unsuitable to protect indigenous peoples' intellectual efforts and creations. However, there may be alternatives in 'soft' IP tools such as collective marks or geographical indications, or even in the use of competition law principles which

Table 7.1 Key international milestones in the protection of TK

UNESCO	Model Provisions for National Laws on the Protection of Expressions of Folklore Against Illicit Exploitation and other Prejudicial Actions (1982).[7]	Offers mainly guidance on elements to develop measures for the protection of cultural expressions (not necessarily 'TK' as defined in more recent times).
FAO	International Undertaking on Plant Genetic Resources (1983).	Recognizes Farmers' Rights and the International Fund to support its implementation (mainly as a compensatory measure).
FAO	International Treaty on Plant Genetic Resources for Food and Agriculture (2001).	Develops the concept of Farmers' Rights, including through the protection of PGRFA related TK.[8]
UNESCO	Convention for the Safeguarding of the Intangible Cultural Heritage (2003).	Recognizes TK as an expression of the national cultural heritage and implements a system of national lists to identify these specific expressions.
United Nations	United Nations Declaration on the Rights of Indigenous Peoples (2007).	Recognizes a series of rights, including intellectual rights and the collective TK of indigenous peoples and the need for its protection.
United Nations	Nagoya Protocol on Access to Genetic Resources (2010).	Recognizes the need to protect TK mainly through PIC principles and biocultural protocols and ensure the equitable sharing of benefits arising from their utilization.
WIPO IGC	Draft provisions for the protection of TK (2014).	Recognizes a series of tools and instruments (PIC, registers, commercial secrets principles, etc.) to protect TK and implement them at the national level.

could, given certain circumstances, provide some forms of protection to indigenous peoples' intellectual creations.

Finally, and one of the most critical but often underestimated aspects in the development of policies and legal frameworks, there are the very general references to 'traditional knowledge' which are almost never a precise definition of the concept. This issue – together with unclear scope – is one potential limitation at the time of implementation. Do these policies and norms refer to TK as an intangible *per se*, or in its more tangible expression (i.e. a technique, a process, a product)? Do they cover only TK which is publicly accessible or do they mostly refer to TK which is still maintained as confidential by communities or specific community members? Without exception, broad definitions and scope facilitate legal drafting, and in fact even the TRIPS Agreement for example does not define 'invention' in its text.[9] However, in the absence of a 'tradition' of dealing with TK, it makes practical implementation of measures often complicated, as is being experienced in many countries and regions.[10]

Shared and dispersed traditional knowledge

Traditional knowledge in its more classic form as an intangible, is in essence *information.* Information related policy, management, institutional and regulatory frameworks have long been studied and abundant literature has been produced regarding its socio-economic role and implications for development, particularly in the context of markets, decision making and intellectual property (Boyle, 2003). Prominent scholars such as Stigler, Akerlof and Stiglitz have reflected in this field and in some cases won Nobel Prizes. The economics of information theories are a good starting point to understand the best policy and legal options available, especially in the context of TK which is shared by and widely distributed among communities or indigenous peoples.[11]

Very simply, and in the context of indigenous peoples, certain types of goods (e.g. a medicinal potage, a food recipe, a cultivating technique or knowledge regarding the venomous feature of a plant) take very considerable time to be produced and developed. These 'intellectual investments' often involve physical effort, passage of time, continued observation, repeated trial and error processes, further testing and validation and the existence of cultural contexts under which they can effectively evolve. They may also involve inputs from a series of collective and cultural, religious and spiritual activities.

At the same time, these categories of goods follow principles which are also applicable to other informational goods. On one hand, they are cheap to replicate. Any member of a community or a stranger can, with some observation, replicate the good at a very low marginal cost or effort. In very exceptional cases, certain information or TK is maintained, closely guarded and known only by a particular member of the community. These goods are also non-rival and non-exclusionary in the sense that a person's consumption or use does not deprive another from potentially exactly the same consumption or use opportunity.

Furthermore, these goods, and more specifically the TK embedded in them, once released by the creator – whether a single individual or a collective – cannot be easily controlled or managed. Information and TK in this case, flow freely, unimpeded and easily (Vogel, 2000). It is almost impossible to keep TK confined to a community cultural context. In practice, TK can become part of the public domain and/or at the very least, become freely accessible to many.[12] This applies to TK which effectively 'escapes' a single person's control and becomes part of the collective knowledge assets which indigenous peoples and communities hold and claim to have rights over – often *collective* rights.

If this holds true, basic underlying economics and IP principles regarding access to and use of informational goods may be applicable. However, classic IP tools may not be fully applicable or effective in the case of TK given some additional variables, including difficulties in determining a right holder or singling out an inventor or creator, a cultural resistance to assigning monopoly-like property rights, plus the collective nature of the innovation process within communities and trans-generational passing on of specific TK, among others.

Indigenous peoples and their communities generally share TK and are also faced with its wide dissemination and distribution outside community contexts.

This includes broad dissemination within and also across national borders, with concerns regarding the policy and legal implications of this 'transboundary' situation. Only recently has the issue and related problems of shared and widely distributed TK and genetic resources, been placed on the international agenda – albeit timidly.[13]

What makes TK different from other forms of shared and disseminated information and knowledge, may lie in the cultural context under which TK develops and the fact that classic IP tools seem unable to offer appropriate protection responses.

The Nagoya Protocol on ABS establishes in article 10 (Global Multilateral Benefit Sharing Mechanism) that:[14]

> Parties shall consider the need for and modalities of a global multilateral benefit-sharing mechanism to address the fair and equitable sharing of benefits derived from the utilization of genetic resources and **traditional knowledge** associated with genetic resources that occur in **transboundary situations** or for which it is **not possible to grant or obtain prior informed consent**. The benefits shared by users of genetic resources and **traditional knowledge** associated with genetic resources through this mechanism shall be used to support the conservation of biological diversity and the sustainable use of its components globally.

There may be situations where TK is so disseminated and shared so extensively, that it is in practice impossible to celebrate a contractual agreement and ensure PIC conditions of any sort. For these situations, a potential protection policy option may be to develop a multilateral funding mechanism (or use existing mechanisms i.e. GEF, The Global Trust Fund under the FAO International Treaty), which ensures some form of benefit sharing for accessing and using genetic resources and TK. A funding option does not curtail the possibility of other regulatory alternatives, but offers – arguably – a low transaction cost mechanism if designed adequately.[15]

The Nagoya Protocol furthermore recognizes in article 11 (Transboundary Cooperation) that:

> 1 In instances where the same genetic resources are found in situ within the territory of more than one Party, those Parties shall endeavour to cooperate, as appropriate, with the involvement of indigenous and local communities concerned, where applicable, with a view to implementing this Protocol.
>
> 2 Where **the same traditional knowledge** associated with genetic resources is **shared** by one or more indigenous and local communities in several Parties, those Parties shall endeavour to cooperate, as appropriate, with the involvement of the indigenous and local communities concerned, with a view to implementing the objective of this Protocol.

This provision leaves it to Parties to develop a cooperation scheme to ensure benefit sharing is realized in situations of shared, similar TK. Some national legal frameworks (including Andean Decision 391, the African Union Model Law, etc.) have also acknowledged the issue of shared resources and TK but have not really overcome a key challenge at the implementation level: how to technically and efficiently guarantee optimum solutions to a situation where subject matter (TK) is disseminated and widely distributed, and where an identifiable right holder is difficult to engage, is still a major challenge.[16]

Table 7.2 Examples of the treatment of shared or widely disseminated TK in some existing legal instruments

Legal instrument	*Provision*
Law 27811, for the protection of collective knowledge in Peru (2001)	**Article 6.** *The indigenous representative organization, whose prior informed consent is sought [representative organization are deemed the legitimate TK negotiating body on behalf of communities], must inform the widest possible number of communities holders of the same knowledge that it is entering into negotiations, and take into account their interests.*
Decision 391 of the Andean Community on ABS (1996)	**First, Final Provision.** *In the negotiation of the terms of access contracts in cases where more than one Member State is country of origin of genetic resources or derived products, as well as in regards to access activities, the National Competent Authority will take into account the interests of the other countries sharing these resources . . .*
Nagoya Protocol on ABS (2010)	**Article 10.** *Parties shall consider the need for and modalities of a global multilateral benefit-sharing mechanism to address the fair and equitable sharing of benefits derived from the utilization of genetic resources and traditional knowledge associated with genetic resources that occur in transboundary situations or for which it is not possible to grant or obtain prior informed consent. The benefits shared by users of genetic resources and traditional knowledge associated with genetic resources through this mechanism shall be used to support the conservation of biological diversity and the sustainable use of its components globally.*
Nagoya Protocol on ABS (2010)	**Article 11.2.** *Where the same traditional knowledge associated with genetic resources is shared by one or more indigenous and local communities in several Parties, those Parties shall endeavour to cooperate, as appropriate, with the involvement of the indigenous and local communities concerned, with a view to implementing the objective of this Protocol.*
Draft Articles on TK protection (WIPO IGC)	**Article 12** (Trans-boundary Cooperation) provides that: *12.1 In instances where the same [protected] traditional knowledge [under article3] is found within the territory of more than one [Member State]/[Contracting Party], those [Member States]/[Contracting Parties] [should]/[shall] endeavour to cooperate, as appropriate, with the involvement of indigenous and local communities concerned, where applicable, with a view of implementing this [instrument]. 12.2 Where the same [protected] traditional knowledge [under article 3] is shared by one or more indigenous and local communities in several [Member States]/[Contracting Parties], those [Member States]/ [Contracting Parties][should]/[shall] endeavour to cooperate, as appropriate, with the involvement of the indigenous and local communities concerned, with a view to implementing the objectives of this [instrument].*

The main limitations affecting bilateral, contractual negotiations when TK is shared are:

a) How can a contract be negotiated (or PIC obtained) when there is no single nor even defined right holder?
b) Even if this was possible to determine, assuming TK is in the public domain or publicly available, is it feasible, and economically viable, to negotiate an advantageous contract?
c) What are the effects of economic pressures (impacting benefit potential) when TK is in practice found and accessible from various sources?

One additional dimension affecting bilateral contracts in regards to shared TK – and any TK for that matter – are the information asymmetries which in these particular circumstances, indigenous peoples and their representatives may face when negotiating an agreement or contract.

The policy and legal options to protect shared and dispersed TK

The issue of shared and widely distributed TK has become part of international agendas rather recently. The IGC early in its process recognized that obtaining PIC in cases where right holders are not defined could be a problem and, more importantly, affect and create tensions between communities sharing similar resources and TK but with different views about what to do with them. This included situations where TK might even be in the public domain and available. These two well-established principles in IP have been strongly contested by indigenous peoples who claim their rights over TK go way beyond whether or not they have lost control or whether their TK is now disseminated, in the public domain or readily available and accessible.

To define the best and most effective policy and legal option regarding the protection of shared and widely disseminated or distributed TK, decisions should bear in mind the following situations.

TK which is maintained confidential among different communities and is shared

Many communities may possess the same or rather share, very similar confidential and/or sacred TK.[17] If this is the case, there may be unlimited options to obtain and access this TK from multiple sources – or at least as many options as communities there are with this TK. This is *if* communities or holders decide to share the TK.

The almost certain scenario, is that only one or a few selected communities involved in a specific bioprospecting or ethnobotanical project will directly participate in the benefits derived from it. Project proponents will enter into negotiations with one, two or a group of well represented communities. Negotiating with each and all communities is, in practice, impossible.

If TK is restricted to one or just very few, closely related and integrated communities, applying a trade secret based approach may be an option to secure a certain level of exclusivity and control in regards to this category of TK.[18] This would require a certain level of agreement and coordination between communities to ensure the trade secret is maintained as guarded and undisclosed over time.

The advantage of this approach is that trade secrets are universally recognized and are present in almost all national legislations. One disadvantage is that there is a certain expertise and capacity required to enter into contractual negotiations which define conditions upon which this TK will be shared and used.

This capacity is almost certainly not available in communities today – at least in most. An unfair competition framework may also be required, which is flexible enough to incorporate and consider indigenous peoples' TK as trade secret. Finally, in the case of shared and confidential TK, there is a potential downside to seeking trade secret protection, if there is the real possibility that the knowledge eventually 'escapes' or is removed – by chance or on purpose – from a confidential context. This may happen when many communities share the TK and are not that closely integrated in a nation, tribe, group or whatever category is used.

The holder of the trade secret may be a representative member of a community(ies) or the specific individual holding the TK, on behalf of the community(ies). In this latter case, it would be expected that some form of internal benefit sharing is foreseen to ensure there is no single, individual beneficiary. But this will depend on customs and practices that socially organize each community(ies).[19]

It should be recognized *a priori*, that there may be inevitable exclusions of communities or individuals who hold similar TK but which – given the specific circumstance – will be unable to engage and participate directly in the project and its benefits.

Apart from trade secrets as a protection tool, national law may develop *ad hoc* tools to support registration of confidential TK and ensure its use only for defensive purposes – through specific action by IP authorities or bodies responsible for preventing misappropriation or misuse of natural resources and TK.[20]

The advantage of this approach is that *all* indigenous people and communities who possess this category of TK may be protected and defended (provided they register their knowledge or are all recognized by the register as holders of TK) against misuse or 'biopiracy'. The protection measure becomes immediately effective (potentially at least) and benefits a general group of indigenous people and communities. The potential shortcoming of this option is whether or not the registered TK – which is confidential – can validly be presented to attack novelty and inventiveness based on action from national authorities or entities managing this register.[21]

Shared TK which has passed on and is already in the public domain

It seems from existing data and information that it can be safely assumed that the most common situation is that TK is broadly and widely shared between communities and has also entered the public domain in its more classic form (i.e. it has been described in publications or available documents or is simply widely used outside rural, community contexts by broader sectors of society).

Traditional knowledge which is widely shared and accessible to the public, in accordance with IP, cannot be strictly protected though certain limitations could be imposed on its use. Three possibilities which may offer a certain degree of 'protection' come into mind. Protection in this context is defined broadly, encompassing exclusive rights, control, compensation and defensive protection.

First, defensive protection measures could be implemented. The registration of TK may support the prevention of misappropriation of TK even if it is in the public domain or rather *because* of this factor.[22] Whether specific TK or a broad set of systematized TK, national authorities may be in the position to use these registers to attack novelty and inventiveness related to patents.[23] This approach has the advantage that it benefits a broad and general set of indigenous peoples and communities, though it does not secure benefit sharing. Defensive protection only defends against wrongful or openly illegal patents over biodiversity innovations and related TK.

Second, though this has not been explored in detail in this chapter, the copyright derived principle of *domaine public payant* may also be potentially useful, in this case in a compensatory context. In simple terms, this principle applies to works, crafts and arts which have lost copyright protection, are in the public domain but for some specific reason are deemed important enough to receive special policy attention which is expressed in the obligation to pay the state (for subsequent distribution) a fee for the use of these works, crafts and arts.[24]

Third, a relatively new tool which is being explored as a means to ensure protection of TK (and other indigenous peoples' values and resources) are 'bio-cultural or community protocols'.[25] These are basically management and planning tools elaborated by specific communities, which determine the conditions upon which their resources and TK may be accessed and utilized.[26]

Protocols are not binding on third parties but do provide *ex ante* guidance regarding what to expect if projects and activities are planned in the lands or territories of indigenous peoples and communities. To some extent, they contribute to legal certainty. The most notable advantage of these tools is that regardless of the category or condition of TK, rules are freely set (including PIC principles in some cases) which offer the guiding pathway for researchers or entrepreneurs who are interested in accessing and using TK.

Finally, even if TK is clearly in the public domain, widely shared, available and known by broader sectors of society, recognition of TK holders and indigenous peoples in general in publications, audio and visual media materials, official campaigns, and other means, can serve an important goal in the re-evaluation of TK and social inclusion processes.[27] This is another variant of the 'protection' concept which has little to do with substantial rights, compensation or control, but rather focuses on the distinct nature and socially valuable feature of TK which has, almost as a historic trend, been overlooked – until recently.

Funds and TK protection

The FAO International Undertaking process of the 1980s offers an important, albeit often overlooked lesson, regarding policy options for addressing widely distributed resources.[28] The FAO Undertaking recognized, early on, that the most practical and effective international measure to implement Farmers' Rights,[29] was to establish an international (compensatory) fund.[30] Though the fund never materialized it is an interesting example of a multilateral mechanism which sought to overcome operational and management complexities regarding Farmers' Rights (i.e. definition of 'farmer', beneficiary, right holder, etc.).

On the other hand, the FAO International Treaty on Plant Genetic Resources for Food and Agriculture (2001) process has also created a funding mechanism which seeks to ensure that benefits derived from seeds or materials accessed from a Multilateral System for *commercial purposes*, are distributed to farmers in countries of origin.[31] The Multilateral System is applied to a set and agreed pool of common genetic resources.

In the case of TK, there is already relatively good and precise data and information regarding where indigenous people and communities are located worldwide. This coincides and overlaps almost perfectly with centres of agrobiodiversity. In this regard, there is a good idea of potential TK holders and beneficiaries of a multilateral, fund-like, TK benefit-sharing scheme.[32]

If in the context of shared or widely disseminated TK, PIC is too complicated to achieve, and most importantly, frictions and tensions begin to appear when certain indigenous groups are *excluded* from negotiations over TK which they claim is 'theirs', there may be a need to revisit contractual options and look at alternative possibilities. Compensatory funds are one of these possibilities.[33]

Paradoxically, funds offer – in comparison – an easier and more practical approach to address situations of shared and widely disseminated TK, at least more so than seeking PIC and agreeing to contracts where no defined right holders can be identified and on the verge of inequity if not all communities or groups who share the TK participate in the benefits being negotiated.[34] Transaction costs are reduced to the operational costs of the system.

Rather than create new taxes or impose additional burdens on the private sector, or seek additional resources from international cooperation and funding agencies, under an international agreement or decision, a small, fixed percentage (i.e. 1 to 2 per cent)[35] of *existing taxes on sales* of a certain category of products (i.e. natural products, biotrade products, natural cosmetics, pharmaceuticals), could contribute to an international compensatory fund which would then direct monies to conservation and sustainable use projects in selected areas in the world (pre-defined and identified by countries in an official list).[36] Ideally, a binding international agreement would provide the basis for implementing this obligation and ensuring companies and profit-making institutions commit to it.

This approach does not run counter to existing options, including those under consideration by the draft text in IGC – which leaves considerable leeway for action at the national level. However, agreeing and developing a list of such products or goods may not be simple. If there is a lesson to be learned from international processes, the FAO International Treaty managed to develop a closed (for the moment) list of plant genetic resources for food and agriculture for which very specific rules and a Multilateral System on ABS applies. It is almost irrelevant where these came from, as long as they are part of this pool of resources for which all contributing countries receive a portion of benefits (both monetary and non-monetary).

The main advantages of a fund/compensatory approach include: there is a *de facto* recognition that access and use of TK will be compensated (regardless of its form or status); all indigenous people and communities are entitled to participate in the benefits derived from the fund (because of the fact that they are recognized as indigenous and communities); there is no need for a negotiation between indigenous peoples and communities and a specific company, project or enterprise, and therefore no transaction costs involved – at the negotiating level; indigenous people and communities are not selling their TK, but merely receiving a compensation due to a legal recognition of its value; TK can continue to flow and develop and benefit the broader society. Indigenous people should have an informed participation and involvement in discussions and debate regarding the formal and substantial operational features of such a mechanism, whether at a national or international level.

Conclusions

In the context of TK protection discussions, there is a widespread belief that TK (in its intangible form) can in fact be controlled and subject to effective use restrictions. Information is one of – if not *the* most – complicated goods to control and protect, especially once it becomes shared and spreads among social structures. Traditional knowledge is, mostly, shared, dispersed and disseminated among communities and indigenous people. This is not to say that there may not be very valuable TK which is still maintained and kept confidential within community structures – but for this category of TK, other alternatives should be kept open.

Except in very few cases where TK is still unique, maintained by single individuals or for some reason has not escaped very confined communal contexts, control and restrictions are difficult (and extremely costly) to put into practice. When TK is part or becomes part of the public domain or is freely accessible, protection possibilities are more limited still.

Intellectual property was designed to protect and legally enclose informational goods. A highly complex albeit effective IP system governs creation and dissemination of ideas and innovation in its many forms. Though the use of certain IP tools for TK protection in particular is very difficult in the best of cases, there may be instances where certain IP tools (i.e. trade secrets, copyright) could serve the purpose of protecting certain intellectual interests of indigenous peoples and communities. However, the actual use and application of these tools requires careful definition of the element of TK for which protection is sought: knowledge, a product or a process.

As in the case with genetic resources, TK is almost invariably shared to some extent or degree. This is not an exceptional situation but, rather, the rule. As a result of this, the complete text of the draft 'Protection of TK' should be understood in a broad sense (i.e. exclusive rights, control, compensation, maintenance of TK, etc.), more like a strategic goal than as an IP exclusive type of right. Depending on the emphasis placed on each element of protection, a specific tool or mechanism might be applied, including in the case where TK is shared.

Recommendations

The IGC process in WIPO and the existing draft text for an international regime for TK protection offers nations a broad range of tools which may serve to protect TK and indigenous peoples' intellectual interests.

In the case of shared knowledge and widely disseminated TK, there seems to be no totally equitable nor fair technical solution to protecting shared and widely disseminated TK in particular. Either control or exclusive rights are impossible to exercise; or certain indigenous peoples or communities will be excluded from the benefits which could be generated from use of the TK; or PIC is impossible to obtain (no defined right holder, or representative group); or tensions and frictions may be created among and between indigenous peoples and communities – some of which participate in benefits and some of which are excluded from them.

If this is the case – and it should so be recognized – there is the need to acknowledge the potential of a set of measures and tools which may include: defensive protection, *domaine public payant* based schemes, biocultural or community protocols or a funding, compensatory mechanism. The advantages of the latter option is that it is inclusive rather than exclusive and may have the potential of effectively implementing measures in practice which generate direct benefits to a broad set of indigenous people groups and communities. These funds – whether at the international level or at the national level – at the very least: secure social recognition of TK; place an economic value on TK; ensure benefits accrue directly to indigenous people and communities and are targeted towards conservation and sustainable use projects.

One recommendation for the ongoing IGC process may be to consider incorporating a draft article which refers to widely disseminated and distributed TK and the need to explore policy and regulatory options which reduce transaction costs, prevent tension and frictions among indigenous people and communities and ensure monetary benefits from the use of TK (even if in the public domain) are equitably shared with these groups.

A suggested draft article for the IGC text on TK protection may read, 'In cases where traditional knowledge is shared across borders, between two or more indigenous peoples or communities, in two or more countries, Member States/ Contracting Parties, will collaborate to identify develop and implement, with participation of indigenous peoples and communities, appropriate, feasible, fair and equitable protection mechanisms'.

A fund or permanent financial mechanism, which engages private sector and commercial actors which directly or indirectly use TK and promotes mandatory contributions according to sales of natural or biodiversity products, could offer a cost-effective way for an 'across the board' solution to compensate indigenous people and communities for their intellectual contributions.

As a pilot phase for this type of scheme, a list of potential endeavours or business which it is accepted use biodiversity and its components and to a lesser or greater extent benefit from TK (past or present) could include, for a start: biotrade activities; natural products development; cosmetics (which use natural products). Their contribution to a fund would not be an additional tax or cost but could be designed as part of express State incentives (reflected in taxes, for example) directed specifically at promoting, preserving and developing biodiversity and TK in particular.

This alternative should not be seen as *replacing other potentially useful national or international approaches*, but could be prioritized as a practical multilateral option which reduces transaction costs and overcomes classic problems related to identifying a right holder, implementing a valid and legitimate PIC procedure, potentially excluding indigenous people and communities from direct benefit-sharing arrangements, etc. It is, as in other cases, a second-best solution to an otherwise technically very challenging and still unresolved problem.

Notes

1 The IGC process was formally launched by the WIPO General Assembly in 2001. The Draft Articles were approved during the 28th session of the IGC held in July 2014, in Geneva, Switzerland. The Draft Articles are available at www.wipo.int/tk/en/news/igc/2014/news_0017.html (accessed 23/1/2017).

2 Not all widely shared and dispersed TK is part of the public domain. This will depend on the extent that the TK has been documented, is described in literature such as in academic journals and books, and is available from sources such as libraries, internet and publications. For a detailed analysis of the issue of 'public domain' see, www.wipo.int/edocs/mdocs/tk/en/wipo_grtkf_ic_17/wipo_grtkf_ic_17_inf_8.pdf (accessed 23/1/2017). Note on the Meanings of the Term 'Public Domain' in the Intellectual Property System, with Special Reference to the Protection of Traditional Knowledge and Traditional Cultural Expressions.

3 Articles 10 and 11 of the Nagoya Protocol need to be interpreted in conjunction. Article 10 on the Global Multilateral Benefit-Sharing Mechanism establishes that: *Parties shall consider the need for and modalities of a global multilateral benefit-sharing mechanism to address the fair and equitable sharing of benefits derived from the utilization of genetic resources and traditional knowledge associated with genetic resources that occur in transboundary situations or for which it is not possible to grant or obtain prior informed consent. The benefits shared by users of genetic resources and traditional knowledge associated with genetic resources through this mechanism shall be used to support the conservation of biological diversity and the sustainable use of its components globally.*

Article 11 on the other hand, on Trans-boundary Cooperation determines that: *1. In instances where the same genetic resources are found in situ within the territory of more than one Party, those Parties shall endeavour to cooperate, as appropriate, with the involvement of indigenous and local communities concerned, where applicable, with a view to implementing this Protocol. 2. Where the same traditional knowledge associated with genetic resources is shared by one or more indigenous and local communities in several Parties, those Parties shall endeavour to cooperate, as appropriate, with the involvement of the indigenous and local communities concerned, with a view to implementing the objective of this Protocol.*

4 Article 12 of the Draft Articles (Trans-boundary Cooperation) provides that. *12.1 In instances where the same [protected] traditional knowledge [under article3] is found within the territory of more than one [Member State]/[Contracting Party], those [Member States]/[Contracting Parties] [should]/[shall] endeavor to cooperate, as appropriate, with the involvement of indigenous and local communities concerned, where applicable, with a view of implementing this [instrument]. 12.2 Where the same [protected] traditional knowledge [under article 3] is shared by one or more indigenous and local communities in several [Member States]/[Contracting Parties], those [Member States]/[Contracting Parties][should]/[shall] endeavor to cooperate, as appropriate, with the involvement of the indigenous and local communities concerned, with a view to implementing the objectives of this [instrument].*

5 Most existing policy and legal initiatives are based on a *sui generis* approach to TK protection which entails the rational combination of different tools (e.g. registers, contracts), mechanisms (e.g. PIC), and procedures (e.g. consultation) which 'protect' TK. The most notable example of this may be Law 27811 of Peru which combines registers, licences, trade secret and competition law principles and defensive protection principles to ensure TK of indigenous people related to biodiversity is safeguarded and protected. See, Venero, Begoña, Mitos y Verdades sobre la Biopiratería y la Propiedad Intelectual. In: *Anuario de Derechos Intelectuales*. Kresalja, Baldo. Editor. Palestra, Lima, 2004. For a non-official translation of Law 27811, see www.wipo.int/edocs/lexdocs/laws/en/pe/pe011en.pdf (accessed 23/1/2017).

6 For an analysis of the implications and features of registration and documentation of TK processes, see WIPO, *Traditional Knowledge Documentation Tool Kit*. Consultation Draft 2012. Available at www.wipo.int/export/sites/www/tk/en/resources/pdf/tk_toolkit_draft.pdf

7 See, http://unesdoc.unesco.org/images/0006/000684/068457mb.pdf (accessed 23/1/2017) for the full text of the Model Provisions.

8 According to article 9.9.2(a), the responsibility to realize Farmers' Rights rests with national governments, and measures to protect these rights may include:[. . .] *(a) Protection of traditional knowledge relevant to plant genetic resources for food and agriculture.* [. . .]

9 The TRIPS Agreement does not define an invention, but establishes criteria upon which an invention is measured: novelty, inventiveness and industrial application. These are technical concepts which over time have been developed in terms of content and their specific boundaries.

10 Just as an example of the potential problems of inexistence of definitions, recently in India, the Supreme Court determined that Novartis would not be awarded a new patent over Gleevec (a cancer treating drug), basically because the 'new' 'invention' was not *deemed significantly different from the original version of the drug*. See, www.nytimes.com/2013/04/05/opinion/the-supreme-court-in-india-clarifies-law-in-novartis-decision.html?_r=0 (accessed 23/1/2017).

11 Joseph Vogel, since 1992, has been an active albeit often lonely critic of how little information economics has been considered and assessed in the context of ABS and TK related negotiations, especially in the CBD. Vogel argues that the most limiting but overlooked flaw in the CBD and the Nagoya Protocol is, precisely, ignoring the *informational* nature of genetic resources and TK and their effects on sovereignty and contractual approaches to ABS. See, Vogel, Joseph. 1994. *Genes for Sale*. New York: Oxford University Press.

12 WIPO makes a useful distinction between TK in the public domain and freely accessible TK. It is often assumed that TK is mostly in the public domain and therefore available for free. However, it cannot be assumed that this TK has no owner or does not belong to an indigenous people, community or indigenous individual. It may be possible to

impose certain conditions and limitations on access to and use of TK which is in fact in the public domain. Furthermore, being in the public domain does not mean that it is in practice freely accessible: on the contrary there may be some very simple situations such as electronic safeguards (in the case of databases), or remote and isolated archives, or very limited editions of certain publications, etc. which make the principle of freely accessible almost impossible to be realized. See, www.wipo.int/edocs/mdocs/tk/en/wipo_grtkf_ic_17/wipo_grtkf_ic_17_inf_8.pdf (accessed 23/1/2017).

13 Vogel argues that even the Nagoya Protocol falls short of adequately addressing the true nature of genetic resources (coded, natural information) and shared and widely distributed TK. See, Vogel, J., Alvarez-Berríos, N., Quiñones-Vilche, J.L., et al. (2001). 'The Economics of Information, Studiously Ignored in the Nagoya Protocol on Access and Benefit Sharing'. 7/1 *Law Environment and Development (LEAD) Journal* (2011), pp. 51–65, available at www.lead-journal.org/content/11052.pdf (accessed 23/1/2017).

14 The rationale for articles 10 and 11 of the Nagoya Protocol is rather fuzzy. What is known is that very late in the negotiations of the Protocol (during COP 10 in Nagoya), the African Group insisted on incorporating provisions addressing genetic resources outside national jurisdictions and shared and widely disseminated resources and TK, among others. It is suspected that some of the ideas Vogel had been advocating since the early nineties finally permeated part of the policy discussions. Another point of interest is that an initial draft text of the Nagoya Protocol, included a provision – now missing – regarding TK in the public domain.

15 A detailed study on whether and to what extent funding mechanisms effectively lower transaction costs is yet to be undertaken. Furthermore, resources – other than donor contributions – have to become sustainable and embedded in the system itself. The funding mechanism under the FAO IT, where commercial use of plant genetic resources have yet to materialize in monetary benefits, offers a useful albeit relatively new experience. What is certain is that existing regulatory and institutional frameworks offer few if any advantages in terms of supporting the realization of benefits and sharing.

16 Andean Decision 391 on a Common Regime on Access to Genetic Resources, 1996, and applicable in Bolivia, Colombia, Ecuador and Peru, includes a specific provision which recognizes that Member States may share resources and when this is the case, each State should take into consideration the *interests of the other states* when negotiating access agreements. The problem has been over time that countries are struggling to implement the Decision and *safeguard their own interest*, with little if any incentive to consider neighbouring countries' interests. On the other hand, article 10 of Law 27811 for the protection of collective knowledge in Peru establishes that collective knowledge *may pertain to various indigenous people*. More interestingly, article 6 establishes that '*. . .the indigenous representative organization, whose prior informed consent is sought [representative organization are deemed the legitimate TK negotiating body on behalf of communities], must inform the widest possible number of communities holders of the same knowledge that it is entering into negotiations, and take into account their interests . . . [. . .]*'. See, www.bnp.gob.pe/documents/marcolegal/ley_27811.pdf (accessed 23/1/2017).

17 Clearly, sharing '*the same*' TK is ontologically impossible. What is shared is very similar or even almost identical TK related to biodiversity or genetic resources uses and applications.

18 In classic IP law, and according to the TRIPS Agreement, to be considered a trade secret information (i.e. secret or confidential TK), must: be secret; have some kind of commercial or potential value and have been subject to reasonable steps to be kept confidential. There must be a conscious act of maintaining TK secret – but available for use in a specific commercial or industrial endeavour.

19 Law 27811 in Peru, acknowledges that for *internal* benefit sharing, within a community or communities, their own customary and traditional practices regarding decision making need to be respected.

20 The Peruvian National Commission for the Prevention of Biopiracy (Law 28216, of 2004), is formally responsible for identifying and taking action in cases where Peruvian biodiversity or its communities' TK have been accessed illegally or misappropriated through granting of wrongful IP and patents in particular. See, www.biopirateria.gob.pe (accessed 23/1/2017).

21 The Traditional Knowledge Digital Library (TKDL) in India, for example, only uses the registered TK for defensive purposes and IP authorities can only access this database for these purposes: to review novelty and inventiveness. There are no other uses allowed, even for research purposes.

22 The case of the Traditional Knowledge Digital Library in India is a case in point. Traditional Indian medicinal knowledge (Ayurvedic, Unani and Sidha formulations), widely shared, disseminated and firmly in the public domain (i.e. it has been documented and published for centuries), is systematized and organized in a database used for defensive protection and to prevent 'biopiracy'. It is used *only* for this purpose. This entails institutional arrangements between the Council for Scientific and Industrial Research (CSIR) in India, and the most important IP offices in the world: the European Patent Office (EPO), the Japanese Patent Office (JPO) and the United States Patent and Trademark Office (USPTO). See, www.tkdl.res.in/tkdl/langdefault/common/Home.asp?GL=Eng (accessed 23/1/2017). Another example is action undertaken by the Peruvian IP office (INDECOPI, which is also responsible for implementing the law for the protection of TK), which *ex officio* as well as upon request from indigenous people, is responsible for registering TK in formal, State recognized databases which may be used for defensive purposes. See, Venero, Begoña. La Protección Jurídica de los Conocimientos Tradicionales en el Perú. In: Kresalja, Baldo. Editor. *Anuario Andino de los Derechos Intelectuales.* No. VI, 2009. Lima, Perú. pp. 86–102 INDECOPI's activities in this regard may be reviewed at www.indecopi.gob.pe (accessed 23/1/2017).

23 Interestingly, in 1996 Peru adopted a national regulation on breeders' rights (Supreme Decree 008–1996-ITINCI, 1996) which included the obligation of applicants to demonstrate the legal origin of genetic resources and TK which may have been used in the development of the new variety as a prerequisite for processing the plant breeders' right application. This extended defensive protection measures to PBR. However, after adopting UPOV 91, as part of an obligation in the FTA with the US, the regulation was derogated and these requirements eliminated, in a clear example of the negative effects of adopting UPOV like regimes – and IP stricter rules in FTAs.

24 See the work of the UNESCO Committee of Non-Governmental experts on the 'Domaine Public Payant', *Copyright Bulletin,* vol. XVI, no. 3, 1982, 49. See also Dietz, Adolf, 'A modern concept for the right of the community of authors (domaine public payant)', *Copyright Bulletin,* 1990, XXIV, no. 4, 13–28. Jerome Reichmann, a notable critic of the 'enclosure movement', has also contributed to this conceptual debate, see Reichman, Jerome and Maskus, Keith, 'The Globalization of Public Knowledge Goods and the Privatization of Global Public Goods', in: Maskus and Reichman (eds.) *Journal of International Economic Law* 7(2): 279–320.

25 For specific examples of biocultural protocols in Africa, South East Asia, Asia and South America, see www.wipo.int/tk/en/databases/creative_heritage/indigenous/ (accessed 23/1/2017).

26 The Nagoya Protocol recognizes community protocols. Article 12 of the Nagoya Protocol establishes that: [. . .] *3. Parties shall endeavor to support, as appropriate, the development by indigenous and local communities, including women within these communities, of: (a) Community protocols in relation to access to traditional knowledge associated with genetic resources and the fair and equitable sharing of benefits arising out of the utilization of such knowledge;* [. . .]

27 A notable example of this form of recognition is happening now in Peru. Through an intense and profoundly inclusive campaign to revalue national gastronomy, a potent side effect has been for very broad sectors of society to recognize, acknowledge and appreciate the value of TK and indigenous peoples' contribution to the conservation, development and maintenance of native crops which are crucial to the 'gastronomic boom' of the country. Peru is one of the three world preferred gastronomy destinations, a feat which has been achieved over a period of a decade or so of close collaboration between the State, the private sector, social actors (including communities) and a group of visionary cooks and chefs which have placed biodiversity, and agro-biodiversity in particular, on the national agenda. Sustainably using resources and TK is one way to secure 'protection'. For an analysis of this process see, Ruiz, Manuel. 2009. *Agrobiodiversity Zones and the Register of Native Crops: Learning from Ourselves*. Genetic Resources Policy Initiative, SPDA, Lima, Peru. pp. 55–86.

28 The International Fund for Plant Genetic Resources was created as the tool to realize Farmers' Rights and compensate them for their contribution to maintenance, conservation, development and dissemination of plant genetic resources (FAO Resolution 5/89).

29 At the time, FAO Resolution 5/89 defined Farmers' Rights as '*(. . . rights arising from the past, present and future contributions of farmers in conserving, improving, and making available plant genetic resources, particularly those in the centers of origin/diversity. These rights are vested in the International Community, as trustee for present and future generations of farmers, for the purpose of ensuring full benefits to farmers, and supporting the continuation of their contributions, as well as the attainment of the overall purposes of the International Undertaking) in order to:*

a) ensure that the need for conservation is globally recognized and that sufficient funds for these purposes will be available;
b) assist farmers and farming communities, in all regions of the world, but especially in the areas of origin/diversity of plant genetic resources, in the protection and conservation of their plant genetic resources, and of the natural biosphere;
c) allow farmers, their communities, and countries in all regions, to participate fully in the benefits derived, at present and in the future, from the improved use of plant genetic resources, through plant breeding and other scientific methods.'

30 FAO Resolution 3/91 established that *'Farmers' Rights will be implemented through an international fund on plant genetic resources which will support plant genetic conservation and utilization programs, particularly, but not exclusively, in the developing countries.'* This fund was to be operated by the donors of genetic resources, funds and technology through the Commission on Plant Genetic Resources.

31 The Multilateral System on Access and Benefit Sharing was developed due to the recognition that countries are interdependent in regards to genetic resources for food and agriculture in particular and that efficient mechanisms are required to ensure continued flows of this special category of resources for breeding, conservation and to sustain agricultural systems worldwide. A Standard Material Transfer Agreement adopted in 2001, is a contractual tool which ensures expeditious access to the resources included in a list of resources – a common pool. For a detailed review of the International Treaty see, Moore, Gerald and Tymowski, Wytold. (2005). *Explanatory Guide to the FAO International Treaty on Plant Genetic Resources for Food and Agriculture*. IUCN Environmental Law Programme. Environmental Law and Policy Paper No. 57. IUCN, Gland.

32 This multilateral 'protection' scheme would focus on the *compensatory element* (benefit sharing) and co-exist with parallel specific positive and defensive protection measures to be implemented at the national level.

33 Graham Dutfield discusses the possibility of establishing and creating national 'private collective management institutions' to monitor use of TK, issue licences to users, and distribute fees to right holders in proportion to the extent to which TK is utilized by researchers, industry, etc. These functions could also be assigned to a public institution, depending on specific national contexts and to what extent indigenous people place their trust in the State. See, Dutfield (2003), p. 7.

34 This chapter does not analyse in depth the viability of an international or national fund. Both may be potential alternatives to promote compensatory measures for access to and use of shared TK.

35 This suggested percentage is a general figure. It is not the result of an in-depth economic analysis. The point to be made is that the percentage should be derived from *existing taxes* and not impose additional costs.

36 This assumes that States will have the political will to agree on this type of scheme, and in parallel leave the option open for existing IP-based protection schemes, contractual negotiations, etc.

References

Boyle, James (2003). *The Second Enclosure Movement: And the Construction of the Public Domain*. Duke University.

Dietz, Adolf (1990). 'A modern concept for the right of the community of authors (domaine public payant)', *Copyright Bulletin,* XXIV, no. 4, 13–28.

Dutfield, Graham (2003). *Protecting Traditional Knowledge and Folklore. A Review of Progress in Diplomacy and Policy Formulation.* UNCTAD-ICTSD. Intellectual Property Rights and Sustainable Development. Issue Paper No. 1. June 2003.

Kresalja, Baldo (ed., 2009). *Anuario Andino de los Derechos Intelectuales.* No. VI, Lima, Perú.

McManis, C. and Terán, Y. (2011). 'Trends and scenarios in the legal protection of traditional knowledge', in Wong, Tzen and Dutfield, Graham (eds.) *Intellectual Property and Human Development: Current Trends and Future Scenarios.* Cambridge University Press, Cambridge, pp. 139–174.

Moore, Gerald and Tymowski, Wytold (2005). *Explanatory Guide to the FAO International Treaty on Plant Genetic Resources for Food and Agriculture*. IUCN Environmental Law Programme. Environmental Law and Policy Paper No. 57. IUCN, Gland.

Reichman, Jerome and Maskus, Keith (2004). 'The globalization of public knowledge goods and the privatization of global public goods', in Maskus, Keith and Reichman, Jerome (eds.) *Journal of International Economic Law* 7(2): 279–320.

Ruiz Muller, Manuel (2009). *Agrobiodiversity Zones and the Register of Native Crops: Learning from Ourselves*. Genetic Resources Policy Initiative, SPDA, Lima, Peru.

UNESCO Committee of Non-Governmental Experts on the 'Domaine Public Payant', *Copyright Bulletin* vol. XVI, no. 3, 1982, 49.

Venero, Begoña (2004). Mitos y verdades sobre la biopiratería y la propiedad intelectual, in Kresalja, Baldo (ed.). *Anuario de Derechos Intelectuales.* Palestra, Lima.

Vogel, Joseph (1994). *Genes for Sale*. New York: Oxford University Press.

Vogel, Joseph (ed., 2000). *The Biodiversity Cartel. Transforming Traditional Knowledge into Trade Secrets*. CARE. Quito, Ecuador.

Vogel, J., Alvarez-Berríos, N., Quiñones-Vilche, J.L., *et al.* (2001). 'The economics of information, studiously ignored in the Nagoya Protocol on access and benefit sharing', 7/1 *Law Environment and Development (LEAD) Journal.*

WIPO IGC, *Traditional Knowledge – Operational Terms and Definitions 11*, WIPO Doc. WIPO/GRTKF/IC/13/9 (2002).

WIPO IGC, *The Protection of Traditional Knowledge: Draft Gap Analysis: Revision 4*. WIPO Doc. WIPO/GRTKF/IC/13/5/(b) Rev. (2008).

WIPO, *Traditional Knowledge Documentation Tool Kit*. Consultation Draft 2012. Available at www.wipo.int/export/sites/www/tk/en/resources/pdf/tk_toolkit_draft.pdf (accessed 23/1/2017).

Zamudio, Teodora, Vogel, Joseph, and Ruiz Muller, Manuel. 'Logic Should Prevail: A New Conceptual and Operational Framework for an International Regime of Access to Genetic Resources'. Research Document. *Initiative for the Prevention of Biopiracy*. SPDA (Peruvian Society for Environmental Law), Year V, No. 13, March, 2010. Lima, Peru.

8 Negotiating the public domain in an international framework for genetic resources, traditional knowledge and traditional cultural expressions

Ruth L. Okediji[1]

Introduction

Competing views of the public domain remain a root cause of the complexity and underlying tension surrounding efforts at the World Intellectual Property Organization (WIPO) to develop an international legal framework for the protection of genetic resources (GRs), traditional knowledge (TK), and traditional cultural expressions (TCEs). As of this writing, the public domain has not yet been a topic of formal negotiations at the WIPO Intergovernmental Committee on Intellectual Property, Genetic Resources, Traditional Knowledge and Folklore ('Committee' or 'IGC'). Nevertheless, the subject looms large over every aspect of the ongoing deliberations – and with good reason. The public domain is a major fault line in prevailing conceptions of the role of property as a foremost tool of economic and social organization. When nations agree to recognize property rights in the intellectual creations of citizens from other countries, they also are setting the terms of their competitive relations (Okediji, 2015, 192).[1] Recognition of new categories of property invariably threatens the interests of existing property owners and could adjust the competitive landscape for the long run.

A neutral way to express concerns about the effect of new property rights on the status quo is to suggest that recognition of new types of property harms the public domain. This has been the approach of some opponents of the entitlement-based claims of indigenous groups and local communities. Whether asserting this claim in the context of the IGC should elicit the same consternation associated with expansive intellectual property rights that do threaten the public domain is, however, a matter for some reflection.

1 William L. Prosser Professor of Law & McKnight Presidential Professor, University of Minnesota Law School. The views expressed here are mine alone and reflect my personal opinions. Nothing in this chapter should be attributed to any government I have represented or assisted in the WIPO IGC. Thanks to Olufunmilayo Arewa, Margo A. Bagley and Wendy Gordon for helpful feedback, and to Jessica Sharpe for excellent research assistance.

There is no single definition of the public domain, but most approaches share an instrumentalist vision: the public domain is a reservoir of resources accessible to the public for creative or consumptive uses. Thus, the public domain is a source of (intellectual) property[2] (Litman, 1990; Boyle, 1997). This view is oddly juxtaposed with the knowledge assets at issue before the IGC, especially TK.

Traditional knowledge consists of 'know-how, skills, innovations, practices, teachings or learnings' (WIPO/GRTKF/IC/31/4, 5).[3] It is 'developed, sustained and passed on from generation to generation within a community, often forming part of its cultural or spiritual identity'[4] (WIPO, Traditional Knowledge). The knowledge systems that produce specific manifestations of TK are not typically organized around utilitarian incentives, nor rights of control and exclusion, that make the public domain indispensable to the proper functioning of the intellectual property realm. Further, the entitlement claims sought by indigenous groups and local communities are not principally a means to secure economic returns for their investments but, rather, a way of regulating relations with third-parties seeking access to, and use of, those knowledge assets.

For countries demanding or supporting protection for GRs and TK (so-called 'demandeurs'), a group that consists primarily of developing countries, the main objective of the Committee's work is to create a multilateral framework establishing minimum conditions under which these knowledge assets may be legitimately accessed and used by others. Accomplishing this goal could also facilitate compliance with the access and benefit-sharing regime of the Nagoya Protocol to the Convention on Biological Diversity (Nagoya Protocol, 2010).[5] For a subset of countries, compliance with the Protocol is more important than achieving recognition of entitlement claims for indigenous groups and local communities.

For their part, indigenous groups and local communities view the misappropriation of GRs, TKs and TCEs as fundamentally at odds with global efforts to address biopiracy. They resist the notion that the public domain is a global commons to which they must contribute their knowledge. Most demandeurs do not believe they should jettison concern over longstanding historical deprivation of rights of self-determination in relation to the scope of any prospective instrument(s) for GRs, TK and TCEs.

Developed and developing countries alike recognize the public domain as an important consideration of the IGC's work. However, it has become clear since 2009 as negotiations intensified that these early acknowledgments of the public domain proceeded from different premises. Some developed countries view the status of the public domain as a threshold issue, resolution of which is required for delimiting the nature and scope of any international framework addressing GRs, TK and TCEs. At the other extreme, a mix of developed and developing countries, view public domain issues that have arisen as mere distraction from the negotiations.

Along this continuum, a few countries acknowledge the public domain as an important principle of intellectual property law, but are ambivalent at best about its relevance in an instrument directed at addressing interests that lie beyond traditional justifications animating patent, copyright or trademark regimes. Still other developing countries refuse to consider that a public domain has any role whatsoever in the IGC process. Accordingly, defending or denying 'the' public domain has become a central theme permeating every aspect of the IGC negotiations.

Safeguarding the public domain is unquestionably an important consideration in the optimal design of property rules. In seeking a multilateral instrument at WIPO, the countries that seek property-type entitlements for GRs and TK have tacitly acknowledged that the public domain must feature in any new regime. Similarly, non-demandeurs, by championing 'the' public domain, also have (perhaps unintentionally) signaled recognition of entitlement protection since the public domain exists directly in relation to property rights. To the extent that recognition of some version of a public domain is required (and I believe it is) in any prospective IGC framework, its role might be quite different from what the public domain usually represents in intellectual property law.

In this chapter discussing the public domain in the IGC context, I eschew the debate over the nature of the public domain that has beleaguered intellectual property scholarship and that now threatens the IGC. Instead, I highlight the fact that intellectual property law envisages many public domains, and each is distinctly constituted. The public domain can be expressed, shaped or designed to suit the particular features and function of the property regime to which it relates. The public domain, after all, is an idea – a legal or perhaps even rhetorical, construct – not a scientific discovery or law of nature.

Moreover, given the peculiar features of GRs, TK and TCEs, the property/public domain divide may represent too narrow a set of options for stakeholders. Setting the public domain in opposition to the entitlement claims of indigenous groups and local communities ignores a rich set of options otherwise available in property law. The deployment of 'public domain' terminology could be re-oriented to correspond with a range of interests that appropriately inform the conditions of public access to GRs, TK and TCEs, allowing access to knowledge goods to occur under specified conditions.

To accomplish this richer outcome, however, discourse about the public domain at the IGC must include considerations of the function and role of the public domain in specific relation to GRs, TK and TCEs. Importantly, how such knowledge assets enter into an appropriately designed realm for use by non-members of the group (for convenience, let us also call this realm the 'public domain') will need to be settled. A useful first step to designing the public domain for the IGC requires that delegations discern the vision of the public domain already deeply embedded in the demand for property rights in GRs, TK and TCEs. The chapter outlines a few possible negotiating options that may be consistent with that vision.

The importance of engaging the public domain at WIPO's IGC

Underlying justifications for the entitlement-like claims at issue at the IGC have roots in the history of unauthorized access, use and uncompensated appropriation of property – real and incorporeal – of indigenous and minority groups dating back to the so-called age of discovery. Even well into the twentieth century, scientists and researchers from developed countries continued to operate under assumptions that the genetic resources and accumulated knowledge in these communities could be freely accessed, used and/or taken (Reichman, Uhlir and Dedeurwaerdere, 2016).[6]

Armed with legal tools such as 'the common heritage of mankind' and 'the public domain', scientists and international institutions facilitated the development of a global knowledge infrastructure for research and innovation, utilizing plant genetic resources and traditional knowledge. International regimes for science and research coalesced around the view that those resources were part of an uncharted global commons that could – indeed *should* – be freely and methodically exploited (Reichman *et al*., 2016; Halewood, 2014).[7] The existence of a substantial body of scientific data, knowledge and discoveries important to research in fields such as microbiology, agriculture and environmental science is at least partially attributable to the unfettered access that scientists had to the plant genetic resources and associated traditional knowledge of indigenous groups and local communities.

Given this sensitive background, focus on the public domain in the early years of the IGC certainly would have been both misplaced and untimely. The concept of the public domain, justifiably, is perceived with deep hostility by developing countries and indigenous groups as shorthand for the continued appropriation of their biodiverse resources and other knowledge assets. Yet, direct engagement with the idea of the public domain in the IGC is both necessary and inevitable. The 2014 Draft Articles for the protection of TK includes references to the public domain liberally inserted throughout. Moreover, during the twenty-seventh session of the IGC (IGC 27), a delegate from one of the developed countries proposed a definition of the public domain during the session on TCEs. According to this definition, the public domain refers to:

> intangible materials that, by their nature, are not or may not be protected by established intellectual property rights or related forms of protection by the legislation in the country where the use of such material is carried out. This could, for example, be the case where the subject matter in question does not fill the prerequisite for intellectual property protection at the national level, or as the case may be, where the term of any previous protection has expired.[8]

This definition of the public domain has not yet been revisited; nonetheless since IGC 27 the idea of the public domain has metastasized in the IGC. It is formally part of the negotiating record, and firmly embedded in the line-drawing exercises in which delegations commonly engage at WIPO. The issue should be explicity addressed at a future IGC session for a number of reasons.

First, unresolved skepticism from IGC demandeurs towards the public domain threatens the ability to usefully contrast the role of intellectual property ('IP') in the production of knowledge assets for economic growth and development in the modern state, and the role of GRs and TK in dynamic knowledge systems integral to the innovative processes of indigenous groups and local communities. The comparison between GRs/TK and IP has drawbacks, certainly, since most advocates for the former are not typically interested in IP analogues. Nonetheless, the kind of rights that TK holders seek have particularly strong parallels in copyright and trademark law. For all practical purposes, any international regime for GRs, TK and TCEs will invariably be shaped both in opposition to, and in coordination with, the universe of existing IP rights in which the public domain plays an integral role. The IGC cannot avoid consideration of an appropriate counterpart to the public domain in any framework for GRs, TK and TCEs.

Second, there is no single public domain. Rather, every type of IP has a differently constituted public domain. In copyright law, for example, the public domain includes unprotectable subject matter (such as ideas or facts) and expired copyrighted works. In some countries, copyright's public domain may also include works of the federal government.[9]

The public domain in patent law similarly comprises ineligible subject matter, expired patents, invalidated patents and prematurely expired patents for which maintenance fees were not paid. And despite the TRIPS mandated 20-year term of protection, in practice countries have different terms of patent protection, with some allowing term extensions and adjustments while others do not. In trademark law, the public domain consists mainly of subject matter that has lost its source-identifying function. Of all the IP categories, trademark law arguably has the narrowest public domain. While all other forms of IP have statutorily imposed term limits, trademarks do not. As long as the mark continues to serve as an indication of source, it remains entitled to all the exclusive rights associated with this form of property.[10]

Third, not only is the public domain dissimilarly constituted in the various IP categories, the public domain(s) serve distinct functions in each of these subject areas. For example, the public domain in patent law serves an explicit role in the innovation ecosystem, evidenced by carefully delineated exclusions to eligibility. Laws of nature, abstract ideas and living things are barred from patentability in many countries on the view that granting exclusive rights to such fundamental knowledge threatens innovation. In other jurisdictions, such as the EU, additional patentable subject matter limits are imposed both to protect innovation and, sometimes, to address broader moral considerations (Bagley, 2009).[11]

Concerns about furthering innovation and creativity are also evident in copyright law's conception of the public domain. In addition to the prohibition on protection for ideas, also ineligible for copyright protection are systems and processes (Samuelson, 2007).[12] Add to this stock phrases, short words and a host of other exclusions based on functionality, and it becomes evident that copyright's public domain serves many purposes, including safeguarding the right to participate in culture. Copyright's public domain polices boundaries between the public interest and private reward, between freedom to create and rights to control the fruit of one's labor, and helps channel creativity into the appropriate IP subject matter fields.

The complex relationship between property rights and the public welfare interest in ensuring that downstream innovators and creators have adequate access to the building blocks of creativity is managed through a dynamic balance between rights, limitations and exceptions (Gordon, 2016; Samuelson and Scotchmer, 2002).[13] The innovation function of the public domain in patent and copyright laws is expressed not only in the statutorily required eligibility criteria, but also in a range of limitations and exceptions (L&Es) for each subject. Resources in the public domain may include protected works that the public can freely access for specifically recognized purposes, including to identify underlying ideas, facts or other unprotectable features. A number of L&Es facilitate competition and the exercise and enjoyment of personal freedoms such as privacy and freedom of expression.

Finally, because determinations of subject matter eligibility, and conditions for maintaining title in IP assets are defined at the national level, the public domain is largely shaped by domestic, not international, law. A resource that exists in the public domain of one country may not exist in the public domain of another. In addition, the consequences of public domain status differ among the different types of IP. Some resources in the public domain might still have certain exclusive rights that attach to them. A good example is an author's moral rights which survive expiration of the economic rights and may last well beyond the author's death (Berne Convention, 1971).[14]

These considerations illustrate the fallacy of any notion of 'the' public domain. It is worth repeating that there is no single public domain. Further, national public domains are not static; their contours will shift depending on the policy objectives of the underlying domestic property regime and they may include resources that are not entirely free from constraints associated with property entitlements. In short, the public domain is largely what each country makes it. As a socio-legal construct, the public domain can be, and is often, structured in view of the goals and objectives we seek to promote within a specific legal order. The public domain is malleable and susceptible to concerns that both embrace and oppose property rights.

How the public domain is structured in the IGC will require a principled consensus about the rights of resource holders to assert control over their knowledge assets and to defend the lifestyles and institutions that produce them. In the current Draft Articles for TK, insertions of the public domain appear principally as efforts to curtail the scope of TK that could be subject to entitlement claims asserted by resource holders (Traditional Knowledge Draft Articles).[15] But simply leveraging the public domain as a tool for denying property rights in GRs, TK and TCEs is both incomplete and inconsistent with the underlying logic of most property regimes, including the IP system. If the current deployment of the public domain in the IGC process is maintained alongside a potentially broad set of limitations and exceptions, there could be very little TK left to protect.

A limited scope of protection for TK may be the desired outcome for those opposed to a multilateral framework for the protection of GRs, TK and TCEs. Certainly, narrow rights in TK appear on the face of it consistent with a liberal view of the public domain. But effectively defining GRs, TK and TCEs *out* of entitlement-based protection by unduly expansive views of the public domain further compromises WIPO's already limited role in the unfolding international framework for the protection of GRs, TK and TCEs. National efforts to implement the Nagoya Protocol will, invariably, collapse the interface between the international GRs/TK/TCEs regime and the international IP system for which WIPO is chief custodian. Diluting WIPO's voice in this rapidly evolving transnational regime for GRs, TK and TCEs is not an ideal outcome for either advocates or opponents of the IGC process.

In sum, confronting the issue of the public domain is important for the longterm viability and practical application of any prospective IGC instrument. Without some coordination among states to define the outer boundaries of the public domain for GRs, TK and TCEs, the benefits of an international instrument for protection may instead be severely compromised by territoriality-driven approaches.

Public domain challenges for a framework for genetic resources and traditional knowledge

Beyond general statements highlighting 'protection of the public domain', there has been no formal scrutiny of the relationship between the public domain and the subjects under debate at the IGC. I have already alluded to the fact that a principal reason for the culture of silence around the public domain at the IGC has been the lasting effect of the historicized view of the status of knowledge assets in indigenous communities. Now that the public domain has come to the fore of international discussions, that view is fully evident.

References to the public domain in relation to the IGC appear to question, at the outset, whether GRs and TK are not *inherently* part of a global public domain. Indeed, the current definition of the public domain in the text clearly displays this view by positing all 'intangible materials that . . . are not or may not be protected by established intellectual property rights' as resources in the public domain. In this environment, the public domain doesn't delimit a set of recognized entitlements but, instead, poses a challenge about whether these bodies of knowledge are entitled to protection as such.

Private property, with its attendant bundle of entitlements, is the social institution typically deployed to regulate control of valuable resources. From the perspective of the resource holders, demands for rights-based protection in international law are not claims for *new* forms of property that imperil the set of policy concerns represented by the public domain; rather, the claims being considered at the IGC, represent efforts to delineate rights and interests that have long existed within culturally distinctive communities. To put it bluntly, demandeur countries and the communities whose resources are at issue seek to defend their *already existing* domestic legal, moral and equitable rights within a legally unassailable international framework.

To carve out meaningful space for a rights-based international framework for GRs, TKs and TCEs, the undifferentiated notion of the public domain currently deployed in the IGC needs to be disciplined. One way to do this is by addressing features of the conventional understanding of the public domain that pose particular challenges for IGC demandeurs. Below, I highlight four of these challenges.

1 The public domain generally describes a realm that comprises intellectual resources that are free for others to access, use and build upon. It has been notably described as the 'opposite of property'[16] or 'the 'outside' of intellectual property' (Boyle, 2003).[17] No one owns the public domain so the classic right of property to exclude is not usually associated with this realm (Boyle, 2003).[18] Put simply, the public domain is not property. Thus, if GRs, TK and TCEs are in the public domain, there can be no property-based protection.
2 At least for IP purposes, public domain resources are typically deemed freely accessible and free to use (Litman, 1990)[19] (cf. Lessig, 2002).[20] The notion that public domain resources are available to third parties at no cost stems from a conception of the public domain as principally the result of expired IP rights. Having granted exclusive rights in the knowledge asset, it becomes a public resource at the end of the prescribed term of protection. This is the essence of the social welfare bargain: IP rights secure a reward for the creator but mandates a return to the public. At a minimum, opponents of entitlement-like protection seek a similar social bargain for any framework for the protection of GRs and TK.
3 An important challenge for indigenous groups and local communities is the desire to maintain cultural and spiritual identity with their knowledge systems and with the goods those systems produce. The risk of dissociation posed by placing GRs and TK in the public domain is a critical concern for the IGC. The fact that works in the public domain are freely accessible encourages use (or overuse) that may accelerate processes of dissociation.
4 As noted earlier, one of the most important functions of the public domain is to protect the basis of future innovation and creativity by excluding raw materials and building blocks of innovation and creativity from enclosure. A long history in international law of advocating free access to plant genetic resources as part of the 'common heritage' of mankind evoked bitter division between the technologically rich developed countries and the biodiversity-rich Global South. Efforts since the 1980s through the conclusion of the Convention on Biological Diversity have largely addressed the controversy surrounding the common heritage principle.[21] Nonetheless, plant genetic resources are the basis of plant production and agriculture. In the IP space, granting exclusive rights to control these raw materials/resources would adversely affect innovation and creativity, certainly in the agricultural sciences (Halewood, 2014).

These features of the public domain – its non-property status, its emphasis on (mostly) free access and use, its disassociation from communities that cultivated the body of knowledge, and finally, the public domain's essential role in ensuring future innovation and creativity – are distinct challenges for demandeurs. But for each of these challenges there are fairly strong defences.

GRs (and associated TK) do not simply exist in an untouched state of nature belonging to no one – *res nullius*. While GRs in particular are products of nature, they also reflect years of ingenuity and skill that has resulted in their current genetic makeup (Oguamanam, 2005, 2014). GRs *qua* GRs are the result of labor and possession, making them quintessentially eligible for property rights under the same Lockean theory that justifies the grant of intellectual property rights (Gordon, 1993).[22] The very possibility that GRs and TK can satisfy the eligibility criteria in corresponding IP subject areas affirms that these are not diffuse resources incapable of ownership; they are not, as some might imply, like the air or like fish in the oceans too disperse for exclusive possession. Such analogies, imperfect as they are, in any event fall short since modern property law recognizes rights of sovereign states to claim ownership over resources such as the air (e.g., capping air use in relation to pollution), and to pass environmental laws regulating certain kinds of fish stock in the ocean (Rose, 1999, 2003).[23]

Whether applied TK manifests in arts, crafts, songs, performances, medicines, or rituals, these goods undoubtedly satisfy copyright's minimum criteria for originality and, where required, fixation. The yearning expressed by indigenous groups and local communities for perpetual association with their knowledge assets is no different from the perpetual association possible between a product or service and the indication of source that trademark law protects. Nor is it that much different from the desire of the French to limit use of 'Champagne' to producers of sparkling alcoholic beverages from Champagne, and the Italians' desire to limit the use of 'Parma' to producers of goods from Parma where, like Champagne, the distinctive *terroir* of these regions has facilitated cultivation of a distinctive product. Indeed, WIPO recently concluded a new international arrangement dedicated to this deeply felt, somewhat mysterious (at least to Americans), quasi-spiritual association between product and place. The Geneva Act of the Lisbon Agreement on Appellations of Origin and Geographical Indications (as adopted on May 20, 2015) is a regime that protects precisely this type of association on far less stringent criteria than trademarks, and also with perpetual provenance (Lisbon Agreement, Art. 2).[24]

But especially under a utilitarian perspective, claimants of GRs, TK and TCEs deserve protection if there is any expectation that they should continue to invest in their creative activities. This is particularly relevant for the IGC because the twin forces of misappropriation and globalization continue to disrupt and undermine the capacity of many traditional systems to sustain plant and animal life. One important reason to conclude a multilateral framework regulating access and use of GRs and TK is to encourage ongoing investment in these bodies of knowledge goods by indigenous groups and local communities themselves. As the United Nations Education, Scientific and Cultural Organization (UNESCO) puts it:

> Local communities . . . possess their own rich understandings about the natural milieu and their own interpretations about how it should be managed.

> When these are ignored by scientists and resource managers, efforts to conserve local ecosystems may falter and local livelihoods may be undermined.[25]

Further, any notion of the public domain in GRs and TK must be differently constructed. Unlike intellectual property, GRs and TK are intimately connected to a depletable resource, namely, land. One of the most influential arguments in support of property rights is the need to prevent overuse that leads to waste and, even, 'the tragedy of the commons'. These concerns have serious implications for GRs and TK where key considerations of protecting biodiversity and sustaining the life and cultures of resource owners has greatly informed the need for an entitlement based regime. Certainly, sustainability considerations permeate the Nagoya Protocol's regime of access and benefit sharing (see Convention on Biological Diversity, Art. 1, 8).[26] So far, public domain assertions in the IGC context have not at all addressed the intersectionality of tangible and intangible property interests reflected in the claims of demandeur countries.

This feature alone, i.e., intersectionality, requires much more nuanced approaches to the idea of the public domain in the IGC. As in the case of real property freely accessible to the public, the danger of overuse or underinvestment requires imposition of some limits. Public access to physical spaces that are especially valuable (or vulnerable, such as forests) requires some degree of oversight. Moreover, public access and private ownership are not incompatible in the realm of real property. Thus, even if GRs and TK were by their nature open to all, as leading property scholar Professor Carol Rose notes:

> the idea of [things belonging to the public and open to the public] works hand in glove with a regime in which most resources are the subject of private property. Indeed, open public access would make little sense without an underlying regime of private property and trade. The openness of trade routes presumes that the users of these routes have their own incentives to trade, and that those incentives come in large part from private ownership.[27]

The same could be said of claimants of GRs and TK who have their own systems of incentives, governance and land management. These systems derive from property rights and values that are internal to the indigenous group and local community and sometimes reinforced by other legal regimes. Existing international law regimes, for example, already recognize property rights in GRs. The United Nations Resolution on Permanent Sovereignty Over Natural Resources[28] makes clear that peoples and countries enjoy permanent sovereign authority over their natural resources, including the right to determine the rules and conditions for the 'exploration, development and disposition' of these resources. The various public domains in IP cannot displace the property rights to GRs recognized by the Resolution and other domestic legal regimes, including sovereign land rights.

Considerations of 'publicness' and 'exclusivity' in real property law should be brought to bear in constructing an appropriate public domain for purposes of the

IGC. GRs and TK cannot be treated in isolation from the physical, cultural and institutional context in which they exist and are created. The constellation of moral and spiritual values, that animate systems of TK protection, are deeply woven into the nature and scope of property entitlements and interests being sought at the IGC. And to the extent some countries consider a multilateral framework for GRs and TK as a means to facilitate implementation of the Nagoya Protocol, limiting the conception of the public domain to what applies in the IP space would fall short of the IGC's mandate. An appropriately designed public domain in the context of the IGC requires consideration of the ways in which the public interest is served in regard to tangible *and* intangible property, both of which are implicated in the IGC's efforts to establish a multilateral framework for GRs, TK and TCEs.

Finally, the notion that the public domain consists of material that is entirely free of ownership interests is misleading. To say that property is free to access or use is not necessarily the same as saying that it has no owners. Take for example, Creative Common's licenses or distributed methods of production such as open source software. Access and use of these works may be free, but there is hardly any question that they have owners and that there are laws – whether IP or other legal rules – that recognize such ownership. Even traditional literary works, such as books, don't lose all incidents of ownership when copyright protection lapses. An author, for example, is still usually known as the creator of a book in which copyright has expired. I previously mentioned moral rights which, in some jurisdictions, extend long after the author's economic rights have lapsed. Certainly in these jurisdictions, a public domain exists in which third parties freely access authorial works but the works still bear a legally protected connection to the author.

The prospect of a custom-made public domain for GRs, TK and TCEs is not exceptional. It is already the rule in the IP space with important precedents in multilateral treaties, including the CBD and related biodiversity agreements. It is therefore not infeasible to construct a public domain that addresses the concerns of resource holders, and that aligns with broader goals to promote innovation and creativity. But it is important first to identify whose interests such a customized public domain will serve.

As currently framed in the various Draft Articles in the IGC, designing the public domain appears to be principally about protecting *existing* beneficiaries of the IP system, and doing so in a way that undermines creativity and innovation in local communities and among indigenous groups. But to be stable, any emerging framework for GRs, TK and TCEs must coordinate the desires of resource holders in a way that coheres with the IP system. For this reason, it is puzzling that non-demandeurs who defend the public domain so stridently are also the ones most opposed to the property rights that the public domain ought to facilitate in the communities that produce the knowledge goods at issue.

In the next section I describe the limits of the so-called 'defensive' protection of GRs and TK proposed by non-demandeurs at the IGC. Defensive protection devolves to treating the public domain merely as a realm outside of IP. This is grossly insufficient. There are a variety of other legal interests – both within and outside the narrow confines of the IP system – that extend protection beyond the set of rights traditionally conferred upon rights holders. Examples include unfair competition, consumer protection, and other interests in tort, privacy or contract. To the extent the public domain is viewed as a device solely to limit or prohibit exclusive rights in GRs, TK and TCEs, it will fall short of the critically important role it should play in any international instrument emerging out of the IGC process.

The current state of the public domain in the IGC negotiations – The rise of defensive 'protection'

Under the mandate for the 2016/2017 biennium, WIPO member states are required to continue work 'with a primary focus on reaching a common understanding on core issues, including the definition of misappropriation, beneficiaries, subject matter, objectives, and what TK/TCEs subject matter is entitled to protection at an international level, including consideration of exceptions and limitations and the relationship with the public domain.' This is the first IGC mandate to explicitly direct the attention of negotiators to the public domain. Just how much the formal mandate will drive the substantive content of IGC negotiations is open to question; yet resolving tension over the role of the public domain is important for the definitional exercise that confronts the IGC's work in the current biennium.

The demand for entitlements in GRs, TK and TCEs, a so-called 'offensive' protection model, has been opposed by critics who instead propose 'defensive' protection. The defensive approach is favored by a small number of developed countries as an alternative to the property rights regime. It refers to 'measures aimed at preventing the acquisition of intellectual property rights over TK or genetic resources by parties other than the customary custodians of the knowledge or resources' (WIPO/GRTKF/IC/6/8, para 2).[29] Defensive measures include strategies 'to ensure that third parties do not gain illegitimate or unfounded IP rights' over specific applications of these forms of knowledge.[30] The defensive model reflects the most hefty criticism of the efforts in the IGC to grant offensive protection on grounds that entitlement type protection will recapture assets that should not be enclosed by a proprietary regime. Property-like protection for TK, skeptics argue, would 'remove what is now in the public domain from that domain' (Munzer and Raustiala, 2009).[31]

Defensive protection has been part of a palette of options for the protection of GRs, TK and TCEs since the inception of the IGC. In early efforts to determine the dimension, scope and focus of the IGC's work, a significant part of the task entailed working out the extent to which (and how) traditional knowledge/traditional cultural expressions might fit within existing intellectual property categories. In particular, substantial attention was devoted to the ways in which IP mechanisms might facilitate the protection of these subjects.[32]

A 2003 report prepared by the Secretariat reviewing the first five sessions of the IGC noted, for example, that protection of the full range of interests underlying the demand for TK protection, would require elements of defensive protection (WIPO/GRTKF/IC/5/12).[33] In its initial formulation, 'defensive protection' was used to highlight means by which resource holders could exert legal rights against violations of their pre-existing rights recognized within domestic or traditional law. The IGC report noted that discussions over the first four IGC sessions considered defensive protection equally important and a necessary supplement to the ability to assert positive rights in TK. The report defined a defensive strategy as one 'aimed at preventing others from gaining or maintaining adverse IP rights'.[34] It went on to add that '[b]oth strategies are typically used in conjunction, in a coordinated manner, and applicable to the interests of any group of TK/TCE holders'. Importantly, the report noted that defensive protection is part of the exercise of IP rights.[35] In sum, the early conception of defensive protection was along a continuum of rights *in* the creative goods and not *apart* from them.[36] Or, put differently, defensive protection referred to *how* rights were used, not *whether* those rights existed.

This rights-based formulation of defensive protection has, however, undergone critical changes. A 2008 gaps analysis by WIPO observed the implementation of defensive protection for traditional knowledge within the international patent system (WIPO/GRTKF/IC/13/5(b)).[37] The analysis highlighted a number of features of the patent system as evidence of such protection. It included, for example, the fact that an inventor's name must be listed on the patent obligation, the expansion of PCT minimum documentation to include a range of traditional knowledge publications, revision of the International Patent Classification in 2006 to extend and focus its coverage of a specific type of traditional knowledge related material, namely 'medicinal preparations of undetermined constitution containing material from algae, lichens, fungi or plants, or derivatives thereof, e.g. traditional herbal medicines' (A61K 36/00).[38]

The WIPO analysis also identified other defensive measures that might safeguard traditional knowledge, namely: the adoption of standards for documenting traditional knowledge, including standards to govern access to and use of such documentation; preparation of special guidelines for examination of patents related to traditional knowledge to 'enhance the likelihood that illegitimate patents would not be granted over traditional knowledge'[39] by, among other things, defining the existing state of art in the field; development of a portal for recorded traditional knowledge that could serve as a resource for patent examiners, and a traditional knowledge database to facilitate determinations about the patentability of the invention that would be available only to patent examiners.[40] In addition to these enhanced administrative measures, WIPO's analysis also highlighted the possibility of heightened disclosure requirements,[41] including disclosure of source of the traditional knowledge or genetic resources used in the claimed invention, as a significant tool for the defensive protection of TK.[42]

A number of the measures described in the gaps analysis have been successfully implemented under WIPO's umbrella charter (WIPO/GRTKF/IC/6/8).[43]

Nonetheless, it is seriously misleading to refer to these initiatives, laudable as they may be, as an effort to protect the knowledge assets of indigenous groups and local communities. Defensive measures (not 'protection') share one thing in common namely, defense of the patent (and conventional IP) system. The measures to prevent the grant of exclusive rights to third parties in GRs, TK and TCEs serve primarily the ultimate purpose of denying entitlement claims, thus effectively transforming GR, TK and TCEs into a commons and justifying anyone's access and use of these resources.

In addition to the defensive measures already incorporated in discrete administrative aspects of the international patents system, a proposal by the United States, Japan, Russia and South Korea seeks the establishment of traditional knowledge databases that will prevent 'the erroneous grant of patents' (WIPO/GRTKF/IC/31/6).[44] The U.S. in particular views prevention of the erroneous grant of patents as the primary utility of any international framework emerging from the work of the IGC with respect to GRs and associated TK. The goal of this particular kind of 'defensive protection' of GRs and associated TK is to strengthen the patent system by facilitating access to 'prior art' by patent examiners.[45] On the face of it, this approach to the IGC mandate view would, in effect, return to the status quo in which GRs and associated TK were treated as part of a global commons and non-appropriable by the communities from which scientists obtained the knowledge.

The focus on defensive protection is closely linked to a number of widely publicized cases in which researchers or firms from developed countries sought (and sometimes received) patents or other intellectual property rights over knowledge developed and held by indigenous groups and local communities (Robinson, 2015).[46] In some cases the knowledge was obtained as a result of unauthorized access to the knowledge custodians; in others the knowledge was freely accessible or shared by the source community. In almost all cases, outrage arose over the inequitable sharing of economic benefits earned from the commercialization of the traditional knowledge. The consequences of granting exclusive rights in knowledge held and developed over many generations by indigenous groups and local communities has also raised particular concern in the field of agriculture.

The measures-based form of defensive protection is unlikely to significantly enhance the public domain or result in better enforced patentability criteria. Patent examiners function under a wide range of constraints of which limited information is only one (Lemley, 2001; Posner, 2012).[47] To the extent that resource constraints and lack of information plague the patent system, mechanisms such as a mandatory disclosure requirement can helpfully improve information asymmetries. However, non-demandeur countries have also opposed a mandatory disclosure requirement for patent applicants.

Access to information about GRs and TK can certainly be helpful as proponents of the measures-based model assert. But as Peter Drahos notes, patent examiners are themselves a distinct epistemic community, with their own internal philosophies, ideas and views about knowledge systems and what counts as a protectable invention (Drahos, 2010; Bagley, 2011).[48] Examiners channel certain values, and they exercise judgment and choices that can differ from office to office and across countries. Simply having more information will not necessarily result in denial of a patent for an invention that covers GRs and TK.

A recent example is illustrative. In 2007, the cosmetics company Mary Kay filed an Australian patent application 200720583 for 'Compositions comprising Kakadu Plum Extract or Acai Berry Extract' which was examined by the Australian patent office from 2008 to 2010. Claim 1 in the application stated:

> A method of firming skin or reducing the appearance of fine lines or wrinkles comprising topically applying, to skin in need thereof, a composition comprising: (a) palmitoyl tetrapeptide 7; and (b) a dermatologically acceptable vehicle comprising: (i) water; (ii) glycerin; (iii) butylene glycol; (iv) propylene glycol; and (v) a chelating agent, wherein topical application of the composition to the skin firms the skin or reduces the appearance of fine lines or wrinkles, and wherein the composition is not a sunless tanning composition.

Like all WTO member states, Australian patent law requires that an invention be novel, non-obvious and have an inventive step (TRIPS Agreement, 1994).[49] Experts testified that Aboriginal people had known of the healing properties of the Kakadu plum for years, and that this knowledge had been well documented. The Mirrar Aboriginal people also voiced concerns to the press. Pointing to a strong record of prior art, the Australian patent examiner noted a strong prior art base and observed in its written report that the invention claimed lacked an inventive step and was likely to be rejected. Following the examiner's report, Mary Kay withdrew the patent application (see Robinson and Raven, 2016).

Proponents of defensive protection point to this as one among numerous examples of successful defensive protection. Proof that the claimed invention did not satisfy the patentability criteria helped ensure that no patent issued in this instance. But this outcome applies only in Australia.

In 2011, shortly before the Australian patent application was withdrawn, Mary Kay filed U.S. patent 8,691,300 which issued on April 8, 2014 for 'Skin Care Formulations'. While the claimed invention was slightly different from the Australian application, it still implicated the genetic resource. Independent claim 1 reads:

> A method of firming skin or reducing the appearance of fine lines or wrinkles comprising topically applying, to skin in need thereof, a composition comprising:
>
> (a) a chemically compatible combination of palmitoyl tetrapeptide 7 and *Euterpe oleracea* fruit extract or *Terminalia ferdinandiana* fruit extract; and

(b) a dermatologically acceptable vehicle comprising:

(i) water;
(ii) glycerin;
(iii) butylene glycol;
(iv) propylene glycol; and
(v) disodium EDTA,

wherein topical application of the composition to the skin firms the skin or reduces the appearance of fine lines or wrinkles, and wherein the composition is not a sunless tanning composition.[50]

The U.S. patent that ultimately issued covers a method for firming or reducing wrinkles which is comprised of palmitoyl tetrapeptide-7 and *Euterpe oleracea* fruit extract or *Terminalia ferdinandiana* fruit extract (The Composition) and a dermatologically acceptable vehicle (claim 1). The claim covers facial creams (claim 2) that are aimed at reducing fine lines and wrinkles (claim 3) and also where the Composition includes a skin lightening agent (claim 4) and where that skin lightening agent is lactobacillus. Are the new elements claimed in the U.S. patent sufficiently different from the knowledge known and used by the Aboriginal people? The answer appears to be 'yes' since the patent ultimately issued in the United States despite the examiner initially rejecting the invention as obvious.[51]

One is left with one of two ways to view this divergence between the patent offices of two developed countries: either the U.S. examiner wrongfully issued the patent under U.S. law as careful analysis would suggest, or the standard of inventiveness under U.S. law is so low that minor variations to the traditional knowledge could circumvent the argument that the invention lacked an inventive step. Both these considerations have been well-advanced in the patent literature. In a world where patent offices hold significant power over the public domain in the patent realm, defensive protection is a mirage; it offers no meaningful defense against the unjust appropriation of traditional knowledge, and it confers no certain value-added to the operation of the patent system. Especially in the case of the U.S., patent applications are reviewed under a presumption of patentability that exerts pressure on the examiner, rather than the applicant, to show patent ineligibility (Seymore, 2013).[52]

With the incorporation of IP in the world trading system, a patent or other IP grant over GRs and TK can have significant implications over opportunities for economic empowerment and growth for indigenous groups and local communities. In the Australian case described above, Kakadu plum is a preferred embodiment; it is within the scope of the claimed invention, because the claims are read and interpreted in light of the specification. If a third party sells a product that meets the elements of the composition in the method claim that has Kakadu plum as the chelating agent, and has instructions to use the composition that contains all of the elements of the method claim, the patent owner can stop that party from selling the product.

The implications of the U.S. Kakadu plum-based patent thus are hugely significant. Aboriginal groups who wish to export the plum and its pulp to the U.S. have been precluded from doing so given the existing patent. The patent has had real life effects on the livelihood of Aboriginal women who sought to establish their own export business for the Kakadu plum (author interview with attorney representing Aboriginal women, 2015). In this example, with information easily available to the U.S. patent office, and with one patent office that had already raised questions about the legitimacy of the invention in light of the TK, it is hard to justify how the U.S. patent was nonetheless issued. My point is that defensive measures safeguard neither the public domain nor TK holders. Available information about TK only means that it is a factor that patent examiners take into account – it is not dispositive of the outcome and may, in fact, supply an opportunity for patent applicants to strategically overcome objections as was the case with this Mary Kay patent.

The focus of defensive measures is *not* to protect systems of traditional knowledge protection, to reward traditional innovation, or to encourage private investment in developing new applications of these knowledge forms. The measures proposed do not recognize independent value in these knowledge assets, but see them only as tools to support administration of the patent system. At their most effective, these defensive measures attempt, in the end, to channel GRs and TK into something that has not previously existed in international intellectual property law – a global public domain comprising of anything not recognized by existing international IP treaties. But even this outcome has significant limits; third-party innovators who choose not to seek protection from the patent system but to, instead, utilize domestic trade secret laws may still obtain the economic benefit derived from GRs, TK and TCEs. Moreover, as the Kakadu plum case suggests, denial of a patent in one country does not mean the invention will be denied a patent elsewhere.

Property rights and the public domain are far more malleable concepts than is currently reflected in the WIPO IGC negotiations. Ultimately, the goal of the public domain is the public good – not just *any* public good but the public good of the societies in which the knowledge is created, used and made available. The notion of the 'public' is not a vague reference to an unidentified group of persons; rather, laws governing creativity, including laws pertaining to the public domain – what constitutes that domain, what goes in or out of that domain (including on what terms) is governed by national law. The public domain of countries can and will differ under all categories of intellectual property law. This reflects aspects of the design of the international IP system which sets only minimum rules for member states.

Possible ways forward for the IGC

The *idea* of the public domain – the conviction that there exists a realm of freely accessible and useable materials that enable creativity to flourish – influences every aspect of the IGC negotiations. The question is whether the public domain for GRs, TK and TCEs is intended to benefit indigenous groups and local communities, or whether it is simply a rhetorical tool to facilitate ongoing uncompensated takings of knowledge goods vital for the well-being of the communities that produce them.

The competing visions of what a possible multilateral framework for GRs, TK and TCEs should look like, and the principles that inform them, are linked squarely to assumptions about the nature and role of the public domain. Specifically, they are linked to assumptions about the public domain as distinctly conceptualized within existing intellectual property subject matter categories. But even the association with the public domain in IP is leveraged incompletely and imprecisely. Inserting a monolithic concept of the public domain in the IGC context doesn't advance progress as much as one would hope. There must still be consideration of the precise elements of TK and GRs that merit access by third parties under prescribed conditions. A customized public domain requires shared understanding of the economic and scientific value of GRs, TK and TCEs, and a rational theory for why that value should be available to communities beyond those that have invested in its cultivation or development.

Below are a few possible ways forward that future IGC sessions might consider when dealing with the public domain.

Embrace the tiered approach with its concept of the public domain

During IGC 27, a tiered approach to TK was introduced reflecting expert discussions and, specifically, interventions by the Africa Group.[53] Building on the work of the IGC Secretariat, this approach sought to provide a structure for delineating between different kinds of TK. Roughly, the tiers are divided along various degrees of publicness – 'secret TK', 'sacred TK', 'closely held TK', and 'widely diffused TK'. The idea behind the tiers is an acknowledgment that certain knowledge is already available to the global population, having been historically appropriated through various means, whether legitimate or contrived. Proposed entitlement claims for the various tiers vary from strong property rights for secret TK (akin to trade secrets) to possible rights of attribution, where possible, including for widely diffused TK.

The tiered approach has been a constructive tool in recent IGC sessions, in large part because it helps define the body of knowledge at issue, and it provides a framework for considering the kind of exclusive rights that should apply. Importantly, the category 'widely diffused TK' acknowledges that there is some knowledge to which the international public has already been exposed that can't be easily enclosed. In other words, a kind of public domain already exists by default. Embracing the tiered approach could offer insights into how to begin constructing a stylized public domain – one in which access may be unfettered but rights of attribution or other forms of acknowledgment may nevertheless apply.

There is both value and practical merit to the tiered approach. Certainly, entitlement claims in traditional knowledge – exclusive rights, moral rights, and neighboring rights – map well onto the various rights recognized in IP, and particularly in copyright law. The tiers of TK facilitate a parallel allocation of rights, at least for purposes of better evaluating what bundle of rights could be considered for TK holders. Even if economic reward is all or mostly what indigenous groups and local communities are after, a strong case could be made for property-type entitlements, given the distinguishing features of traditional knowledge and the parallels in modern copyright law.

But for societies organized around these distinctive inter-generational knowledge systems, such a policy choice between economic and moral rights would contravene the organizational principles around which knowledge goods are produced, traded, consumed and shared. To put it differently, the rights being negotiated at the IGC are rights that enable the exercise of sovereignty over the substance and modalities of relations among indigenous groups and local communities and third-parties. In this regard, an alternative conception of property as a 'social system' comprising entitlements which 'shape the contours of social relationships' is especially relevant to the claims of indigenous groups and local communities in their GRs, TK and TCEs[54] (Singer, 2000, 1993). Even within this conception, which more suitably maps onto the cultural frames of most indigenous societies, property has limits and the tiered approach explicitly acknowledges this.

Move the locus of public domain discussions to limitations and exceptions

There is no question in the IP space, that limitations on property rights, such as fair use, freedom of speech or reverse engineering, impose positive externalities on society. These externalities also serve as justifications for the public domain, but they do so without impairing the basic claim to property status.

An ideal locus of arguments in favor of the benefits of unauthorized access to TK and GRs is not the definitional exercises that invoke the public domain to question the property claims of indigenous groups and local communities but, instead, in negotiations over L&Es. Moving the public domain discussion to consideration of appropriate L&Es makes clear that parties seek to create boundaries that enhance utilization of GRs and TK to enhance human welfare. Appropriately constructed L&Es, particularly those framed as standards that allow a court or administrative agency to focus on the reason for the access and use by third parties, may better bridge the gap between the rights of indigenous groups and local communities and broader welfare considerations that should be served by an appropriately designed public domain.

Non-demandeur countries have raised specific concerns regarding possible adverse effects on agricultural research and innovation, including for biotechnology if property rights are granted in GRs and associated TK. However, other international regimes already manage a vast research and innovation complex for genetic resources.[55] The IGC would not be operating on a blank canvas in attempting to balance these interests. The legal framework established by a constellation of treaties such as the Food and Agriculture Organization's International Undertaking on Plant Genetic Resources, the CBD, and even the Nagoya Protocol, is supplemented by a network of privately constructed research commons that have global reach (Reichman *et al.*, 2016). Considered in this rich context, there is little likelihood that any minimum standards framework emerging from the IGC will injure prospects for research and innovation in GRs.

One might argue that where the IGC situates the question of access to GRs and TKs is irrelevant, and that the real challenge is to ensure resources holders have absolute rights of ownership which includes the absolute power to control.[56] Perhaps. But no system of property – entitlement-like protection – is ever absolute. Property rights exist in some considerable tension; one person's property right in inventions for genetically modified (GMO) seeds could (and has) interfered with the property rights of farmers to grow their crops without fear of being contaminated by GMO seeds.[57] A copyright owner's rights in literary and artistic works can conflict with the privacy or free speech interests of others. In seeking entitlement-like protection at WIPO, resource holders will necessarily face limits like every property regime, including IP. Indigenous groups and local communities must confront the reality that their property rights will be subject to some limits. '[I]t is not possible for property rights to be absolute and also to comprise [a] complete bundle of [the entitlements typically associated with property regimes]. . . . Absolute property rights are self-defeating.'[58]

Similarly, insistence by non-demandeurs that any IGC instruments address issues of the public domain implicitly recognizes some type of (pre) existing entitlement for TK and GRs. Put differently, the protection of the public domain is inherently at odds with the premise of the defensive approach. In my view, L&Es provide the ideal structure to confirm entitlement rights while acknowledging the necessary limits that attend to those rights. L&Es actually allow recognition of public domain status on aspects of protected works so it is quite possible that certain works remain protected even as they contribute to the public domain (Okediji, 2016).[59] The question for the IGC is how best to construe, and where to situate, those limits in a manner consistent with the goals and interests of the creative knowledge systems from which TK and GRs emerge.

Explicitly acknowledge the national provenance of the public domain

As I have already observed, the starting part for what ends up in the public domain is determined solely at the national level where implementation of TRIPS standards occurs. The public domain is first and foremost a national construct. For this reason, while a definition supplied in an international instrument may have political or ideological import, it is often of little practical value. A better approach is to acknowledge the idea of the public domain and to acknowledge also, as we do in other areas of IP, that limits to those rights must be carefully circumscribed to advance clearly exceptional national goals. In this regard, a three-step test formulation appropriately modified for TK could be appropriate.[60] Using the three-step test mechanism would be coherent with the international framework that currently governs knowledge goods in the global economy.

Acknowledge and integrate traditional knowledge holders' public domain

The existence and importance of the public domain is hardly in doubt in IP law. Yet for as often as it is invoked, the contours of the public domain are also sharply contested. Over the years in intellectual property circles the public domain has been at the epicenter of important scholarly debate (Samuelson, 2003). Questions about what constitutes the public domain, whether and how it can be preserved against encroachment through contracts, digital tools and numerous other para-intellectual property schemes and, above all, the degree to which it plays an equal, or perhaps greater, role than property entitlements in fostering creative industries, have been important subjects of academic attention.

Defining the public domain for IGC purposes solely in reference to intellectual property is especially problematic. When viewed primarily as a product of intellectual property, the public domain requires line drawing between protected subject matter and unprotected public domain that is infeasible for GRs and TK. Insistence on such bright line distinctions makes sense for an intellectual property-driven public domain in which the primary source of regular sustenance are works that have been stripped of their property status. This happens most commonly through expiration of the term of protection for patents and copyrighted works, or through genericide for trademarks.

Within systems of knowledge that have their own ideas of 'public' and well-defined notions of use and ownership, the task of the public domain is different. The IGC should recognize the public domain inherent within traditional knowledge systems, and ideally it should hoist this conception as a starting point for what definition of the public domain is included in an international instrument. I am not advocating for theoretical precision in defining the public domain to suit the peculiar nature of TK; my point is that there *is* a universe of goods and resources that indigenous peoples and local communities already recognize as available to the public, or as unprotected within their own systems of property. This public domain is, at a minimum, a starting point for further discussion about what else belongs to the TK public domain and why. What is not going to be feasible, however, is conflating the public domain in IP and the public domain in TK.

A measures-based approach with teeth

A proposal by Professor William Fisher that counts as 'measures based' (because it doesn't directly vest rights in GRs and TK) and that, with minor adjustments, could potentially satisfy the interests of both demandeurs and non-demandeurs is worth consideration as an alternative way forward for the IGC. Professor Fisher suggests that instead of a harmonized global regime for GRs and TK holders with rights 'whose contours are determined by national laws' what is needed is 'a complex and more variegated system of norms' (Fisher, 2007).[61] He proposes three parallel provisions:

(a) *It shall be a defense to a claim of patent infringement that the inventor(s), in developing the protected product or process, relied substantially upon materials or knowledge taken from a member country in violation of that country's laws.*

(b) *It shall be a defense to a claim of trademark infringement that the trademark holder, or the original developer of the mark, relied substantially upon materials or knowledge taken from a member country in violation of that country's laws.*

(c) *It shall be a defense to a claim of copyright infringement that the work in which copyright is claimed constitutes a reproduction of a work registered within a member country, and that reproduction is not authorized by a license from the country in question.*

The central idea of these provisions is to preclude enforcement of IP rights in instances where those rights derive from unauthorized access and use of GRs and TK.

> The effect . . . would be to increase the leverage of countries in determining the terms on which flora, fauna, medicinal knowledge, folklore, and traditional art forms are exploited by others. Many countries already have . . . laws that deal with [genetic resources and traditional knowledge], but those laws rarely figure significantly in the debate . . . because it is so easy to violate them with impunity. Adding the three provisions . . . would give the local laws teeth, not by penalizing violations directly, but by exposing violators to the economically devastating sanction of the forfeiture of their own intellectual-property rights.
>
> The countries in which traditional knowledge is currently concentrated could be expected to exercise their enhanced powers in various ways. Some would likely demand greater compensation from individuals and firms using their materials. Others would insist upon attribution. Still others would insist that the production of goods (drugs, clothing, et cetera) based upon traditional knowledge be done in the country where that knowledge originated. Finally, some would forbid the use of traditional knowledge altogether.[62]

Professor Fisher acknowledges the extreme difficulty of adding these provisions to the TRIPS Agreement; however, adding them to a new international framework specifically designed to address legal protection for GRs and TK is something entirely conceivable. The beauty of this proposal for IGC purposes is its simplicity and effectiveness at addressing, in one blow, the problems of misappropriation, the erroneous grant of patents and meaningful enforcement of the variety of interests of owners of GRs and TK. This is something that so far has proven elusive in practice, not to mention challenging to devise at the multilateral level. Importantly, the proposal has a public domain 'kicker': forfeiture of IP rights for misappropriation or unauthorized access. This outcome could greatly enhance the public domain in all three subject areas. Moreover, the disincentive to engage in unauthorized use and access of GRs and TK caused by the threat of forfeiture, while also encouraging investment in GRs and TK by indigenous groups and local communities, is welfare improving.

Recognition of rights in GRs and TK can and should occur without abandoning the logic underlying longstanding conceptions of the public domain. Moreover, designing an international framework for the protection of cultural knowledge assets created by indigenous groups and local communities is possible without embracing the view that only the strongest property rights can foster optimal levels of innovation. Both these perspectives are inconsistent with the values that foster dynamic production of traditional knowledge and stewardship of the genetic resources with which they are associated. Indeed, within many local communities and indigenous groups, the co-existence and mutual reinforcement between property rights and their public domain is central to the processes that engender creativity and innovation. These groups would hardly have survived without this symbiosis.

One final point should be made. The easiest and most effective way to protect the concept of the public domain is to stop the unremitting expansion of IP rights. It simply is improbable to assume that the growing thicket of private rights that empower cross-border legal measures against domestic users of knowledge will not generate defensive reactions as IP rights increasingly encroach upon, or more directly conflict with, the knowledge systems of other societies. As one commentator has shrewdly observed, property begets property (Safrin, 2007).

Conclusion

The orthodoxy of the public domain is intimately bound with intangible resources. Its application to a context in which claims are principally directed at the intersectionality of the physical and intellectual realms will not be a straightforward exercise. Moreover, because the rhetoric of the public domain is most often deployed as a defense to expanding intellectual property rights, it might seem that asserting the public domain in the IGC context is, indeed, consistent with creativity and innovation.

I have suggested instead that in the context of the Committee's work, IP's public domain cannot support the continued production of TK, nor the cultivation and protection of GRs, in a manner necessary to sustain these resources for the long term. In this sense, the public domain as asserted in the IGC is inconsistent with the very policy objectives with which it has been most associated – to enhance creative output in specific communities.

The issues under consideration at the WIPO IGC are not entirely, or even mostly, about the right to exclude others from use of knowledge goods created by indigenous groups and local communities. Rather, the claims of indigenous groups and local communities address a more fundamental concern – one that turns directly on the question of whose sovereign prerogative it is to define the terms and conditions under which creative activity flourishes within specific territories and cultures.

The purpose of the public domain is not to deny property rights to creative works, but rather to establish a legitimate source of raw materials to foster creativity; the public domain serves creators – it doesn't steal from them. At a minimum, the public domain should not be a tool to transfer knowledge from one community to another as has been the case with TK. The public domain cannot provide a basis for the unauthorized access, taking and use of the creative content belonging to other creators. Re-using, reformatting or otherwise transforming TK also cannot strip TK creators of rights in the resulting derivative work, any more than it does for copyright owners.

As copyright law in most jurisdictions makes clear, subject to limitations and exceptions, an author retains the rights to control many downstream uses of her work. Taking creative content from the public domain in one jurisdiction to re-appropriate it in another is arguably a foreseeable and allowable outcome of differences among the copyright laws of nation states. However, appropriation in whole of content owned by one set of creators by another set of creators is consistent with neither intellectual property law, its various public domains, nor the global public interest.

Notes

1 Ruth L. Okediji, 'Legal Innovation in International Intellectual Property Relations: Revisiting Twenty-One Years of the TRIPS Agreement', 36 *University of Pennsylvania Journal of International Law* 191, 192 (2015) ('. . . the TRIPS Agreement sought to reshape the conditions of *future* global competition. . .')

2 Jessica Litman, 'The Public Domain', 39 *Emory Law Journal*, 965 (1990); *see also*, James Boyle (1997) *Shamans, Software, and Spleens: Law and the Construction of the Information Society*. Harvard University Press.

3 WIPO/GRTKF/IC/31/4, Annex, p. 5.

4 *See Traditional Knowledge*, WIPO, www.wipo.int/tk/en/tk/ (last visited Oct. 1, 2016).

5 Nagoya Protocol on Access to Genetic Resources and the Fair and Equitable Sharing of Benefits Arising from their Utilization to the Convention on Biological Diversity, Oct. 29, 2010 [hereinafter Nagoya Protocol].

6 *See* Jerome H. Reichman, Paul F. Uhlir, and Tom Dedeurwaerdere (2016) *Governing Digitally Integrated Genetic Resources, Data, and Literature: Global Intellectual Property Strategies for a Redesigned Microbial Research Commons.* Cambridge University Press (noting that the scientific norms and practices supporting free access by researchers to explore biodiversity-rich environments in former colonies/developing countries for purposes of discovering, isolating and collecting genetic resources for deposit in ex-situ repositories were well-established by the 1950s).

7 For example, the International Undertaking on Plant Genetic Resources (1983) was based on 'the universally accepted principle that plant genetic resources are a heritage of mankind and consequently should be available without restriction' (Article 1); The Preamble to the Recommendation for the Protection of Movable Cultural Property (1964) states '[m]ovable cultural property representing the different cultures forms part of the common heritage of mankind'. Similarly, the Preamble to Recommendation on the Safeguarding of Traditional Culture and Folklore (1989) states 'folklore forms part of the universal heritage of humanity' as did the Preamble to the UNESCO-WIPO Model Provisions for National Laws on the Protection of Expressions of Folklore for Illicit Exploitation and Other Prejudicial Actions (1982) ('folklore represents an important part of the living cultural heritage of the nation'). *See also*, Michael Halewood, 'International Efforts to Pool and Conserve Crop Genetic Resources in Times of Radical Legal Change', in Mario Cimoli, Giovanni Dosi, Keith Maskus, Ruth L. Okediji, Jerome H. Reichman and Joseph E. Stiglitz (2014) *Intellectual*

Property Rights: Legal and Economic Challenges for Development, p. 288. ('Since the earliest days of settled agriculture, crop domestication and improvement efforts have generally taken place within open systems of innovation. . .The maintenance of collaborative systems to collect, conserve, and provide access to plant genetic resources in support of crop science and plant breeding distinguishes agricultural innovation systems from those associated with pharmaceutical, cosmetic and industrial research and development.')

8 *See* Traditional Knowledge Draft Articles, Use of Terms.

9 *See, e.g.*, 17 U.S.C. § 105 (2012) ('Copyright protection under this title is not available for any work of the United States Government. . .); *but see*, Ruth L. Okediji, 'Government as Owner of Intellectual Property? Considerations for Public Welfare in the Era of Big Data', 18 *Vanderbilt Journal of Entertainment and Technology Law* 331, 335 (2016) ('The public domain status of federal government works is a deliberate policy choice justified in reference to the public interest although. . .there are important exceptions to the rule').

10 *See, e.g.*, Nu-Enamel Corp. of IL v. Armstrong Paint & Varnish Works, 95 F.2d 448, 450 (7th Circ. 1938) ('a manufacturer of goods is entitled to the reputation he has established and the public has the right to rely upon his distinctive means of distinguishing between his and other goods. Courts will not unduly extend monopolies under trademarks, but where a clear property right is shown and it is apparent that the right is being invaded and trespassed upon as in the present case, we shall not hesitate to grant relief.'); Dreyfus Fund, Inc. v. Royal Bank of Canada, 525 F. Supp. 1108, 1115 (S.D.N.Y. 1981) ('A trademark owner is not required to maintain any particular level or type of advertising to avoid the abandonment or weakening of its mark. The owner must only continue to use the mark in a manner that preserves its consumer recognition or distinctiveness.')

11 Margo A. Bagley, 'The New Invention Creation Activity Boundary in Patent Law', 51 *William and Mary Law Review* 577 (2009).

12 *See* Pamela Samuelson, 'Why Copyright Excludes Systems and Processes From the Scope of Its Protection', 85 *Texas Law Review* 1921 (2007); TRIPS Agreement, art. 9(2) (Copyright protection shall extend to expressions and not to ideas, procedures, methods of operation or mathematical concepts as such.') See also, Feist Publ'ns, Inc. v. Rural Tel. Serv. Co., 499 U.S. 340 (1991).

13 *See* Wendy J. Gordon (2017) 'How Oracle Erred: The Use/Explanation Distinction and the Future of Computer Copyright', in Ruth L. Okediji (ed.) *Copyright Law in an Age of Limitations and Exceptions*. Cambridge University Press; *see also*, Pamela Samuelson and Suzanne Scotchmer, 'The Law and Economics of Reverse Engineering', 111 *Yale Law Journal* 1575 (2002).

14 Berne Convention for the Protection of Literary and Artistic Works, art. 15(4), July 24, 1971, 25 U.S.T. 1341, 828 U.N.T.S. 221, art. 6bis.

15 *See, e.g.*, Traditional Knowledge Draft Articles, ([Protection does not extend to traditional knowledge that is widely known or used outside the community of the beneficiaries as defined in Article 2.1, [for a reasonable period of time], in the public domain, protected by an intellectual property right or the application of principles, rules, skills, know-how, practices, and learning normally and generally well-known]); (Policy Objectives '(xviii) promote access to knowledge and safeguard the public domain.']; General Guiding Principles (k) [Principle of recognizing that knowledge that is in the public domain is the common heritage of mankind]; (l)[Principle of protecting, preserving and expanding the public domain]).

16 James Boyle, Foreword: 'The Opposite of Property', 66 *Law and Contemporary Problems* (2003).

17 Ibid.

18 Ibid.

19 Jessica Litman, 'The Public Domain', 39 *Emory Law Journal* 965 (1990).

20 Laurence Lessig, 'The Architecture of Innovation', 51 *Duke Law Journal* 1784 (2002).
21 *See, e.g.*, 'Interpretation of the International Undertaking on Plant Genetic Resources', U.N. Doc. C89/24 (Nov. 11–30, 1989).
22 Wendy J. Gordon, 'A Property Right in Self-Expression: Equality and Individualism in the Natural Law of Intellectual Property', 102 *Yale Law Journal* 1533 (1993).
23 Carol M. Rose, 'Expanding the Choices for the Global Commons: Comparing Newfangled Tradable Allowance Schemes to Old-Fashioned Common Property Regimes', 10 *Duke Environmental Law and Policy Forum* 45, 51 (1999); Carol M. Rose, 'Romans, Roads, and Romantic Creators: Traditions of Public Property in the Information Age', 66 *Law and Contemporary Problems* 89, 93–94 (Winter 2003). Available at: http://scholarship.law.duke.edu/lcp/vol66/iss1/4 (accessed 21 December 2016).
24 Article 2 of the Lisbon Agreement provides that the treaty applies to '(i) any denomination protected in the Contracting Party of Origin consisting of or containing the name of a geographical area, or another denomination known as referring to such area, which serves to designate a good as originating in that geographical area, where the quality or characteristics of the good are *due exclusively or essentially to the geographical environment, including natural and human factors*, and which has given the good its reputation; (ii) any indication protected in the Contracting Party of Origin consisting of or containing the name of a geographical area, or another indication known as referring to such area, which identifies a good as originating in that geographical area, where a given quality, reputation or other characteristic of the good is essentially attributable to its geographical origin' (emphasis added).
25 *See* Traditional Knowledge, UNESCO, available at www.unesco.org/bpi/pdf/memobpi48_tradknowledge_en.pdf (accessed 21 December 2016).
26 *See* Convention on Biological Diversity, art. 1, 8(j), June 5, 1992, 31 I.L.M. 822 [hereinafter CBD].
27 Rose (2003), *supra* note 23 at 100.
28 *See* Permanent Sovereignty Over Natural Resources, G.A. Res. 1803 (XVII), U.N. Doc. A/RES/1803 (Dec. 14, 1962).
29 *See* WIPO/GRTKF/IC/6/8 'Defensive Protection Measures Relating to Intellectual Property, Genetic Resources and Traditional Knowledge: An Update', Document prepared by the Secretariat for the IGC 6, March 15–19, 2004.
30 *See* WIPO/GRTKF/IC/5/12, para 28–31. See also http://www.wipo.int/tk/en/tk/ (accessed 21 December 2016).
31 Stephen R. Munzer and Kal Raustiala, 'The Uneasy Case for Intellectual Property Rights in Traditional Knowledge', 27 *Cardozo Arts and Entertainment Law Journal* 37, 41 (2009).
32 www.wipo.int/edocs/mdocs/tk/en/wipo_grtkf_ic_5/wipo_grtkf_ic_5_12.pdf, para. 28, at p. 11.
33 *See* Overview of Activities and Outcomes of the Intergovernmental Committee, WIPO/GRTKF/IC/5/12, available at http://www.wipo.int/edocs/mdocs/tk/en/wipo_grtkf_ic_5/wipo_grtkf_ic_5_12.pdf (accessed 21 December 2016).
34 Ibid. at para. 28.
35 Ibid.
36 Examples used in the report include 'commercial defensive publication services, a practice of filing patent applications not with a view to gain patent rights but to ensure later patents on the same subject matter are not granted, defensive trade mark registers, specific register of prohibited material, and other defensive publication strategies'. *See Ibid.*
37 *See* WIPO Gaps Analysis, WIPO/GRTKF/IC/13/5(b) Rev.
38 WIPO/GRTKF/IC/13/5(b) Rev., Annex I, para. 31 at p. 13.
39 Ibid. at para. 33.
40 Ibid. at para. 34.

41 Ibid.
42 Ibid. at para. 36.
43 *See* 'Practical Mechanisms for the Defensive Protection of Traditional Knowledge and Genetic Resources within the Patent System', available at www.wipo.int/edocs/mdocs/tk/en/wipo_grtkf_ic_5/wipo_grtkf_ic_5_6.pdf; and 'Defensive Protection Measures Relating to Intellectual Property, Genetic Resources and Traditional Knowledge: An Update', WIPO/GRTKF/IC/6/8 December 15, 2003, available at www.wipo.int/edocs/mdocs/tk/en/wipo_grtkf_ic_6/wipo_grtkf_ic_6_8.pdf (accessed 21 December 2016) noting that the IGC 'has developed and implemented a range of practical mechanisms for the defensive protection of traditional knowledge (TK) and genetic resources. It has also referred proposals to other WIPO bodies which have taken up defensive protection measures relating to TK.'
44 *See* 'Joint Recommendation on the Use of Databases for the Defensive Protection of Genetic Resources and Traditional Knowledge Associated with Genetic Resources', Document submitted by the Delegations of Canada, Japan, the Republic of Korea and the United States of America, WIPO/GRTKF/IC/31/6.
45 Ibid.
46 *See* Daniel Robinson (2015) *Biodiversity, Access and Benefit-Sharing: Global Case Studies*. Routledge/Earthscan, Oxon.
47 Mark A. Lemley, 'Rational Ignorance at the Patent Office', 95 *Northwestern University Law Review* 1495 (2001); Richard A. Posner, 'Why There Are Too Many Patents in America', *The Atlantic*, July 12, 2012, available at www.theatlantic.com/business/archive/2012/07/why-there-are-too-many-patents-in-america/259725/; Melissa Wasserman and Michael Frakes, 'Does the U.S. Patent & Trademark Office Grant Too Many Patents? Evidence from a Quasi Experiment', 67 *Stanford Law Review* 613 (2015).
48 *See* Peter Drahos (2010) *The Global Governance of Knowledge: Patent Offices and their Clients.* Cambridge University Press. *See also*, Margo A. Bagley, 'Review of the Global Governance of Knowledge: Patent Offices and Their Clients, by Peter Drahos' (September 1, 2011). *The IP Law Book Review* Vol. 2, No. 1, pp. 15–22, September 2011; Virginia Public Law and Legal Theory Research Paper No. 2011–38, available at SSRN: http://ssrn.com/abstract=1934853.
49 Agreement on Trade-Related Aspects of Intellectual Property Rights, Apr. 15, 1994, Marrakesh Agreement Establishing the World Trade Organization, Annex 1C, 1869 U.N.T.S. 299, 33 I.L.M. 1197, art. 27 [hereinafter TRIPS Agreement].
50 U.S. Pat. No. 8,691,300 (issued April 8, 2014).
51 I assigned the Kakadu plum case to a team of students in my Spring 2016 International Intellectual Property class at Harvard Law School. The team concluded that, based on their findings of prior art that was available to the patent examiner, independent claim 1 of the patent did not meet the requirements for non-obviousness. Some of the prior art the students identified that should have precluded the patent issuance includes *Skin Aging Handbook: An Integrated Approach to Biochemistry and Product Development*, edited by Nava Dayan, June 2008. See also www.bulkactives.com/product/product/pal-ghk-and-pal-gqpr-liposome-emulsion.html
52 *See* Sean Seymore, 'The Presumption of Patentability', 97 *Minnesota Law Review* 990 (2013).
53 IGC Report, Document 27/10, paras. 95 and 226.
54 J.W. Singer (2000) 'Property and Social Relations: From Title to Entitlement', in Charles Gesler (ed.) *Property and Values: Striking an Equitable Balance of Public and Private Interests.* Island Press: Washington, pp. 3–20; J. W. Singer and J. M. Beermann, 'The social origins of property', 6 *The Canadian Journal of Law and Jurisprudence* 217–248 (1993).
55 International Undertaking on Plant Genetic Resources art. 9, Nov. 3, 2001–Nov. 4, 2002, U.S.T. LEXIS 137 [hereinafter International Undertaking], www.fao.org/Ag/

cgrfa/iu.htm (last visited Oct. 19, 2015); Agreed Interpretation of the International Undertaking, Res. 4/89, FAO Conference, 25th Sess. (1989); Farmers' Rights, Res. 5/89, FAO Conference, 25th Sess. (1989); UPOV; CBD.

56 Joseph William Singer (1998) 'Property', in D. Kairys (ed.) *The Politics of Law: A Progressive Critique.* Third edition, Basic Books, NY, pp. 240–258.

57 Monsanto Canada Inc. v. Schmeiser, [2004] 1 S.C.R. 902 (Can.); Organic Seed Growers & Trade Ass'n v. Monsanto Co., 718 F.3d 1350 (Fed. Cir. 2013).

58 Ibid. at 241.

59 *See* Ruth L. Okediji, 'When Should we Invent International Intellectual Property Rights?' (2016 Working paper on file with the author).

60 There are now several different variations of the three-step test but all have the same basic design and effect. *See, e.g.*, Berne Convention, Art. 9(2); TRIPS, Art. 13; WIPO Copyright Treaty Art. 10(1); WIPO Copyright Treaty, Art. 10(2).

61 William Fisher, 'Two Thoughts About Traditional Knowledge', 70 *Law and Contemporary Problems* 131–134 (Spring 2007), available at: http://scholarship.law.duke.edu/lcp/vol70/iss2/8 (accessed 21 December 2016).

62 Ibid. at 133.

Bibliography

Agreement on Trade-Related Aspects of Intellectual Property Rights, Apr. 15, 1994.

Angelopoulos, Christina. (2012) 'The Myth of European Term Harmonisation: 27 Public Domains for the 27 Member States,' *Int'l Rev. Intell. Prop. & Competition L.* (43) 567.

Bagley, Margo A. (2009) 'The New Invention Creation Activity Boundary in Patent Law,' *William and Mary Law Review* (51): 577.

Bagley, Margo A. (2011) 'Review of the Global Governance of Knowledge: Patent Offices and Their Clients, by Peter Drahos', *The IP Law Book Review* (2): 15–22, Virginia Public Law and Legal Theory Research Paper No. 2011–38. Available at SSRN:http://ssrn.com/abstract=1934853

Beerman, Jack. (1993) 'The Social Origins of Property,' *Canadian Journal of Law and Jurisprudence* (6): 217.

Berne Convention for the Protection of Literary and Artistic Works, Art. 15(4), July 24, 1971, 25 U.S.T. 1341, 828 U.N.T.S. 221, Art. 6bis.

Berne Convention, Art. 9(2).

Boyle, James. (1997) *Shamans, Software, and Spleens: Law and the Construction of the Information Society*. Harvard University Press.

Boyle, James. (2003) 'Foreword: The Opposite of Property?' *Law and Contemporary Problems* (66).

Boyle, James. (2008) *The Public Domain: Enclosing the Commons of the Mind.* Yale University Press.

Chander, Anupam and Sunder, Madhavi. (2004) 'The Romance of the Public Domain', *Cal. L. Rev.* (92): 1331.

Convention on Biological Diversity. 1992, Art. 1, 8(j), 31 I.L.M. 822.

Defensive Protection Measures Relating to Intellectual Property, Genetic Resources and Traditional Knowledge: An Update, WIPO/GRTKF/IC/6/8 December 15, 2003, available at www.wipo.int/edocs/mdocs/tk/en/wipo_grtkf_ic_6/wipo_grtkf_ic_6_8.pdf (accessed December 2016).

Documents prepared by the Group of Countries of Latin America and the Caribbean, (GRULAC). 2000. *Traditional Knowledge and the Need to Give it Adequate Intellectual*

Property Protection. Available at www.wipo.int/edocs/mdocs/govbody/en/wo_ga_26/wo_ga_26_9.pdf.

Drahos, Peter. (2010) *The Global Governance of Knowledge: Patent Offices and their Clients*. Cambridge University Press, Cambridge.

Drahos, Peter. (2014) *Intellectual Property, Indigenous People and Their Knowledge*. Cambridge University Press, Cambridge.

Dutfield, Graham. (2004) *Intellectual Property, Biogenetic Resources, and Traditional Knowledge* 5–6. 2nd ed.

Fisher, William. (2007) 'Two Thoughts About Traditional Knowledge', *Law and Contemporary Problems* (70): 131–134. Available at: http://scholarship.law.duke.edu/lcp/vol70/iss2/8.

G.A. Res. 1803 (XVII), U.N. Doc. A/RES/1803 (Dec. 14, 1962) available at www.un.org/ga/search/view_doc.asp?symbol=A/RES/1803(XVII)

Gana, Ruth. (1995) 'Has Creativity Died in the Third World? Some Implications of the Internationalization of Intellectual Property', *Denver Journal of International Law and Policy* (24): 109.

Geisler, Charles (ed.) (2000) 'Property and Social Relations: From Title to Entitlement', in *Property and Values: Striking an Equitable Balance of Public and Private Interests*.

Gordon, Wendy J. (1993) 'A Property Right in Self-Expression: Equality and Individualism in the Natural Law of Intellectual Property', *Yale Law Journal* (102): 1533.

Gordon, Wendy J. (2017) 'How Oracle Erred: The Use/Explanation Distinction and the Future of Computer Copyright', in Ruth Okediji, (ed.) *Copyright Law in an Age of Limitations and Exceptions*. Cambridge University Press.

Halewood, Michael. (2014) 'International Efforts to Pool and Conserve Crop Genetic Resources in Times of Radical Legal Change', in Mario Cimoli, Giovanni Dosi, Keith Maskus, Ruth L. Okediji, Jerome H. Reichman, and Joseph E. Stiglitz (eds) *Intellectual Property Rights: Legal and Economic Challenges for Development.*

Hardin, Garrett. (1968) 'The Tragedy of the Commons', *Science* (162): 1243. http://scholarship.law.duke.edu/lcp/vol66/iss1/4 (accessed 21 December 2016).

IGC Report, Document 27/10, paras. 95 and 226.

Intergovernmental Committee on Intellectual Property and Genetic Resources, Traditional Knowledge and Folklore, Twenty-Seventh Session, Geneva, March 24 to April 4, 2014, The Protection of Traditional Knowledge: Draft Articles, WIPO/GRTKF/IC/27/4, January 23, 2014.

Interpretation of the International Undertaking on Plant Genetic Resources. 1989. U.N. Doc. C89/24.

'Joint Recommendation on the Use of Databases for the Defensive Protection of Genetic Resources and Traditional Knowledge Associated with Genetic Resources', Document submitted by the Delegations of Canada, Japan, the Republic of Korea and the United States of America, WIPO/GRTKF/IC/31/6.

Lemley, Mark. (2001) 'Rational Ignorance at the Patent Office', *Northwestern University Law Review* (95): 1495.

Litman, Jessica. (1990) 'The Public Domain', *Emory Law Journal* (39): 965.

Marrakesh Agreement Establishing the World Trade Organization, Annex 1C, 1869 U.N.T.S. 299, 33 I.L.M. 1197, art. 27.

Moy, Carl. (2011) Moy's Walker on Patents § 9:27.

Munzer, Stephen R. and Raustiala, Kal. (2009) 'The Uneasy Case for Intellectual Property Rights in Traditional Knowledge', *Cardozo Arts & Entertainment Law Journal* (27): 37.

Nagoya Protocol on Access to Genetic Resources and the Fair and Equitable Sharing of Benefits Arising from their Utilization to the Convention on Biological Diversity, Oct. 29, 2010

Oguamanam, Chidi. (2005) 'Genetic Use Restriction (or Terminator) Technologies (Gurts) in Agricultural Biotechnology: The Limits of Technological Alternatives to Intellectual Property', *Canadian Journal of Law and Technology.*

Oguamanam, Chidi. (2014) 'Farmers' Rights and the Intellectual Property Dynamic in Agriculture,' in Matthew David and Deborah Halbert (eds). *Sage Handbook of Intellectual Property.* Sage, London.

Okediji, Ruth L. (2015) 'Legal Innovation in International Intellectual Property Relations: Revisiting Twenty-One Years of the TRIPS Agreement', *University of Pennsylvania Journal of International Law* (36): 191, 192.

Okediji, Ruth L. (2016) 'Government as Owner of Intellectual Property? Considerations for Public Welfare in the Era of Big Data', *Vanderbilt Journal of Entertainment and Technology Law* (18): 331, 335.

Okediji, Ruth L. (2016) *When Should we Invent International Intellectual Property Rights?* Working paper on file with the author.

Overview of Activities and Outcomes of the Intergovernmental Committee, WIPO/GRTKF/IC/5/12, available at www.wipo.int/edocs/mdocs/tk/en/wipo_grtkf_ic_5/wipo_grtkf_ic_5_12.pdf (accessed December 2016).

Permanent Sovereignty Over Natural Resources, G.A. Res. 1803 (XVII), U.N. Doc. A/RES/1803 (Dec. 14, 1962).

Posner, Richard A. (2012) 'Why There Are Too Many Patents in America,' *The Atlantic*, July 12. available at www.theatlantic.com/business/archive/2012/07/why-there-are-too-many-patents-in-america/259725/

Practical Mechanisms for the Defensive Protection of Traditional Knowledge and Genetic Resources within the Patent System, available at www.wipo.int/edocs/mdocs/tk/en/wipo_grtkf_ic_5/wipo_grtkf_ic_5_6.pdf

Preamble to Recommendation on the Safeguarding of Traditional Culture and Folklore. 1989.

Preamble to the UNESCO-WIPO Model Provisions for National Laws on the Protection of Expressions of Folklore for Illicit Exploitation and Other Prejudicial Actions. 1982.

Reichman, Jerome H., Uhlir, Paul F. and Dedeurwaerdere, Tom. (2016) *Governing Digitally Integrated Genetic Resources, Data, and Literature: Global Intellectual Property Strategies for a Redesigned Microbial Research Commons.* Cambridge University Press.

Ricketson, Sam and Ginsburg, Jane. (2006) *International Copyright and Neighboring Rights: The Berne Convention and Beyond.* Oxford University Press.

Robinson, Daniel. (2015) *Biodiversity, Access and Benefit-Sharing: Global Case Studies.* Routledge/Earthscan, Oxon.

Robinson, Daniel and Raven, M. (2016) 'Identifying and Preventing Biopiracy in Australia: patent landscapes and legal geographies for plants with Indigenous Australian uses', *Australian Geographer*, Online, forthcoming issue.

Rose, Carol. (1999) 'Expanding the Choices for the Global Commons: Comparing Newfangled Tradable Allowance Schemes to Old-Fashioned Common Property Regimes', *Duke Envtl. L. & Pol'y F.* (10): 45, 51.

Rose, Carol. (2003) 'Romans, Roads, and Romantic Creators: Traditions of Public Property in the Information Age', *Law and Contemporary Problems* (66): 89, 93–94.

Safrin, Sabrina. (2007) 'Chain Reaction: How Property Begets Property', *Notre Dame Law Review* 82: 1917.

Samuelson, Pamela and Scotchmer, Suzanne. (2002) 'The Law and Economics of Reverse Engineering'. *Yale Law Journal* (111): 1575.
Samuelson, Pamela. (2003) 'Mapping the Digital Public Domain', *Law and Contemporary Problems* (66): 148.
Samuelson, Pamela. (2007) 'Why Copyright Excludes Systems and Processes From the Scope of Its Protection', *Texas Law Review* (85): 1921.
Seymore, Sean. (2013) 'The Presumption of Patentability', *Minnesota Law Review* (97): 990.
Singer, Joseph William. (1998) 'Property', in D. Kairys (ed.) *The Politics of Law: A Progressive Critique.* 3rd edn, Basic Books, NY.
Skin Aging Handbook: An Integrated Approach to Biochemistry and Product Development, 2008. Edited by Dayan, Nava.
The Preamble to the Recommendation for the Protection of Movable Cultural Property (1964).
Traditional Knowledge, UNESCO, available at www.unesco.org/bpi/pdf/memobpi48_tradknowledge_en.pdf
Traditional Knowledge, WIPO, www.wipo.int/tk/en/tk/ (last visited Oct. 1, 2016).
US Copyright Office Circular 34, *Copyright Protection Not Available for Names, Titles, or Short Phrases*, available at www.copyright.gov/circs/circ34.pdf
Wasserman, Melissa and Frakes, Michael. (2015) 'Does the U.S. Patent & Trademark Office Grant Too Many Patents? Evidence from a Quasi Experiment', *Stanford Law Review* (67): 613.
World Intellectual Property Organization Committee on Intellectual Property and Genetic Resources, Traditional Knowledge and Folklore (IGC), www.wipo.int/tk/en/igc/ (accessed December 2016).

Statutes and treaties

WIPO Copyright Treaty, Art. 10(2)
WIPO/GRTKF/IC/13/5(b)
WIPO/GRTKF/IC/31/4
WIPO/GRTKF/IC/5/12
WIPO/GRTKF/IC/5/6
WIPO/GRTKF/IC/6/8
TRIPS Agreement, Art. 9(2)
U.S. Pat. No. 8,691,300 (issued April 8, 2014)
17 U.S.C. § 105 (2012)
17 U.S.C. § 102(b) (2012)
35 U.S.C. § 103 (2012)
35 U.S.C. § 154(a)(2) (2012)
37 C.F.R. § 1.362 (2015)

Judicial Cases

Alice Corp. Pty. Ltd. v. CLS Bank Int'l, 134 S. Ct. 2347 (2014).
Ass'n for Molecular Pathology v. Myriad Genetics, Inc., 133 S. Ct. 2107 (2013).
Bayer Co. v. United Drug Co., 272 F. 505 (S.D.N.Y. 1921).
Bilski v. Kappos, 561 U.S. 593 (2010).
Bonito Boats, Inc. v. Thunder Craft Boats, Inc., 489 U.S. 141, 150 (1989).

Dreyfus Fund, Inc. v. Royal Bank of Canada, 525 F. Supp. 1108, 1115 (S.D.N.Y. 198).
Feist Publ'ns, Inc. v. Rural Tel. Serv. Co., 499 U.S. 340 (1991).
Harper & Row Publishers, Inc. v. Nation Enterprises, 471 U.S. 539, 556 (1985).
Intellectual Ventures I LLC v. Symantec Corp., No. 2015–1771, 2016 U.S. App. LEXIS 17695, at *26 (Fed. Cir. Sep. 30, 2016).
Mayo Collaborative Servs. v. Prometheus Labs., Inc., 132 S. Ct. 1289 (2012).
Monsanto Canada Inc. v. Schmeiser, [2004] 1 S.C.R. 902 (Can.).
Nu-Enamel Corp. of IL v. Armstrong Paint & Varnish Works, 95 F.2d 448, 450 (7th Circ. 1938).
Organic Seed Growers & Trade Ass'n v. Monsanto Co., 718 F.3d 1350 (Fed. Cir. 2013).
Sony Corp. of America v. Universal City Studios, Inc., 464 U.S. 417, 429 (1984).
United States v. Harvey Steel Co., 196 U.S. 310, 315 (1905).

9 Traditional cultural expressions

Marc Perlman

'Traditional cultural expression' (TCE) is not a well-defined concept, but a stage in a decades-long attempt to construct an international legal framework around the cultural politics of identity.[1] It is the latest addition to a vocabulary of partially overlapping terms used in everyday and specialist discourses (custom, heritage, tradition, folklore, *Volkskunde, arts populaires*, etc.).

In this chapter I survey the many proposals that have been advanced over the decades for the protection of folklore/TCEs. I preface my survey with a characterization of the problems that were perceived to require *sui generis* legislation, and I conclude with an analysis of the major difficulties that beset the enterprise. My purpose is not to advocate for any point of view, but to offer a panorama of the variety of solutions that have been proposed at one time or another.[2]

Background: the problem

TCEs were thought to be vulnerable to several perceived threats. One was posed by the mass media, said to have a corrosive effect on the integrity and authenticity of national folklore. Worse, the influx of films, recordings, radio, and television – especially bearing the products of foreign entertainment industries – could become an existential threat, endangering the very survival of folklore by distracting the people from their own culture.

Apart from the threat of eradication, the culture industries of the developed world posed a threat of misappropriation. Foreigners could 'free ride' on what they considered 'public domain' expressions (see Chapter 8 by Okediji, this volume). A solution to this problem would be doubly useful, because it could also solve a balance of trade problem: developing countries tended to be net importers of intangibles. If they could turn their folklore into a source of remuneration by capturing some of the foreigners' profits, they might be able to offset the outward flow of payments to the developed world.

We can identify three analytically distinguishable aspects of these perceived problems: the existential, the moral and the economic. A group may worry that its traditions can't survive in the modern world, it may feel injured because it has been insulted or denied proper recognition, or it may feel deprived of opportunities for economic gain. The agent held responsible for the problem is generally either a distant, powerful, alien actor (e.g. the developed world), or else a nearby one, such as a neighbouring country whose culture too closely resembles that of the threatened group.

Why was the IP regime thought to be incapable of solving these problems? TCEs had at least four features that were inconsistent with the fundamentals of IP law: TCEs required perpetual protection; rights over TCEs were collective rights; originality was not a prerequisite; and the beneficiary might be an informal, amorphous, loosely defined collective.

This problem was brought to the attention of the international legal community as early as 1952, and UNESCO and WIPO devoted several years to it in the 1970s and 1980s. Thus the soil was already well ploughed by the time the IGC arrived on the scene. Despite this long history of discussion, as of the time of this writing (February 2017), the IGC has yet to produce a draft convention or a consensus on holding a Diplomatic Conference. The document that has emerged after thirty-two meetings has bracketed alternatives in almost every sentence.

Rather than attempt a definitive reading of this fractured text, I will try to map its universe of discourse. Drawing on the voluminous body of drafts, reports, studies, position papers, analyses, policy statements, lists of objectives and principles, 'non-papers', scholarly publications, and other documents produced by these decades of negotiation, I survey the universe of possible solutions. To some extent I am following the example of the WIPO Secretariat, whose Traditional Knowledge Division has extracted a 'menu of options' from the splintered mass of provisions under consideration. I have tried to add value by ranging over a wider swath of proposals, digging deeper into the past, and using for my sources not just municipal law and plurilateral agreements, but also the writings of legal scholars and other commentators. I have engaged in a certain degree of interpretation: when there is an option for the protection of traditional knowledge that could plausibly be advanced (*mutatis mutandis*) for TCEs, I include it here. I have tried to add analytic depth by summarizing the various arguments that have been advanced for and against each option.

My ambition was to make this survey comprehensive, but there will inevitably be gaps. The reader who would like a truly complete, authoritative reading of the IGC texts should consult the stocktaking documents periodically issued by the WIPO Secretariat.

Survey of proposed provisions

I organize these proposals under nine headings, roughly following the recent practice of the IGC: the rationale for protection; the subject of protection; the beneficiary, term, and scope of protection; exceptions and limitations; mechanisms of protection; the institutional framework (variously called the Agency, Board, Commission, or Registration Office); and transboundary expressions. In the interest of brevity, I don't address other topics of discussion, such as sanctions and penalties, formalities, national treatment, transitional measures or capacity building.

Rationale for protection

We may classify the rationales for protection into a few broad and overlapping categories: to prevent moral or existential injury to the group; to prevent damage to its material interests; and to protect the TCE 'itself'.

Traditional cultural expressions are often said to be crucial to the *survival* of communities. While this could refer to the revenue that communities earn from selling crafts or staging performances, it is sometimes meant in a deeper sense. For example, in some indigenous belief systems, misuse of a community's TCEs by outsiders is thought to cause spiritual harm to the custodians (*Tulalip* 2003).

The expressions for which protection is sought are often described as linked to a group's *identity* (Wendland 2008:170), though the nature of this linkage is not usually made explicit. 'Identity' can refer to self-identity, and indeed some writers use psychological terms that imply that folklore is an identity-marker for the group itself, a way it can enhance its cohesion and foster a positive self-image. 'Identity' can also refer to the image of the group in the minds of outsiders. In this regard, the practice of traditions by outsiders can lead to *misrepresentation*: third parties may come to identify those traditions with the outsiders rather than with the community of origin.

The right to control expressions of traditional culture is said to be necessary in order to prevent or halt uses of them that the source community finds *offensive.* Even when outsiders do not imitate expressions exactly but create new expressions in the same style, the community's own works in that style could be tarnished by association.

Often the propertization of tradition is justified by pointing to the *revenue* that could be harvested from users of folklore. Sometimes it is suggested that this revenue could be used to *safeguard* the tradition.

Often writers protest the 'illicit exploitation' of traditional culture without indicating exactly what makes exploitation licit or illicit. Some mention *commercialization* in ways that suggest it is diagnostic of illicit exploitation. On the latter interpretation, illicit exploitation is *unjust enrichment*: outsiders reap benefits from traditional culture without sharing them with the community. Furthermore, the practice of traditions by outsiders may conflict with the community's own efforts to present itself to a wider audience, and hence represent *unfair competition.*

Finally, the right to prevent a tradition from being practised outside its community of origin may be necessary to safeguard its *authenticity*, guarantee its *quality* and protect it against *deterioration*, *debasement* and *distortion*.

It is often said that allowing outsiders unrestricted access to these expressions will lead to the *extinction of the tradition*, though the reasons for this are rarely stated. Sometimes it is suggested that the threat is specifically from the *commercial* exploitation by outsiders, which is said to remove traditional expressions from their 'natural environment', to lead to the standardization and deformation of those expressions, and indeed to be alien to the (presumably non-commercial) nature of folklore itself.

While unique, tangible expressions can be literally 'transferred' from one environment to another, this is not true of *intangible* expressions. How is it possible, then, for unfair competition to damage a tradition as it is practised in its original community? Two answers can be gleaned from the documents. First, if a community markets a traditional product, and an outsider can duplicate it at a lower price, the community – finding it can no longer sustain itself by selling that product – will turn to other pursuits.

What, then, of expressions that are not made commercially available by the community? It isn't possible to undercut the price of something that is not for sale. This question is seldom addressed, but some authors posit a sort of cultural self-immolation. Even if tradition-bearers have no desire or need to profit from their tradition, seeing it commercialized by others may dishearten them so much that they will destroy it in order to save it: that is, they may refuse to pass the tradition on to their own children, preferring that it die out rather than be exploited.[3]

The subject of protection

Protected TCEs have been variously described as the 'characteristic', 'unique', 'genuine', 'authentic' or 'indicative' heritage of a community, including all expressions 'generally recognized as representing a communal identity' (*Revised Provisions* 2005:12) or 'linked' or 'affiliated' with a community.[4] TCEs are the products of 'creative intellectual activity' which embody 'traditional culture and knowledge'; they are recognized by the 'tacit consensus of the community' as 'authentic' (*Policy Objectives* 2004:17). These expressions must have been passed down from generation to generation (for at least three – or possibly two – generations, 'whether consecutively or not'), and still be 'maintained, developed or used by the community or its individual members' as 'part of their cultural or social identity or heritage' (*Draft Articles* 2014).

Most of the items listed as eligible for protection as TCEs belong to genres that would be eligible for copyright protection (if they were the original creations of an individual and had not outlived her long enough to fall into the public domain): (oral) literature, music, dance and drama. However, there are also others that might not be eligible, but could qualify as intangible cultural heritage (such as rhythms, ceremonial processions, rituals, sacred places, languages, etc.).

The beneficiary of protection

One of the most controversial issues concerns the identity of the beneficiary. A wide array of entities has been mentioned to fill this role: individuals, families, communities, local communities, cultural communities, nations, peoples, indigenous peoples, minorities, the State. (None of these terms is defined.)

On one view, the beneficiaries of protection would be the communities in whom 'the custody, care and safeguarding of the TCEs/EoF are entrusted in accordance with the customary laws and practices of the communities' (see Chapter 10 by Tobin on customary law, this volume). The communities themselves should issue authorizations and receive benefits. Above all, the State must not be allowed to represent the community, since their interests might conflict.

Sometimes this position is qualified to take into account the 'borrowing' of TCEs between communities. Since a traditional expression may be encountered in a community that did not originate it, but that acquired it from some other community, we must distinguish the 'proximate' and 'ultimate' origins of an expression. Which one(s) should be the beneficiary/ies? On one view, only the community that developed the expression – the community of ultimate origin – should be able to authorize its use. Otherwise, a pirate who was rebuffed by a group that carefully guarded its TCEs could approach other, more agreeable, neighbouring communities that had 'borrowed' the expression he sought. In this way the pirate could 'launder' his illicitly acquired TCE.[5]

Suppose, however, that the TCE died out in the 'ultimate source' but continued to be practised at a 'proximate source' – in other words, suppose no single community *both* originated *and* maintained the TCE. Which one, if either, should be the beneficiary? Moreover, when expressions migrate they sometimes achieve distinctive new forms, as with Canadian pizza or Trobriand Island cricket. Would the 'proximate source' get to claim it, or only the 'ultimate source'? Or both?

In opposition to the claim that only communities can be beneficiaries of TCEs, some argue that there are many circumstances in which a State should be the beneficiary. Some States claim to be so small, or their population so homogeneous, that they have no distinguishable communities within them. In larger or more diverse States, there may be TCEs that are not clearly attributable to any one community; in such cases, too, the State should be the beneficiary. The same would be true of a TCE that was common to many communities.

But even when a TCE is clearly associated with one and only one community, some insist that the State should still be the beneficiary. This is because benefits that accrue to only one community within a State can have a destabilizing effect, weakening national unity. Moreover, it is important to respect 'the principle of national sovereignty of States over their resources' (Iran in *Draft Report* 2004:97), although the idea that cultural practices can be considered as resources (on a par with forests, rivers or petroleum deposits) could be questioned.

One other possibility deserves mention, that of 'negative' or 'defensive' protection. In this case the beneficiary would be everyone (or no one), since the TCE would not become the object of exclusive proprietary rights.

The term of protection

This is probably the least controversial of the topics. There have been occasional suggestions that protection could be limited in time (at least for published TCEs, or in 'cases that are only of commercial importance'), in order to balance the interests of the community (remuneration) with those of international society (access).[6] But I have not found any specific suggestions as to how the duration of protection could be calculated; one alternative in the 2014 Draft Articles allows the Contracting Parties to limit the term of protection of TCEs 'at least as regards their economic aspects', although another provision allows the Contracting Parties to protect TCEs against distortion or mutilation for an indefinite period.

More frequently it is suggested that protection shall be 'not limited in time' (*Model Provisions* 1985:22). In some versions of this option the Parties would be allowed to institute a term of prescription (*Working Group* 1981:4). Another variant posits an indefinite period of protection, but conditions it on the maintenance of an active link between the TCE and the community: protection would last 'as long as the traditional cultural expressions fulfill/satisfy the [criteria of eligibility for protection]' (*Draft Articles* 2014:14). That is, 'once the community that the TCE is characteristic of no longer uses the TCE or no longer exists as a distinct entity (analogous to abandonment of a trademark, or a trademark becoming generic), protection for the TCE would lapse' (*Revised Provisions* 2005:30).

Scope of protection

The uses of TCEs regulated by the instrument are to include: 'fixation; reproduction; public performance; translation or adaptation; making available or communicating to the public; distribution; any use for commercial purposes, other than their traditional use; and the acquisition or exercise of intellectual property rights.'

TCEs can be protected through any or all of the following three options: the establishment of exclusive property rights; of rights to equitable remuneration (a 'liability regime'); and of dignitary rights.

A property right would require a prospective user to obtain authorization for any use. Such authorization could include prior informed consent and mutually agreed terms.

Under a liability regime, the user would not need authorization, but would have to fulfill some specified requirement. Usually that involves payment of a predetermined sum (or percentage of revenue), though conceivably it could be a simple registration.

Dignitary rights could include an attribution requirement, a prohibition of false endorsement, and/or an integrity requirement (prohibiting use that distorts or mutilates the TCE, or that is derogatory or offensive to the beneficiaries, or that diminishes the TCE's cultural significance for the beneficiaries). They might also bind the user to respect the beneficiary's cultural norms and practices.

The type of protection can be tailored to the nature of the TCE. The most recent IGC documents recognize three major categories:

- '[sacred], [secret] or [otherwise known only] [closely held]'
- 'publicly available [but neither widely known, [sacred], nor [secret]]'
- 'publicly available, widely known [and in the public domain]'

The three types of protection can be applied to the three types of TCEs in various combinations. All could be made the subject of dignitary rights, or property rights, or a liability regime. Dignitary rights could apply to all three, but property rights to only one or two. Or each category could have a unique combination of rights: dignitary rights could apply to all, property rights to sacred/secret TCEs, and remuneration rights to publicly available but not widely known TCEs.

Note that the inventory of restricted uses follows quite closely the uses that usually require authorization under copyright law. Thus non-profit private performance and limited distribution would not need to be authorized. The prohibition on acquiring or exercising IP rights is not self-explanatory, but presumably these rights would not include the *sui generis* rights recognized under the envisioned legal instrument. Would an assignment of the latter rights be possible? It often seems taken for granted that they are inalienable, but some have suggested that they could be temporarily or 'conditionally' alienated.[7] If so, it would seem that TCE holders would be able to license them to corporations, which could exploit them for periods far beyond any current or imaginable copyright term. An entertainment corporation could license a folk tale for a few centuries and exploit it without ever having to lobby lawmakers for extended copyright terms (perhaps the holders of the tale would even accept a one-time payment up front).

The rights recognized can be interpreted more or less broadly. For example, besides requiring authorization for the literal copying of traditional expressions, authorization could also be mandatory for the production of new works in the *style* of traditional expressions. (We can draw an analogy with the idea-expression continuum in copyright. Reproducing the TCE literally is copying the expression; reproducing its style is copying the idea.)

This is a controversial point. Nigeria in 2008 declared itself opposed to the protection of style, comparing it to unprotectable ideas in copyright (*Initial Draft Report* 2008:55). However, for those who consider TCEs to be expressions of traditional knowledge (Perlman 2011:127), then insofar as style is a kind of knowledge, it too should be protected (Palethorpe and Verhulst 2000:8).

One major disagreement concerns expressions that have long been widely distributed outside of the relevant community. On one side is the view that such expressions are not eligible for protection. By contrast, others would extend protection to such TCEs, reasoning that when widespread distribution has taken place without the authorization and compensation of the community, it should not deprive those expressions of protection. Some would introduce a time limit for distribution. For example, expressions whose wide distribution occurred within a certain number of years before protection is instituted (e.g. 50 years) should be protected; expressions freely circulated before that date would not be protected.[8]

All of these proposals agree that commercial use would not need to be authorized if it was done in the TCE's traditional context. There have, however, been voices urging that all commercial use should be subject to authorization, even in a traditional context. Thus Sierra Leone claimed that leaving such use unregulated 'could (in most cases) lead to abuse of the expressions of folklore' (*National Experiences* 2002:11). Since it would presumably be the community itself committing the abuse, requiring community authorization in this case would be otiose. This alternative presupposes that authorization is handled by a separate institution, perhaps a State body (discussed *infra* as the Agency).

There have also been idiosyncratic attempts to demarcate the universe of protectable TCEs using criteria other than those given above. For example, Elizabeth Coleman has suggested that the only expressions that warrant protection are those that serve as insignia (distinguishing marks or emblems 'that are institutional facts . . . maintained by rules concerning who may use signs, and when'; Coleman 2004:22). Not all expressions linked to a communal identity will necessarily serve this function.

Exceptions and limitations

With regard to exceptions and limitations, most of them are familiar from copyright statutes, but there are also some that go beyond any of the conventional exceptions.

An exception for prior use would allow users of a TCE from outside the TCE-holding community to continue in its use (Panama, in *Composite Study* 2003:36). Alternatively, prior users would be treated like any other user, but those unable to obtain authorization for continued use would be allowed a transition period before ceasing to use the TCE.

An exception for teaching, learning, and library preservation could allow public archives and libraries to make reproductions of TCEs 'for non-commercial safeguarding purposes only' (*Revised Provisions* 2005:27). Similarly, there could be an exception for 'non-commercial research' on traditional expressions (2005:26).

There could be an exception for incidental use of publicly displayed expressions. Images of 'expressions of folklore which are permanently located in a place where they can be viewed by the public' may be used 'in a photograph, in a film or in a television broadcast' (*Model Provisions* 1985:11). Alternatively, such incidental use could be disallowed, since it might 'undermine community interests and customary rights' (*Revised Provisions* 2005:27–28).

There could also be various enumerated exceptions (such as for folkloric dance groups, small artisans, public entities, and so forth).

The most controversial exception has been that for original works inspired by folklore (derivative works). This would allow the 'borrowing of expressions of folklore for creating an original work of an author or authors' (*Model Provisions* 1985:10). (While a 'work inspired by folklore' is usually understood to be one that mixes original elements with those drawn from the community's cultural heritage, in at least one case it has been defined as a work composed *exclusively* of the latter.[9]) While this exception was a prominent feature of the 'Model Provisions' of the 1980s, it has received less attention from the IGC.

Mechanisms of protection

Most of the discussion at the IGC presupposes that TCE protection will be realized through *sui generis* rights. However, there have been occasional reminders that some already-existing forms of protection (such as trademarks, certification marks or geographical indications) can be extended to cover some uses of traditional expressions.

Another option that is immediately available is non-legal norm setting. Holders and users of TCEs can establish protocols, guidelines, codes of ethics, 'best practice' statements and suchlike (see Chapter 6 by Robinson and Chiarolla, this volume). These codes, if approved by governments or by national or international standards bodies, or if adopted by professional associations of researchers, curators, archivists and others, could create an informal baseline that commercial actors would need to follow to maintain their reputations. (For an extremely useful survey of such protocols, guidelines and statements, see Torsen and Anderson 2010).

There is also a legal mechanism that is not presently available: customary law. Some indigenous peoples hold that the use of their traditional expressions is regulated by customary law. If the beneficiary of protection is to be determined by a community's customary law, such law would have to be acknowledged internationally by means of reciprocity agreements or mutual recognition between States (see Chapter 10 by Tobin on customary law, this volume).

The Agency/Board/Commission/Registration Office

Most proposals prescribe or allow for an institution (variously called the 'competent authority', 'competent office or organization', 'Governmental authority', 'Agency', or 'Commissioner') to assume certain tasks related to the negotiation and enforcement of rights in TCEs. (I will refer to it as the Agency.)

The Agency can be an existing body (such as a collecting society or regional IP organization) or newly established for the purpose. The officers of the Agency are not usually specified, though one proposal holds that they should include prominent practitioners of the State's cultural traditions (*Music in Common* 2006). The number of agencies is not mentioned, though many proposals seem to assume that each State would have one Agency. Alternatively, multiple Agencies could be organized in each State 'to help ensure that local communities have choice in negotiating bodies' (Kansa *et al.* 2005:303).

Aside from its chief mission, in some proposals the Agency would have subsidiary functions, such as education and awareness-raising (*Model Provisions* 1982:24). It could also document and archive the masterworks of the tradition (*Music in Common* 2006).

But the main purpose of the Agency would be to interface with users of TCEs. As to its precise role, there are several possibilities. It could itself negotiate with prospective users, or else put them in contact with the TCE holders and monitor the negotiations (to guard the best interests of the latter). In some proposals, the Agency 'would have exclusive rights to negotiate royalties . . . thereby eliminating the risk of a race to the bottom' (Kansa *et al.* 2005:303). Other proposals emphasize rather that the Agency can act only 'at the request, and on behalf of, the community', and 'in appropriate consultation' with it. It could assist the community with registrations (for those TCEs that had to be registered), with monitoring compliance and with 'instituting civil, criminal and administrative proceedings' in cases of noncompliance.

Besides looking after the interests of the community, the Agency could provide some legal protection for the user. The Agency could issue clearances that would establish a rebuttable presumption that 'the proposed "use" would not constitute an offence. The effect of a clearance would be to constitute a defence to any prosecution for an offence that any proposed "use" or existing use was of sacred-secret material or was debasing, mutilating or destructive' (*Report* 1981:52).

In cases where the Agency sets royalty rates, it would distribute the revenue among the holders in any way it considered 'to be equitable and most likely to produce amity. It is possible that there might be a substantial number of traditional owners fairly entitled to claim in respect of the commercial use of a given item. Whilst this would affect the amount each claimant might receive, it should not affect the total payment by the user. That would depend only on the nature of the item and the nature of the use' (*Report* 1981:36).

Finally, in cases where more than one community lays claim to the same traditional expression, the Agency could determine which one(s) should be considered the beneficiary of protection, 'using customary laws and processes, alternative dispute resolution (ADR) and existing cultural resources, such as cultural heritage inventories, as far as possible' (*Revised Provisions* 2005:32). If the identity of the beneficiary cannot be determined, the Agency could be the beneficiary.[10]

It might also be possible to replace this system of national Agencies with a single global Agency. It could carry out most of the same functions, but as a centralized institution would be easier to supervise and audit, thus serving the interest of transparency while minimizing transaction and enforcement costs.[11]

Transboundary expressions

The problem of transboundary TCEs received little attention until relatively late in the Committee's work, and there have been correspondingly few options proposed to deal with it. It has often been discussed under the rubric of 'regional TCEs'. For example, when communities in several States share an expression, prospective users, who are not citizens of any of those States, could be required to apply to the Agencies of all of the respective States. If the user receives divergent answers from the Agencies, they would consult with each other to reach agreement (Ngombe 2004:440). Alternatively, if a regional arbitration authority exists, composed of members from the various State Agencies, it could make a determination (Ngombe 2004:441).

This characterization of the problem as 'regional' is quite limiting, however. It presumes that shared TCEs only arise because of groups whose territory straddles a State boundary, or because of 'culture areas' that encompass several neighbouring countries. It ignores the problem of global migration (for example, there are substantial numbers of Japanese in Brazil, and there are Indian diasporas in South Africa and the Caribbean).

For such cases a resort to arbitration or mediation has been proposed. WIPO, which handles much commercial arbitration, could study the 'possibilities of offering alternative dispute resolution services . . . which were particularly appropriate for the problems involving intellectual property issues related to traditional knowledge and folklore' (India, in *Report* 2001:8). This, however, is controversial, and many parties object to WIPO participation. Often customary law is mentioned as being especially suitable for use in mediation (though it would probably not be applicable in the examples given above).

Finally, there is the option of a UN body establishing a global Agency (*Music in Common* 2006). This could function in much the same way as the national or regional Agencies, determining the identities of beneficiaries, facilitating negotiations and distributing royalties. However, it could also maintain a fund similar to that of the Multilateral System of the International Treaty on Plant Genetic Resources for Food and Agriculture, or the Global Multilateral Benefit-Sharing Mechanism contemplated in Article 10 of the Nagoya Protocol. That is, member States could agree to pool certain TCEs or genres of TCEs that would be available for all to use, with fees or royalties payable to a global Benefit-Sharing Fund.

Commentary

Definition of TCE

The concept of TCE has been described in fairly consistent terms from the 1980s to the present, though the terms leave a great deal of room for interpretation. In the 1982 *Model Provisions*, a protected 'expression of folklore' must consist of '"characteristic elements" of the traditional artistic heritage' (where 'characteristic' is glossed as 'linkage with a community's cultural and social identity and cultural heritage'). This identity-bearing role is also emphasized in the *Revised Provisions*: the expressions must be 'generally recognized as representing a communal identity' (p. 12); '. . . at the heart of TCEs/EoF is that they are characteristic of and identify a community' (p. 30).

In recent working documents, TCEs are described as:

> [the unique product of] [directly] [linked with]/[distinctively associated with] the cultural [and]/[or] social identity and cultural heritage of indigenous [peoples] and local communities [or nations] . . .

'Unique product' would seem to prescribe a clear, objective condition: the expression would have to be used in a single community, and no other. 'Distinctively associated with' could be given a similar interpretation, that a TCE co-occurs with a certain cultural heritage and not others. But it could also mean that the expression is associated with the group *by outsiders*. The requirement that it be 'generally recognized as representing a communal identity' could be read as implying this. Further, the definition of 'publicly available' – which in this context we might consider an antonym of TCE – is an expression which 'has lost its distinctive association with any indigenous community and that as such has become generic or stock knowledge, notwithstanding that its historic origin may be known to the public' (*Draft Articles* 2014:5). This, too, could be naturally interpreted to mean that it is outsiders who must make the association. In other words, what makes a TCE a TCE is the association the global public makes between the expression and its community.

If so, how widespread must the association be? If a TCE had to be *widely* associated with a particular community, then it would seem that any expression held by a small community that had little visibility on the world stage could be freely misappropriated.

Perhaps the necessary link could be merely generic: one could recognize a plaid textile design as a tartan without knowing which (if any) Scottish clan it signified. In such cases we can imagine a pattern that had a 'tartan look' but was claimed by nobody in particular. But if that general 'look' was enough to qualify a pattern as a TCE, then in effect an entire style could be protected – indeed, *must* be protected.

Putting aside for now the issue of style and returning to the question of recognition, we might want to consider the possibility that the perceptions of outsiders are irrelevant to the status of TCEs, that it is only the perceptions of insiders that matter. In other words, we could define TCEs the same way that indigenous heritage is defined, as the 'practices, representations, expressions . . . that indigenous peoples and individuals recognize as part of their cultural heritage'.[12]

Perhaps all that is required, then, is that a community identifies itself with the TCE. Would it be enough if only a small segment of the community so identified itself? That might be a necessary condition for a sacred expression kept secret from all but a group of initiated elders. However, that would mean that a TCE need not be 'generally recognized' *even by its own community.*

If self-recognition is the criterion, then the actual origin of the expression would be irrelevant. The concept of TCE would cover 'both the expressions originating in the community concerned and those originating elsewhere but having been adopted, further developed or maintained through generations by that community' (*Model Provisions* 1985:16). If Bolivians believe that the song 'El cóndor pasa' is a Bolivian folk song, then it wouldn't matter that it was composed by a Peruvian.

But the nature of the association would still need to be clarified. Would potential TCEs have to have identity-signalling as their primary function (as with flags and anthems – what Coleman calls 'insignia')? Or would mere differential association satisfy the criterion? That is, would it be enough if community members knew the expression was a feature of their culture, and did not know of any other culture in which it was present? But if the status of the expression depends on the state of the group's knowledge, would a change in that knowledge change the nature of the expression?

Perhaps the criterion should instead be community members' beliefs about the expression's *origin.* Bolivians could be aware that 'El cóndor pasa' is sung in Peru, but if they believed that the song was originally a Bolivian folk song, that would forge the proper identity linkage, and it would be their TCE. In that case, 'El cóndor pasa' would be a TCE for Bolivians who didn't know about its Peruvian origin. But would knowledge of this origin make a difference? In 2009, the Bolivian Minister of Culture, Pablo Groux, officially recognized 'El cóndor pasa' as a Peruvian composition and part of Peru's national heritage. If the song had been a Bolivian TCE before, would it then have ceased to be one?

Or perhaps it is not beliefs (about the origin or distribution of an expression) that matter, but feelings. A strong emotional connection with an expression on the part of its holders would be required for it to be considered a TCE. On this interpretation it would be important to decide how deeply felt the emotional connection had to be, and how widely shared within the community, for an expression to qualify as a TCE. Emotionally powerful cultural expressions can be promiscuous, however, and different people in different societies can bond with the same expression. This is true even of antagonists in war, as when during the Second World War both German and Allied troops became enchanted with the sentimental German song 'Lili Marlene'.

Alternatively, it could be that the crucial identity linkage is neither epistemic nor emotional, but rather established by customary law. Of course, this option would be available only for societies where a body of customary law is recognized, and it would raise all of the well-known problems of establishing in court the terms of a society's unwritten customary law (especially if that society were undergoing rapid change).[13] And since it is possible for the same expression to be found in two communities, such an expression could be governed by different customary laws in each one.

Any definition of TCEs that rested on the self-identification of the holding community could increase the uncertainty of potential users. How would they be put on notice about the status of the expression they wanted to use? A registration system would be one solution (as long as it could somehow deal with transboundary expressions); short of that, innocent infringement could be established by the user's *ex ante* efforts to determine if there was a holder. This would require a standard of due diligence to which users would be bound. Finally, in some of the proposals, the Agency is tasked with determining the beneficiary for any TCE, and/or for issuing clearances that would serve as a defence against charges of derogatory treatment. Possibly a finding by the Agency that there was no known holder of an expression could function as a similar defence.

Definition of community

The concept of 'community' in IGC discourse is even more complex than that of TCE. This is because, beyond the notorious difficulty of 'community' itself, the contested notion of 'indigenous community' plays a large role in the negotiations.

Early in the Indigenous Peoples' Movement, indigeneity was used to refer to historically prior inhabitants of a territory later invaded and settled by foreigners. These inhabitants, especially if they maintained a culture and lifestyle distinctly different from that of the settlers, and especially if they had been stripped of their sovereignty, dispossessed, dislocated and oppressed by the latter, were identified as indigenous. The term was applied mostly with regard to the settler societies of the New World, Australia and New Zealand.

However, the term was also adopted by groups (mostly in Africa and Asia) who were either not the first inhabitants of their land or who had no special claim to priority of occupation compared with their neighbours. This second, broader, sense of indigeneity has been accepted by some and rejected by others. The latter insist that the historical experience of dispossession is the defining factor, while the former plead for a more 'flexible' concept of indigeneity (Kingsbury 1998). Both uses crop up in IGC discussions. The idea of indigenous peoples as first occupants is often used by Member States that deny the presence of any indigenous peoples on their territory, or who claim that their entire population is indigenous. But in most other contexts it seems to be the broader notion of indigeneity that is relevant.

Indeed, indigenous peoples seem to have moved to the forefront as 'the key stakeholders in these negotiations' (Australia in *Draft Report* 2013:61). A certain image of them seems to be presupposed in much of the discussion, an image that has come to represent the paradigmatic holder of TCEs: a minority group with well-defined criteria of membership, strong consciousness of its distinctness from the larger society (perhaps reinforced by a history of persecution), unified under its traditional governance and regulated by a robust system of customary law, possibly maintaining a way of life regarded as 'close to Nature' and/or espousing beliefs in spiritual beings or forces that are unrecognized in the Western scientific world-view, and privileging a communal orientation over the individualistic values usually associated with the modern Western liberal polity. So pervasive is the presumption of indigeneity that occasionally one hears a Member State reminding the others that they must not concern themselves *solely* with indigenous peoples.[14]

A focus on indigenous TCEs would not be problematic if there were no meaningful differences between them and TCEs in general, but this does not seem to be the case (Chaudhuri 2009:195). Laws that are framed under the assumption that each TCE belongs to a group with well-delineated membership and its own system of traditional governance, a group that regards outside uses of its TCEs as inherently threatening or dangerous, will be ill-suited to situations where these assumptions do not hold.

The other communities of concern to the IGC are described with various terms appended to 'indigenous peoples'. One often hears 'indigenous peoples and local communities' (a modification of CBD, Art. 8(j)), but there is also a longer form: 'indigenous peoples and traditional and other cultural communities' (*Revised Provisions* 2005:16). This is usually accompanied by a note that 'the term "cultural communities" is intended to be broad enough to include also the nationals of an entire country' (2005:17). It is a carefully detailed, elaborately specified definition, yet seems to end up being all-inclusive.

This definition is frequently criticized for its vagueness. The problem is caused by the enormous semantic gulf between 'indigenous communities' and 'states'; they seem to share no characteristics that aren't also shared with most other types of community.

The difficulty of conceptualizing subnational groups is not fortuitous – it is built into the structure of the forum. WIPO is organized on Westphalian lines: its members are treated as States with distinct, clearly demarcated borders, territorial sovereignty, and sovereign equality, living under an imperative of nonintervention. As is well known, non-State actors are achieving increasing recognition in international law, and the IGC has become more and more receptive to the (informal) participation of NGOs representing indigenous peoples.

But how are these subnational groups to be understood, if not as miniature Westphalian entities? The Westphalian perspective is just one example of the general phenomenon Rogers Brubaker calls 'groupism': 'the tendency to take discrete, sharply differentiated, internally homogeneous and externally bounded groups as basic constituents of social life, chief protagonists of social conflicts, and fundamental units of social analysis' (Brubaker 2002:164). Arguably, the IGC needs to view all stakeholders as such 'unitary collective actors' (as Brubaker terms them). If indigenous peoples are the subnational communities that best fit this description, they will naturally gravitate to the centre of discussion. Other, less-well-defined, communities will tend to be neglected.

The problem of embracing indigenous peoples and States under a single term may be insoluble, because it is not so much a conceptual problem as an example of constructive ambiguity, designed to hold together a coalition. Many *demandeurs* insist that States should be beneficiaries, while most indigenous delegates and several developed countries want to guard against that possibility. It is yet to be seen if this phrase will be allowed to remain its equivocal self, or if it will eventually be forced to reveal its true colours.

Looking forward

The amount of progress the IGC has achieved so far remains controversial. Some optimistically point to the texts, now couched in treaty language, becoming gradually more streamlined. Others claim to see polite deadlock in the bracket-infested Draft Articles. It is common to hear delegates, absent from the negotiations for several years, claim on their return to find nothing changed.

Meanwhile, if the IGC has made dubious progress since 2001, the facts on the ground may slowly be becoming more hopeful. Professionals and institutions outside the legal field – scientists, anthropologists, folklorists, activists, librarians, archivists, museum curators – are becoming increasingly sensitive to the issue of TCEs. With the recent publication by the Traditional Knowledge Division of a survey of best practices (Torsen and Anderson 2010), this trend may grow in strength. The latest reported examples of exploitation and malfeasance seem on the whole less brazen than the much-discussed classic cases – some of them almost four decades old now – that are fast receding into history.

If this trend is not illusory, we can envision a very different endgame. If consciousness-raising brings global practices into reasonable congruence with the best ethical guidelines, even a non-binding declaration might function as the *opinio juris* necessary to establish customary international law.

Meanwhile, what is the prognosis for the IGC? It's impossible to even guess. If States are ever to commit themselves to vague new norms – especially ones that seem inconsistent with existing legal principles – they will probably only do so in an atmosphere of great trust (Forrest 2006:237). Trust is something that seems conspicuously absent from the IGC nowadays. Can it be restored? Perhaps; but it might not happen until the world becomes a less unequal place.

Notes

1 My thanks to Wend Wendland, Francis Gurry, Manon Ress and Thiru Balasubramaniam for their invaluable intellectual and logistical help. The research for this chapter was supported by a New Directions Fellowship from the Andrew W. Mellon Foundation.
2 Due to my own limited expertise in comparative law, when I present legal examples they will be mostly drawn from United States law. When I present factual examples of TCEs, they will mostly be musical examples, because of my greater familiarity with that artistic domain.
3 Erica-Irene Daes, cited in Barsh (1999:42).
4 Traditional Knowledge is defined in similar terms, as 'distinctively associated with' or 'integral to the cultural identity of' an indigenous people or local community (*Draft Articles* 2011:2).
5 The term 'laundering' was used – in the context of traditional knowledge – by Pedro Garcia (reported in Tobin 2001:54). The terms 'proximate origin' (for the 'immediate source' of an expression) and 'ultimate origin' (for the place of its 'historical conception') are from *International Protection* (1985:41).
6 *Report* (2004:21).
7 Consent 'is always temporary and revocable: heritage can never be alienated, surrendered or sold, except for conditional use' (Daes 1993:9).
8 Ms. Lilyclaire Bellamy made this suggestion at the first Inter-sessional Working Group of the WIPO IGCGRTKF, Thursday, 22 July 2010.
9 In Senegal's Copyright Act of 1973 a 'work inspired by folklore' is defined as a 'work composed *exclusively* of elements borrowed from the Senegalese traditional cultural heritage' (*Updated Draft Outline* 2006, Annex I, p. 18; emphasis added). It is not clear how such a work would differ from an expression of folklore *tout court.* Nor is it clear whether a work mixing traditional with original elements – the usual definition of a 'work inspired by folklore' – would fall outside the scope of the 1973 Act.
10 '. . . the Pacific Regional Model, 2002 provides for prior informed consent and for the establishment of a "Cultural Authority" . . . If no "traditional owners" can be found or there is no agreement as to ownership, the cultural authority can be determined to be the traditional owner' (*Updated Draft Outline* 2006, Annex I, pp. 36–37).
11 Peter Drahos envisions a global collecting society of this type for rights in traditional knowledge (2000).
12 Cf. the revised 'Principles and Guidelines for the Protection of the Heritage of Indigenous Peoples' of the United Nations Working Group on Indigenous Populations (as cited in *Updated Draft Outline* 2006, Annex I, p. 18). Here I gloss over a possible ambiguity in this formulation. 'Recognize as part of their heritage' can bear at least two readings: it could mean that the criterion is entirely subjective (that a community's heritage is whatever they say it is), or that it is defined by a certain objectively determinable relationship – a relation of recognition – between the members of the community and the expressions concerned. Of course, in any given case both interpretations might yield an identical result.
13 The difficulty of establishing customary law to the satisfaction of a court was raised by the Arts Law Centre of Australia, which brought up the possibility that this requirement could be a burden on plaintiffs (WIPO/GRTKF/IC/11/15 Prov. and WIPO/GRTKF/IC/11/4 (a) Add. 2, reproduced in *Extraction* 2008:66). For a wide-ranging reflection on this difficulty and others (including the problem of distinguishing between customary law and custom), see Kuruk (2002).
14 Thus at one point the Iranian delegation expressed its concern 'over the importance placed on indigenous traditional knowledge. The Delegation was concerned that the traditional knowledge held in countries without indigenous populations may be overlooked' (*Cooperation* 1999:27).

References

Barsh, Russel L. 1999. 'How Do You Patent a Landscape? The Perils of Dichotomizing Cultural and Intellectual Property'. *International Journal of Cultural Property* 8(1):14–47.

Brubaker, Rogers. 2002. 'Ethnicity Without Groups.' *Archives européennes de sociologie* 42(2):163–189

Chaudhuri, Shubha. 2009. 'Who Is the "Holder" of Intangible Cultural Heritage?: Revisiting the Concept of Community from an Intangible Cultural Heritage Perspective.' Pp. 189–210 in Toshiyuki Kono (ed.), *Intangible Cultural Heritage and Intellectual Property: Communities, Cultural Diversity and Sustainable Development.* Antwerp: Intersentia.

Coleman, Elizabeth Burns. 2004. 'Aboriginal Art and Identity: Crossing the Border of Law's Imagination.' *Journal of Political Philosophy* 12(1):20–40.

Composite Study 2003. 'Intergovernmental Committee on Intellectual Property and Genetic Resources, Traditional Knowledge and Folklore, Fifth Session (Geneva, July 7 to 15, 2003): Composite Study on The Protection of Traditional Knowledge.' WIPO document WIPO/GRTKF/IC/5/8.

Cooperation 1999. 'Report of the Permanent Committee on Cooperation for Development Related to Intellectual Property, First Session, Geneva, May 31 to June 4, 1999.' PCIPD/1112.

Daes, Erica-Irene. 1993. 'Discrimination Against Indigenous Peoples: Study on the Protection of the Cultural and Intellectual Property of Indigenous Peoples.' UNESCO Commission on Human Rights. Sub-Commission on Prevention of Discrimination and Protection of Minorities. 28 July 1993. E/CN.4/Sub.2/1993/28.

Draft Articles 2011. 'Intergovernmental Committee on Intellectual Property and Genetic Resources, Traditional Knowledge and Folklore, Nineteenth Session (Geneva, July 18 to 22, 2011): The Protection of Traditional Knowledge: Draft Articles.' WIPO document WIPO/GRTKF/IC/19/5, Annex.

Draft Articles 2014. 'The Protection of Traditional Cultural Expressions: Draft Articles,' July 2014. WIPO/GRTKF/IC/28/6, Annex.

Draft Report 2004. 'Intergovernmental Committee on Intellectual Property and Genetic Resources, Traditional Knowledge and Folklore, Seventh Session (Geneva, July 1 to 5, 2004): Draft Report.' WIPO document WIPO/GRTKF/IC/7/15 prov. 2.

Draft Report 2013. 'Intergovernmental Committee on Intellectual Property and Genetic Resources, Traditional Knowledge and Folklore, Twenty-Fifth Session (Geneva, July 15 to 24, 2013): Draft Report.' WIPO document WIPO/GRTKF/IC/25/8 prov. 2.

Drahos, Peter. 2000. 'Indigenous Knowledge, Intellectual Property and Biopiracy: Is a Global Bio-Collecting Society the Answer?' *European Intellectual Property Review* 22:245–250.

Extraction 2008. 'Intergovernmental Committee on Intellectual Property and Genetic Resources, Traditional Knowledge and Folklore, Twelfth Session (Geneva, February 25 to 29, 2008). The Protection of Traditional Cultural Expressions/Expressions of Folklore: Factual Extraction.' WIPO document WIPO/GRTKF/IC/12/4(b).

Forrest, Scott. 2006. 'Indigenous Self-determination in Finland: A Case Study in Normative Change'. *Polar Record* 42(222):229–238.

Initial Draft Report 2008. 'Intergovernmental Committee on Intellectual Property and Genetic Resources, Traditional Knowledge and Folklore, Thirteenth Session (Geneva, October 13 to 17, 2008): Initial Draft Report.' WIPO document WIPO/GRTKF/IC/13/11 Prov.

International Protection. 1985. 'Group of Experts on the International Protection of Expressions of Folklore by Intellectual Property (Paris, December 10 to 14, 1984): Report.' *Copyright* 21(2):40–60.

Kansa, E. C., J. Schultz, and A. N. Bissell. 2005. 'Protecting Traditional Knowledge and Expanding Access to Scientific Data: Juxtaposing Intellectual Property Agendas Via a "Some Rights Reserved" Model.' *International Journal of Cultural Property* 12(03): 285–314.

Kingsbury, Benedict. 1998. '"Indigenous Peoples" in International Law: A Constructivist Approach to the Asian Controversy.' *American Journal of International Law* 92: 414–457.

Kuruk, Paul. 2002. 'African customary law and the protection of folklore.' *Copyright Bulletin* 36(2):4–32.

Model Provisions 1985. 'Model Provisions for National Laws on the Protection of Expressions of Folklore Against Illicit Exploitation and Other Prejudicial Actions'. Paris and Geneva: UNESCO and WIPO.

Music In Common 2006. 'Music In Common Proposal.' Available at www.matcallahan.com/downloads/words/Music_In_Common_Proposal.pdf (accessed 23/1/2017).

National Experiences 2002. 'Final Report on National Experiences with the Legal Protection of Expressions of Folklore. Intergovernmental Committee on Genetic Resources, Traditional Knowledge and Folklore, Third Session (Geneva, 13–21 June, 2002).' WIPO/GRTKF/IC/3/10.

Ngombe, Laurier. 2004. 'Protection of African Folklore by Copyright Law: Questions That Are Raised in Practice.' *Journal of the Copyright Society of the U.S.A.* 51:437–452.

Palethorpe, Stephen, and Stefaan Verhulst. 2000. 'Report on the International Protection of Expressions of Folklore Under Intellectual Property Law.' (Study Commissioned by the European Commission's Internal Market Directorate-General), October 2000. Contract Number ETD/2000/B5–3001/E/04. Available at http://ec.europa.eu/internal_market/copyright/studies/studies_en.htm (accessed 23/1/2017).

Perlman, Marc. 2011. 'From "Folklore" to "Knowledge" in Global Governance: On the Metamorphoses of the Unauthored.' Pp. 115–132 in Mario Biagioli, Peter Jaszi, and Martha Woodmansee (eds.), *Making and Unmaking Intellectual Property*. Chicago: University of Chicago Press.

Policy Objectives 2004. 'Intergovernmental Committee on Intellectual Property and Genetic Resources, Traditional Knowledge and Folklore, Seventh Session (Geneva, July 1 to 5, 2004): Draft Policy Objectives and Core Principles: Background Information and Discussion.' WIPO document WIPO/GRTKF/IC/7/3, Annex II.

Report 1981. 'Report of the Working Party on the Protection of Aboriginal Folklore.' Canberra: Department of Home Affairs and Environment.

Report 2001. 'Intergovernmental Committee on Intellectual Property and Genetic Resources, Traditional Knowledge and Folklore, Second Session (Geneva, December 10 to 14, 2001): Report.' WIPO/GRTKF/IC/2/16.

Report 2004. 'Intergovernmental Committee on Intellectual Property and Genetic Resources, Traditional Knowledge and Folklore, Sixth Session (Geneva, March 15 to 19, 2004): Report.' WIPO/GRTKF/IC/6/14.

Revised Provisions 2005. 'Revised Provisions for the Protection of Traditional Cultural Expressions/Expressions of Folklore.' WIPO/GRTKF/IC/8/4, Annex.

Tobin, Brendan. 2001. 'Redefining Perspectives in the Search for Protection of Traditional Knowledge: A Case Study from Peru.' *RECIEL* 10(1):47–64.

Torsen, Molly, and Jane Anderson. 2010. *Intellectual Property and the Safeguarding of Traditional Cultures: Legal Issues and Practical Options for Museums, Libraries and Archives.* Geneva: WIPO.

Tulalip 2003. 'Statement by the Tulalip Tribes of Washington on Folklore, Indigenous Knowledge, and the Public Domain.' Presented to the Fifth Session of the Intergovernmental Committee on Intellectual Property and Genetic Resources, Traditional Knowledge and Folklore, 9 July 2003. Available at www.wipo.int/export/sites/www/tk/en/igc/ngo/tulaliptribes.pdf (accessed December 2016).

Updated Draft Outline 2006. 'Intergovernmental Committee on Intellectual Property and Genetic Resources, Traditional Knowledge and Folklore, Twelfth Session (Geneva, April 24 to 28, 2006). The Protection of Traditional Cultural Expressions/Expressions of Folklore: Updated Draft Outline of Policy Options and Legal Mechanisms.' WIPO/GRTKF/IC/9/INF/4.

Wendland, Wend. 2008. '"It's a Small World (After All)": Some Reflections on Intellectual Property and Traditional Cultural Expressions.' Pp. 150–181 in Christoph Beat Graber and Mira Burri-Nenova (eds.), *Intellectual Property and Traditional Cultural Expressions in a Digital Environment.* Cheltenham, UK: Edward Elgar.

10 Now you see it now you don't

The rise and fall of customary law in the IGC

Brendan Tobin

> Attempting to alter intellectual property law so that it accommodates traditional knowledge, knowledge that is completely different in essence, is reminiscent of the proverb, 'You can't fit a round peg in a square hole.' No matter how one tries, it just does not fit. It is for this reason that a completely new and customized approach is needed.
>
> Michael Dodson

Introduction

From the outset of the IGC's work it was clear that customary law had a central role to play in protection of traditional knowledge. Indigenous peoples and local communities have consistently argued that protection of their rights to their genetic resources, traditional knowledge and traditional cultural expressions cannot be effectively achieved without due respect and recognition for their own laws, customs and traditions[1] (commonly referred to as their customary laws).[2] Many states have supported this view and international human rights law and constitutional law in more than one hundred countries already recognises a role for Indigenous peoples and/or local communities' customary law in regulation of their rights. A background paper for the first session of the IGC identified the interfaces between customary law and intellectual property as a key issue for the design of appropriate measures for protection of traditional knowledge.[3]

The draft instruments prepared by the IGC have, since the start, consistently recognised the importance of customary law for, among other things, the identification of traditional knowledge and traditional cultural expressions, ascertainment of beneficiaries, benefit sharing, dispute resolution, and protection of traditional customary use. By 2014, however, most references to customary law had been deleted or replaced with references to 'cultural norms and practices', a term unknown to international law and undefined in the extensive glossary of terms prepared by WIPO for the IGC (WIPO 2016).[4] This reflects a move away from a *sui generis* regime based on customary law toward what may be seen as a hybrid intellectual property system. It also evidences a growing resistance by some states to the increasing recognition of Indigenous peoples and local communities' legal regimes in international legal instruments.

This chapter analyses the treatment by the IGC of Indigenous peoples and local communities' laws, in particular their customary laws and protocols, in the development of international instruments for the protection of traditional knowledge and traditional cultural expressions. It begins with an analysis of the relationship between traditional knowledge, customary law and intellectual property rights. It goes on to examine the legal status of customary law and its role in protection of traditional knowledge. It then examines the manner in which customary law has been addressed in the IGC process and its treatment in the IGC draft instruments. Finally it considers the role of biocultural protocols and intercultural justice in the development of law and policy for the protection of traditional knowledge. The chapter concludes that effective and culturally appropriate protection of traditional knowledge rights cannot be secured without due respect and recognition of Indigenous peoples and local communities' laws and legal traditions.

Use of terms

In this chapter, the term 'Indigenous peoples and local communities' will be used, except when referring to specific text employing the term 'indigenous and local communities' in the Convention on Biological Diversity and its Nagoya Protocol on Access to Genetic Resources and the Fair and Equitable Sharing of Benefits Arising from their Utilization. This is in line with the recognition of indigenous peoples as 'peoples' in international human rights law and the use of this term in the current texts under negotiation at the IGC.

The terms 'customary law' and the 'laws, customs and traditions' of Indigenous peoples and local communities are both used throughout the document. The IGC documents and draft instruments and the majority of commentaries on these issues predominantly refer to customary law. For that reason, and to aid the reader, the term customary law is widely used here. However, the United Nations Declaration on the Rights of Indigenous Peoples refers to laws, customs and traditions of indigenous peoples, which is wider than mere customary law. The frequent use of the term here is to remind readers of this wider aspect of Indigenous peoples and local communities' legal regimes, and the importance that they be taken into account in the development of law and policy for the protection of traditional knowledge.

Indigenous law, traditional knowledge and intellectual property

Traditional knowledge plays a fundamental role in the cultural life, subsistence and development strategies, of Indigenous peoples and local communities (Coombe 2005: 115). The struggle by indigenous peoples for recognition of their rights over traditional knowledge is intimately linked to their struggle for realisation of their rights to self-determination (Tobin 2015: 565) and its protection has been described as being 'just as important as the struggle for self-determination'.[5] Traditional knowledge informs community decision-making processes, dispute resolution, and the performance of cultural and spiritual ceremonies and practices. As such, it is both integral to and a defining factor of Indigenous peoples and local communities' legal regimes. Any attempt to regulate traditional knowledge without due respect for the legal regimes that traditionally define, mould and govern its development and use, threatens to distort the nature, legal status and potentially the content, of such knowledge. It also threatens to disrupt the dynamic nature of knowledge development and its customary transmission.

Indigenous peoples and local communities' legal regimes are influenced by their cosmovision (worldview) and distinct epistemologies, which are rooted in relationships between people, community, land, resources, spirituality and culture (Posey 1999). This is reflected in the Navajo word for 'law' *beehaz'aanii*, which according to the Honourable Robert Yazzie, former Chief Justice of the Navajo Nation, means 'the essence of life', something fundamental and absolute existing from the beginning of time (Yazzie 1994: 175). The interrelationship between law and nature is made clear in a statement by Shamans meeting in Brazil, which says '[o]ur knowledge on biodiversity is not separate from our identities, our laws, our institutions, our system of values and our cosmological view as indigenous peoples'.[6]

The notion of law as being an interwoven aspect of cultural and spiritual existence is very distinct from the more mechanistic vision of the law in Western societies. The nature of Indigenous peoples and local communities' legal traditions is also distinct from dominant legal regimes, giving greater primacy to collective rather than individualistic rights and leaning more towards the restoration of community well-being rather than retribution.[7] This focus on restorative justice means Indigenous peoples and local communities' legal regimes tend to be structural and procedural, rather than substantive and rule-oriented, giving customary law a flexibility and dynamism which enables individual cases to be addressed on their merits (Zion 1988).

Existing intellectual property regimes are widely seen as inadequate to protect Indigenous peoples and local communities' cultural and intellectual property (Janke 1998; Dutfield 2000). Not least, because, 'non-indigenous notions of intellectual property differ significantly from Indigenous beliefs' (Janke 1998).

In a statement to the United Nations Permanent Forum on Indigenous Issues, indigenous peoples have expressed their concerns that imposing intellectual property rights over traditional knowledge and resources 'will transform our knowledge and resources into individually owned, alienable commodities'.[8] The statement goes on to stress that:

> [o]ur existing protection systems are legitimate on their own right and any new mechanism for protection, preservation and maintenance of traditional knowledge and associated biological resources must respect and be complementary to such existing systems and not undermine or replace them.[9]

Customary law and traditional knowledge governance

The relationship between customary law and protection of traditional knowledge was first recognised by the international community in Agenda 21, adopted in 1992, which recommends governments '[a]dopt or strengthen appropriate policies and/or legal instruments that will protect indigenous intellectual and cultural property and the right to preserve customary and administrative systems and practices'.[10] In 2002 the Conference of the Parties (COP) to the Convention on Biological Diversity prompted Parties, with the approval and involvement of 'indigenous and local communities':

> to develop and implement strategies to protect traditional knowledge, innovations and practices based on a combination of appropriate approaches, respecting customary laws and practices, including the use of existing intellectual property mechanisms, sui generis systems, and customary law.[11]

More recently, COP adopted the Nagoya Protocol on Access to Genetic Resources and the Fair and Equitable Sharing of Benefits Arising from their Utilization. The Protocol, which entered into force in October 2014, creates obligations for Parties – both as countries home to 'indigenous and local communities' and as countries into which their resources and knowledge may be imported – to adopt measures with the aim of ensuring that access to the genetic resources and traditional knowledge of Indigenous peoples and local communities has their prior informed consent and is the subject of mutually agreed terms.[12] In implementing the Protocol's provisions on traditional knowledge, Parties are obliged under Article 12 to 'take into consideration the customary laws and community protocols and practices' of Indigenous peoples and local communities. [13] This is one of the most important and ground-breaking provisions of the Protocol, recognising as it does the existence of an international dimension and jurisdiction for the laws of Indigenous peoples and local communities where their traditional knowledge is concerned.

Legislation implementing the Nagoya Protocol adopted by the European Union[14] and Switzerland[15] requires users of traditional knowledge to demonstrate due diligence in complying with relevant national access laws in the countries in which the custodians of traditional knowledge are located. Users will, therefore, need to exercise due diligence to ascertain: (a) whether relevant national or international law empowers the custodians of traditional knowledge to regulate their own affairs and/or traditional knowledge in accordance with their own laws, customs and traditions; and, if so, (b) ensure compliance with such laws, customs and traditions. A limitation of the European Union law is that it does not provide any guidance for member states on recognition of customary law, leaving individual member states and their courts to address this issue alone. Even more problematic is the limitation of obligations to take due diligence to traditional knowledge sourced in countries with access laws, and then only with regard to traditional knowledge defined in an access agreement.[16] The result is that protection is only given for traditional knowledge that has been legally accessed, which in essence amounts to a green light for biopiracy (Tobin 2014a).

The limitations of the European Union's law may in part be mitigated by ensuring the recognition of Indigenous peoples and local communities' own laws regulating traditional knowledge, in the countries in which they are located and in the countries into which their knowledge is imported. Although only a handful of countries have adopted specific national legislation on access to traditional knowledge, Indigenous peoples and local communities' own legal regimes are well placed to provide the basis for protection of their traditional knowledge and traditional cultural expressions.

The status of customary law and its role in regulation of traditional knowledge

Indigenous peoples' rights to regulate their affairs in accordance with their own laws, customs and traditions, is grounded in common article 1 of the 1966 United Nations International Covenants on Civil and Political Rights and Economic Social and Cultural Rights, which recognises the rights of all 'peoples' to self-determination. The status of indigenous peoples as 'peoples' is recognised in the United Nations Declaration on the Rights of Indigenous Peoples (UN Declaration) adopted in 2007. At the heart of the UN Declaration is a recognition of indigenous peoples' right of self-determination[17] by virtue of which they are entitled to autonomy or self-government over their internal affairs.[18] This includes the right to maintain and strengthen their distinct political, legal, economic, social and cultural institutions,[19] and the right to promote, develop and maintain their institutional structures and their distinctive customs, spirituality, traditions, procedures, practices and, in the cases where they exist, juridical systems or customs, in accordance with international human rights.[20] Article 31 of the UN Declaration gives specific recognition to indigenous peoples' rights over genetic resources, seeds, medicines, and knowledge of the properties of fauna and flora, as well as over their cultural heritage and intellectual property.[21] Under the Declaration states are obliged to provide prompt, just and fair procedures to resolve disputes between indigenous peoples and states or other parties in a manner that gives due regard to their own laws and to international human rights.[22]

Although highly influential, the UN Declaration is not of itself legally binding. By contrast International Labour Organization (ILO) Convention 169 on Indigenous and Tribal Peoples in Independent Countries, which has been ratified by 22 states,[23] is legally binding on those states. The Convention obliges states to secure Indigenous and Tribal peoples' participation in decision-making processes affecting them[24] and to give due regard to their customs or customary laws when applying national laws and regulations to them.[25] It specifically provides that Indigenous and Tribal peoples 'shall have the right to retain their own customs and institutions, where these are not incompatible with fundamental rights defined by the national legal system and with internationally recognized human rights'.[26] Likewise, ILO Convention 107 on Indigenous and Tribal Populations recognises the rights of Indigenous and Tribal populations to retain their own customs and institutions and obliges states to give regard to their customary laws in defining their rights and duties. Although Convention 107 has been largely replaced by Convention 169 it remains in force in 17 countries,[27] including Bangladesh, India and Pakistan, that together are home to a vast percentage of the world's Indigenous peoples and local communities.

Article 12 of the Nagoya Protocol which obliges states to take into consideration customary law in the implementation of its provisions on traditional knowledge, may be seen as an attempt to give substance to the obligations under international human rights law requiring due respect and recognition for Indigenous and Tribal peoples' laws, customs and traditions, including their customary laws and protocols. This clearly includes their customary laws and protocols relating to traditional knowledge.

At the regional level the Andean Community of Nations, the Organization of African Unity, the African Intellectual Property Organization (OAPI) and the Pacific Islands Forum Secretariat have all adopted measures which recognise an important role for Indigenous peoples and local communities' own laws and decision-making processes in the regulation of traditional knowledge.[28] At the national level the constitutions of at least 115 countries give recognition to the legal regimes of Indigenous peoples and/or local communities (Cuskelly 2011: 6). The most common form of recognition is found in rights to culture, and cultural integrity (Tobin 2014b: 57–8).

Constitutional recognition of Indigenous peoples and local communities' rights to their laws, customs and practices supports recognition of subsisting legal rights based on their own legal regimes (Williams 1990). The common law doctrine of continuity recognises the continuing application of the laws of colonised peoples who have not been the subject of conquest (Gilbert 2006: 18). Where relevant customary laws of indigenous peoples have not been explicitly extinguished by the state, the doctrine of continuity would require that rights over traditional knowledge be defined through deference to those laws (Halewood 2005: 358–60).[29] Following this line of reasoning, customary legal regimes, except to the extent they have been lawfully, intentionally and explicitly overridden, by national law, provide the legal basis for protection of rights over traditional knowledge (Halewood forthcoming). This view finds support in the dissenting opinion of Kirby J., in *Western Australia v Ward*,[30] where he held that:

> recognition of native title rights to protect cultural knowledge (which includes traditional knowledge) is consistent with the aims and objectives of the Native Title Act and is consistent with international norms found in treaties to which Australia is a party.[31]

At the national level, Section 35 of the Philippines Indigenous Peoples' Rights Act of 1997 provides that access to indigenous peoples' knowledge requires their free and prior informed consent, obtained in accordance with their customary laws. The Act includes a right of restitution for any cultural, intellectual, religious or spiritual property taken in violation of indigenous laws, customs and traditions.[32] It provides that customary laws and practices should be used to resolve disputes,[33] and it requires that customary law remedies must be exhausted before any disputes among communities may be brought to the national courts.[34] Panama' s Law No. 20 empowers indigenous communities to govern under their own regulations the rights of use and commercialisation of the art, crafts and other traditional cultural expressions based on their traditions.[35] Peruvian legislation provides that where conflicts arise among indigenous peoples with regards to contracts for access to and/or use of their traditional knowledge these may be resolved with recourse to their own customary laws.[36] In Kenya the Protection of Traditional Knowledge and Traditional Cultural Expressions Bill, 2015[37] recognises a role for customary law in the identification of traditional knowledge, holders of knowledge and beneficiaries, distribution of benefits and resolution of disputes. Notably, it provides for recognition of customary law of both Kenyan and non-Kenyan holders of traditional knowledge.[38]

Even where there is a lack of specific statutory recognition, access to or use of the traditional knowledge of Indigenous peoples and local communities may still be subject to obligations derived from their laws and legal regimes. These obligations may derive directly from customary law itself or from national or international law, regulating rights to land, resources, culture, environment, education, health, and self-determination.

In order to ensure traditional knowledge sourced from Indigenous peoples and local communities within their territories is protected in foreign jurisdictions, states will need to establish national traditional knowledge regimes and/or give specific recognition to, or acknowledge the existence of, Indigenous peoples and local communities' rights to govern access to their traditional knowledge in accordance with their own laws and legal regimes (Tobin 2016). In order to put users on notice of relevant customary law states may inform the Convention on Biological Diversity Access and Benefit-sharing Clearing-House of the existence of Indigenous peoples and local communities' rights to regulate their affairs and traditional knowledge in accordance with their own laws, customs and traditions.

Treatment of customary law within the IGC

Prior to the commencement of the IGC process, WIPO allocated funds in its Budget for 2000–2001 for 'a study on customary law and regulatory systems that apply to the protection of knowledge, innovations and creativity in local and traditional communities, including conclusions relevant for the formal intellectual property system'.[39] This early acknowledgement of the importance of customary law has been reflected throughout the IGC process in its research agenda, in the interventions of representatives of indigenous peoples, local communities and states, and most importantly in the IGC's draft instruments.

Perspectives of Indigenous peoples and local communities

Representatives of the Indigenous peoples and local communities taking part in the IGC process have consistently argued that protection of their traditional knowledge should be based upon respect and recognition of customary law.[40] Participants in the WIPO Panel on Indigenous and Local Communities' Concerns, which, since 2005, has been held immediately prior to each IGC meeting, have continually highlighted the need for recognition of customary law as the primary basis for protection of traditional knowledge.[41] Alphonse Kambu, speaking on the first Panel, stressed that the legal concepts, principles and doctrines of Indigenous peoples and local communities should serve as the 'foundation' for protecting and regulating traditional knowledge and related genetic resources, and 'only when there are gaps then should principles, doctrines and concepts of Western and other legal systems be borrowed to fill in the deficiencies' (Kambu 2005).

At the IGC's ninth meeting the representative of the Kaska Dena Council disputed the presumption made by some delegations that national law, including intellectual property law, supersedes indigenous customary law and proposed, that an objective of any regime should be to:

> [m]eet the actual needs of communities . . . be guided by the aspirations and expectations expressed directly by indigenous peoples and local communities, respect their rights under customary law, including indigenous national, regional and international law. . .[42]

At the same meeting the representative of the Tulalip Tribes made the point that international intellectual property law 'had been largely negotiated at a time when the recognition and respect for indigenous customary law, legal traditions and rights were not taken into account'.[43] This is a key issue for indigenous peoples due to the difficulties posed by applying western notions of the public domain to traditional knowledge whose public disclosure did not signify the extinguishment of customary law based rights.[44] The notion of the public domain is therefore a direct challenge to customary legal rules governing Indigenous peoples and local communities' knowledge-sharing practices, which often place strict controls on the use of traditional knowledge which may be freely shared but not freely used (Raven 2012) (see also Chapter 8 by Okediji in this volume).

State perspectives

Many states from all around the world and with very distinct legal regimes have thrown their weight behind calls for respect and recognition of customary law.[45]

China has argued that customary laws and practices for the protection of folklore should be recognised in legislation regulating traditional cultural expressions.[46] It has proposed that where disputes arise regarding the exercise of rights relating to traditional knowledge these should, as much as possible, be resolved through customary law or alternative dispute resolution mechanisms.[47]

The United States delegation has argued that no government regime could ensure that traditional knowledge was protected.[48] Therefore access to genetic resources and traditional knowledge ultimately should be determined by the traditional knowledge holders and should take into consideration existing customary law, *sui generis* systems and community-based norms.[49] They have also stressed that in recognition and respect for traditional knowledge systems the IGC must acknowledge and support the diverse and numerous customary law regimes that exist.[50]

The African Group have proposed that customary law should play a role in helping to define/identify traditional cultural expressions that should be protected.[51] Nigeria claims that most communities have traditional mechanisms, including customary laws, practices and protocols for the protection of traditional cultural expressions.[52] South Africa has argued that legal certainty and the survival of indigenous cultures requires that any regime recognise the 'customary laws and protocols that govern the creation, transmission, reproduction and utilization of [Indigenous knowledge], the repatriation of cultural heritage and the recording, maintenance, protection and promotion of oral tradition.[53] It takes the view that the IGC is not creating new rights but respecting and recognising existing rights based on customary law.[54]

Similarly, New Zealand has expressed the opinion that *sui generis* laws mean new models of protection emerge, not from existing intellectual property concepts or structures, but rather:

> from the need and aspirations of the holders of [traditional knowledge and traditional cultural expressions] and their communities, from the customary laws, values and protocols associated with [them], and from the concerns . . . expressed by indigenous peoples and local communities about [their] inappropriate use.[55]

This approach was also central to the development of Peru's *sui generis* regime for the protection of traditional knowledge.[56]

The Like Minded Group of countries have given support for the adoption of a regime that requires contracting parties to ensure that protection of rights over traditional knowledge take into account relevant customary laws and practices.[57] Brazil has proposed that any regime for the protection of traditional knowledge should promote its conservation, maintenance, application and wider use in accordance with relevant customary practices, norms, laws and understandings, of traditional knowledge holders.[58] It has also argued that 'the application, interpretation and enforcement of protection against misappropriation of traditional knowledge, including determination of equitable sharing and distribution of benefits, should be guided, as far as possible and appropriate, by respect for the customary practices, norms, laws and understandings' of the holders of the traditional knowledge.[59] The Indian delegation has expressed the view that customary protocols, understandings, laws and practices of local communities and indigenous peoples should as far as possible guide decision making on entitlements to the benefit of protection.[60]

Support for increased recognition of customary law has not been unanimous. Australia, for example, has argued that recognition of customary law should be limited to the extent it conflicts with established international and national laws and policies.[61] Japan has claimed that although, 'there are some existing rights, under customary laws or legal practices, which merit respect' the customary laws and practices which are recognised in some countries or regions, are not necessarily considered legal rights in foreign jurisdictions.[62] Canada has opposed what it describes as 'an unqualified and extraterritorial application of indigenous customary laws and protocols',[63] which they have argued could have significant legal implications for Canada, beyond the role of intellectual property law and policy.[64]

In considering the foregoing views it is important to recognise that while the legal status of customary law in relation to traditional knowledge is primarily determined by national law, it may also derive from obligations under international law. The Nagoya Protocol, for example, requires both states in which Indigenous peoples and local communities are located and the countries into which their knowledge is imported to give 'due consideration' to customary law in the adoption of measures to ensure prior informed and mutually agreed terms exist for access and use.

The differing positions of states on these matters highlights the importance of research on the customary law and its current and potential role for regulating traditional knowledge at the local, national and international levels.

Research on customary law

When setting up the IGC the WIPO General Assembly identified the need for a study of customary laws and regulatory systems that apply to traditional knowledge and of certain conceptual issues, such as collectivity of creation, innovation and ownership and cultural understandings of property rights (WIPO 2000). A paper for the IGC's opening session identified the interfaces between customary law and intellectual property as one of four key issues that needed to be addressed in order to make measures for protection of traditional knowledge clear, practical and accessible to traditional knowledge holders (WIPO 2001a).

In 2001 WIPO published a report on the needs and expectations of holders of traditional knowledge (WIPO 2001b). Based on nine fact-finding missions involving visits to twenty-eight countries, between May 1998 and November 1999, the report identified two main sets of needs and concerns of traditional knowledge holders.

(i) the wish of some to benefit from commercialisation of traditional knowledge and cultural expressions, i.e. a desire for 'positive protection'
(ii) concerns by some at the cultural, social and psychological harm caused by unauthorised use and a wish to control and even prevent altogether any external use, i.e. a desire for 'defensive protection'.[65]

The report concluded that in order to address these needs a study of customary laws and protocols relating to the use, development, transmission and protection of traditional knowledge, and their relationship with intellectual property rights standards was needed.[66]

At its third meeting the IGC approved the preparation of a practical study on the relationship between customary laws and protocols and the formal intellectual property system.[67] A report by the IGC secretariat, submitted at the IGC's fourth session, summarised the desired objectives of Indigenous peoples and traditional communities as including 'the desire for legal recognition of the right to own and control access to and the disclosure and use of traditional cultural expressions in accordance with customary laws and protocols.'[68] The report highlighted the limitations of intellectual property law to achieve this objective noting, for example, that 'the conception of "ownership" in copyright law is incompatible with customary laws and systems'.[69] This incompatibility was apparent in the distinctions between exclusive private property rights under copyright law and the complex 'rules, regulations and responsibilities, more akin to usage or management rights' found in customary law.[70]

In 2006 WIPO commissioned a report on customary law and traditional knowledge in Andean Countries.[71] The report concluded that the focus of any *sui generis* system for the protection of knowledge, innovations and traditional practices should be based on customary law and not through access and commercialisation as privatised resources using intellectual property rights.[72] That same year, WIPO circulated an issues paper on customary law (WIPO 2006) for comment. The paper describes numerous ways in which customary law has already been recognised in intellectual property related cases and national legislation. These include: establishing collective legal standing; settlement of community disputes; asserting an equitable interest in intellectual property; sustaining a claim of breach of confidence; conferring collective ownership of an intellectual property right; as the basis of a general right over biological resources and traditional knowledge; protecting continuing customary use; determining entitlement for damages; and the status of a claimant as a member of an indigenous or other traditional community, or to establish a specific indigenous or aboriginal right (WIPO 2006; Taubman 2005: 531–32).

The WIPO issues paper provides important evidence of the role customary law plays in the protection of traditional knowledge rights. It also demonstrates the potential for significant expansion of this role through effective development of the capacity of national courts and administrative bodies to receive evidence of customary law based rights and duties, and provide adequate recognition and support for their enforcement. Although widely viewed as a useful study, WIPO's research has been criticised by the New Zealand Delegation for focusing on 'existing legal frameworks rather than on the holistic needs of indigenous and local communities'. As a result, Indigenous peoples and local communities' 'priorities, customary laws, protocols and values associated with their TK and TCEs were not the focus or main premises of the analyses'.[73] A study focusing on the actual role of customary legal regimes in the governance of access to genetic resources and traditional knowledge in Andean and Pacific Island countries was prepared for the United Nations University, Institute of Advanced Studies (UNU-IAS) and WIPO.[74] Based upon a series of participative regional and sub-regional Workshops, involving representatives of Indigenous peoples and local communities, governments and non-governmental organisations, the study identified the central role customary law plays in the regulation of traditional knowledge at the local and national levels and the need for increased research from the bottom up into its role in national and international protection of traditional knowledge (Tobin 2008).

Despite frequent calls by both developed and developing countries for WIPO to provide research on customary law, neither the WIPO issues paper or the UNU-IAS/WIPO Andean and Pacific Islands Report were ever formally submitted for information or consideration in the IGC. On the contrary, WIPO withheld the UNU-IAS/WIPO Report for five years before eventually releasing it unedited along with the WIPO issues paper as online publications on its website. By the time of their release in 2013, immediately prior to IGC 24, almost all references to customary law in the IGC draft instruments had already been deleted or replaced with the vague and legally meaningless term 'cultural norms'. Their delayed publication limited their utility to inform the IGC debates and, if their eventual publication was intended as a show of compliance with requests by the IGC for research on customary law, it is quite clearly a case of too little too late.[75]

Customary law in the IGC draft instruments

From early on, the IGC has clearly leant towards the development of measures for the protection of traditional knowledge grounded on customary law.[76] Customary law has been recognised as one of a range of measures available for legal protection for traditional knowledge.[77] It has been seen as having a role to play in identifying traditional knowledge,[78] ensuring equitable benefit sharing[79] and defining cases of misappropriation.[80] The importance of securing the customary usage of traditional knowledge was also firmly established.[81] By making prior informed consent a condition for access to and use of traditional knowledge, the IGC drafts have, in essence, acknowledged the rights of Indigenous peoples and local communities to exercise control over their knowledge in accordance with their customary law and practices. Key terms such as unfair trade were to be defined with attention to concepts of unfair under customary law.[82] A revised draft in 2011 presented a regime built upon concepts of misappropriation[83] in which:

> [t]he application, interpretation and enforcement of protection against misappropriation of traditional knowledge, including determination of equitable benefit sharing and distribution of benefits, should be guided, as far as possible and appropriate, by respect for the customary practices, norms, laws and understandings of the holder of the knowledge, including the spiritual, sacred or ceremonial characteristics of the traditional origin of the knowledge.[84]

It also provided that protection should be extended to traditional knowledge which is:

> integral to the cultural identity of an indigenous or traditional community or people which is recognized as holding the knowledge through a form of custodianship, guardianship, collective ownership or cultural responsibility. This relationship may be expressed formally or informally by customary or traditional practices, protocols or laws.[85]

The recognition of principles of stewardship in the identification of rights was a significant move towards establishment of a system of protection grounded on Indigenous peoples and local communities' own legal principles. This is clear in the definition of 'custodian' as:

> those communities, peoples, individuals and other entities, which, according to customary laws and other practices, maintain, use and develop the traditional knowledge.[86]

This expresses a notion that is different from 'ownership', conveying a sense of responsibility to ensure that traditional knowledge is used in a way that is consistent with community values and customary law.[87]

Taubman, who was Director of the Traditional Knowledge Division of WIPO until 2006, has argued that traditional knowledge should be considered integral to a community based upon factors that may be described by customary law (Taubman 2005). These include: definition of custodianship or the nature of community ownership; the rights and responsibilities associated with custody, access rights, means of dissemination and preservation of knowledge; and the customary mode of defining modalities for prior informed consent, benefit-sharing mechanisms, dispute settlement, and sanctions for infringement of customary law (Taubman 2005: 938). There is, he says, a role for tribal elders in identifying traditional knowledge as against some 'objective, culturally neutral conception of traditional knowledge', which is needed at the level of national or international law if the system is to be workable.[88]

The establishment of a test for the identification of traditional knowledge based on it being 'integral to the cultural identity' of its holders as demonstrated by customary law is not without problems. In Canada an 'integral to a distinctive culture' test, developed in the case of *Van der Peet*,[89] has proved a significant hurdle for aboriginal peoples seeking to secure recognition of their resource rights (Tobin 2014b: 165). The key question here is who should do the deciding on what is integral to cultural identity. If this falls to be decided by state courts, Indigenous peoples and local communities may find themselves facing an 'impossible standard of authentic traditional culture' (Barcham 2007) in order to show continuity of customary law and its continuous application to the relevant traditional knowledge since before colonisation. It is vital therefore that any regime which includes customary law as a condition for identification and enforcement of rights, should be informed by experts in customary law and allow for dynamism and case-by-case analysis.

As the IGC advanced in its efforts to develop legal instruments, developing countries argued these should be binding in nature; this was, however, firmly opposed by many developed countries. The inability of states to broker a compromise at the WIPO General Assembly in 2014 led to the suspension of the IGC process from September 2014. The inability of the international community to find common ground on protection of traditional knowledge over more than fourteen years stands in sharp contrast to the ability of the developed nations to push through the Trade Related Aspects of Intellectual Property Agreement in just a couple of years. What is clearly missing is a lack of political will to recognise and protect indigenous peoples' rights by developed nations, intent on protecting the research and commercial sectors' interest in maintaining as much traditional knowledge as possible within a freely accessible public domain.

From a **sui generis** *regime to an intellectual property hybrid*

In an effort to find a way through the outstanding differences at the IGC the chair, Ambassador Wayne McCook of Jamaica, proposed an innovative, but problematic, proposal in 2013.[90] The proposal describes a three-tier system in which sacred, secret and closely held traditional knowledge would be given the highest level of protection, requiring prior informed consent and mutually agreed terms for access and use. A second tier addresses knowledge that is not secret or sacred and is publically available but not widely distributed. At this level, fair and equitable benefit sharing based on mutually agreed terms is required. It makes no provision, however, for prior informed consent, making it unclear how Indigenous peoples and local communities might defend their interests against a user that refuses to negotiate fair sharing of benefits. At the lowest tier, widely shared traditional knowledge and traditional cultural expressions would be freely usable subject to protection of moral rights, with any benefit sharing to be done by way of payment of user fees to a state fund.[91] This proposal forms the central plank of the draft instruments on both traditional knowledge and traditional cultural expressions submitted to the WIPO General Assembly in 2014.[92]

The proposed three-tiered IGC system reflects in some ways Argumedo's proposal, from the early 1990s, for a stoplight system to regulate access to traditional knowledge held in databases.[93] Under his proposal, free access (green light) would be available to traditional knowledge which was not considered sensitive or economically valuable; limited access (orange light) would apply to knowledge of value which could be accessed subject to prior informed consent; and restricted access (red light) would apply to traditional knowledge that would be available only to members of the relevant indigenous people. The distinction is that under the IGC proposal the state would define under which tier traditional knowledge fell, while under Argumedo's proposal Indigenous people and local communities would always hold responsibility for categorising under which level specific traditional knowledge might fall.

Although the IGC proposal has some *sui generis* aspects about it – it allows, for example, for an indefinite period of protection as long as traditional knowledge and traditional cultural expressions conform to the terms of eligibility set out in the respective draft instruments[94] – it is now leaning more closely towards an intellectual property rights system, where states confer rights 'in accordance with national law'. Both the Draft Articles on traditional knowledge and those on traditional cultural expressions create exemptions for any act permissible under the national law of a Contracting Party for: (a) knowledge protected by patent and trade secrecy laws in the case of traditional knowledge[95] and (b) for material protected by copyright law in the case of traditional cultural expressions.[96] This clearly subordinates indigenous peoples' laws and customs relating to their traditional knowledge to intellectual property law. Both sets of Draft Articles do, however, contain a common proviso limiting their application in cases where it is necessary to protect secret traditional knowledge from disclosure. This is a small but important concession recognising the possibility for customary laws to override intellectual property law.

Customary law vs. cultural norms

Considering the wide support and growing recognition of the importance of customary law for protection of traditional knowledge within the IGC it is amazing that the IGC is working on draft instruments from which the term customary law has been almost completely erased. The only remaining direct references to customary law are found in heavily bracketed text requiring that states' (dangerously paternalistic) rights to establish competent authorities to administer the rights of traditional knowledge be carried out in accordance with their customary laws and protocols[97] and a requirement that states should endeavour to establish databases of traditional knowledge in accordance with customary law.[98] It seems far from coincidental that the only remaining references to customary law are found in two provisions which run directly counter to the expressed wishes of Indigenous peoples and local communities. This includes widespread resistance to the recompilation of their knowledge in databases beyond their control (Alexander *et al.* 2003) and to the notion of states administering their rights over their traditional knowledge. In both cases, the reference to customary law seems intended to avert strident opposition from Indigenous peoples and local communities to these provisions. These references to customary law do, however, serve one positive purpose. They make the continuing use of the term 'cultural norms' in other parts of the draft IGC texts less defensible.

All three tiers of the proposed IGC system would require users to use knowledge in a manner that respects the 'cultural norms and practices' of holders of traditional knowledge. The apparent intention is to make users comply with the laws, customs and traditions of Indigenous peoples and local communities with regard to their moral rights/cultural integrity. While this is a worthy aim, it falls far short of the recognition of a role for Indigenous peoples and local communities' laws, customs, traditions and protocols in the regulation and enforcement of their rights, beyond the issue of moral rights.

As noted earlier the term 'cultural norms' is not defined in the draft instruments or in the extensive glossary prepared by WIPO for the IGC. It appears to be a sociological rather than a legal term and as such is unlikely to be given any significant weight by courts in any case that may arise regarding access to and use of traditional knowledge. The use of the term does not mean that the IGC as a whole has abandoned the recognition of a role for the laws of Indigenous peoples and local communities in regulation of traditional knowledge. It does, however, at best, demonstrate a lack of understanding on the part of IGC member states and Secretariat that the term, unless precisely defined in the text of the draft instruments to cover the laws, customs and traditions of Indigenous peoples and local communities' legal regimes, and at least their customary laws, will have virtually no legal significance. At worst, the use of the term may demonstrate an attempt to beguile participants in the IGC into believing that references to 'cultural norms' are legally equivalent to and interchangeable with specific recognition of Indigenous peoples and local communities' laws and protocols. It is not and its use would appear to neutralise the influence of Indigenous peoples and local communities' customary and other laws on the development and future implementation of the IGC instruments.

Customary law, biocultural protocols and intercultural justice

One of the drawbacks to seeking recognition of traditional knowledge rights based on title as defined by state law is the potential loss of 'epistemological sovereignty' where indigenous laws and customs are subjugated under the categories imposed by an external dominant legal regime (Graber 2012: 14). Overcoming superficial and potentially self-serving state-controlled identification, application and restriction of customary law requires a deep form of legal pluralism based upon recognition and enforcement of extensive rights to self-determination, land, resources and culture (Forsyth forthcoming). Nascent but highly significant moves in this direction are facilitating a resurgence of indigenous law (Borrows 2002), and the emergence of a growing body of law and jurisprudence drawing on indigenous legal principles in the development of what Tsosie describes as intercultural justice (Tsosie 2007: 408 et seq.). This is not the integration of indigenous law within the dominant legal regime as much as an acculturation of the law influenced by a wide range of legal sources and recognising and respecting the continuing and increasingly interrelated role of multiple legal regimes (Tobin 2014b: 180).

Indigenous peoples are recognised as lawmakers under international human rights law and the national law of many countries. Around the world indigenous peoples are exercising this right in new ways through the development of written tribal laws, bylaws and protocols, in many cases directly based upon and subject to their existing customary laws (Fletcher 2007: 61). Protocols are widely in use to regulate, among other things, research activities, commercialisation of artworks and the making of films (Bowrey 2006). Biocultural protocols are increasingly being used to describe the rights of Indigenous peoples and local communities over their lands, resources and cultural heritage and the parameters, elements and modalities for seeking access to and use of genetic resources and traditional knowledge (Swiderska *et al.* 2012).

The development by Indigenous peoples and local communities of protocols as tools for the regulation of their rights to their lands, resources and cultural heritage is an important step in the exercise of rights to self-determination. By taking the initiative Indigenous peoples and local communities play a significant role in defining the interfaces between their own legal regimes and state and international law. In some ways the development of written protocols may be seen as reverse images of the processes leading to the inclusion of indigenous legal principles in state constitutions and statute. Both may be seen as processes for the development of intercultural law drawing on the best of global juridiversity.

The Nagoya Protocol requires states to support the development of community protocols.[99] At the international level, the United Nations Permanent Forum on Indigenous Issues, WIPO, the governing bodies of the Convention on Biological Diversity, Nagoya Protocol and the International Treaty on Plant Genetic Resources for Food and Agriculture might usefully work, together and individually, to support the development by Indigenous peoples and local communities of local, people-wide, regional or even global protocols. Their development may be informed by existing laws, declarations, submissions and draft legislative proposals, made by or in collaboration with Indigenous peoples and local communities as well as by their own customary and other laws and protocols. Protocols of this nature would help to provide a clear indication of the nature, scope and form of protection of traditional knowledge considered appropriate by indigenous peoples. This would serve as an important input for the development of national, regional and international law. Protocols could also help ensure compliance with international human rights law on participation and consultation, secure more effective consideration of customary law and increase legitimacy and likelihood of successful law making at all levels.

Conclusions

After years of leaning towards recognition of the fundamental role of customary law in the development of a *sui generis* regime for protection of traditional knowledge, the IGC has veered sharply away towards a hybridised intellectual property style system devoid of any significant role for customary law. In doing so, the draft IGC instruments signify a backward step in the progressive recognition of customary law and its role in the protection of traditional knowledge under international human rights law, the Nagoya Protocol, as well as regional and national law.

In the end the outcome of negotiations towards development, in a truly good faith process, of a regime for protection of traditional knowledge may not prove so very different whether it starts from a system based on intellectual property or customary law. There are only so many permutations and mechanisms that are available to develop a functional system of protection for the tangible and intangible aspects of traditional knowledge and traditional cultural expressions, especially where these are already fairly readily accessible beyond the control of their traditional holders. The point of departure is, however, likely to prove fundamental to the eventual impact of any regime on the protection of traditional knowledge, not only in terms of monetary values and prevention of misappropriation, but even more crucially in terms of safeguarding the cultural, spiritual and social significance of traditional knowledge and traditional knowledge systems for Indigenous peoples and local communities. Seen from this perspective, the role of customary law in the regulation of traditional knowledge is vital to the protection of cultural integrity and the realisation by Indigenous peoples and local communities of their rights to decide their own development paths and to exercise their human rights to self-determination.

In the interests of securing effective and appropriate protection of traditional knowledge and traditional cultural expressions, the IGC may find it beneficial to prioritise necessary research on the current and potential role of Indigenous peoples and local communities' own legal regimes for their protection. Research is also required to investigate modalities for the recognition, application and enforcement of indigenous peoples' laws and protocols in their regulation of traditional knowledge in foreign jurisdictions. This research and its early provision to IGC members and stakeholders is crucial if the IGC is to work in good faith with Indigenous peoples and local communities to develop instruments for the protection of their traditional knowledge and traditional cultural expressions, in a manner which respects, builds upon and complements, rather than undermines, their own legal regimes.

Notes

1 See for example Mataatua Declaration on Cultural and Intellectual Property Rights of Indigenous Peoples (1992) and the Julayinbul Statement on Indigenous Intellectual Property Rights (1993) in Posey, Darrell (ed.) (1999) *Cultural and Spiritual Values of Biodiversity: A Complementary Contribution to the Global Biodiversity Assessment*, Intermediate Technology Publications. See also Australian Copyright Council (ACC) (September 1998) 'Protecting Indigenous Intellectual Property. A Discussion Paper', Redfern: ACC, 11; WIPO (2013) 'Customary Law, Traditional Knowledge and Intellectual Property: An Outline of the Issues', *World Intellectual Property Organization*, 5.

2 The legal regimes of Indigenous peoples and local communities are largely oral in nature and for this reason there is a tendency to refer to them as customary law. Their regimes are however much broader than merely custom and may include tribal statute, sacred and declaratory law. Borrows, J. (2010) *Canada's Indigenous Constitution*, University of Toronto Press, Toronto, 12.

3 'Matters concerning Intellectual Property and Genetic Resources, Traditional Knowledge and Folklore'. Document prepared by the Secretariat, WIPO/GRTKF/IC/1/3, 21.

4 'Glossary of Key Terms Related to Intellectual Property and Genetic Resources, Traditional Knowledge and Traditional Cultural Expressions', WIPO, Geneva, WIPO/GRTKF/IC/30/INF/7.

5 UNDP Consultation on the Protection and Conservation of Indigenous Knowledge, Sabah, East Malaysia (1995) Basic points of agreement on the issues faced by the indigenous peoples of Asia. Reprinted in Posey, D. (1999), 575.

6 WIPO/GRTKF/IC/2/14.
7 Mayens cited in Reynolds, L. (2006).
8 IPBC (2004) 'Collective Statement of Indigenous Peoples on the Protection of Indigenous Knowledge'. United Nations Permanent Forum on Indigenous Issues, Third Session, New York, 10–12 May 2004. Agenda item 49(e): Culture, cited in Gibson J. (2005), 8.
9 Ibid.
10 Agenda 21, chapter 26.4.
11 See COP Decision VI/10F. 33, See also 'Bonn Guidelines on Access to Genetic Resources and Fair and Equitable Sharing of the Benefits Arising out of their Utilization', Secretariat of the Convention on Biological Diversity, Montreal, 2002, Para. 31.
12 Nagoya Protocol Articles 6.2 (genetic resources) and 7 (traditional knowledge).
13 The Nagoya Protocol does not refer directly to indigenous peoples but instead uses the term 'indigenous and local communities'. This runs counter to the advances made in international human rights law, which recognises indigenous peoples' status as 'peoples' and as such entitled to rights of self-determination. For the purposes of this work the term Indigenous peoples and local communities will be utilised except where referring to specific provisions of the Protocol and Convention on Biological Diversity.
14 Regulation of the European Parliament and of the Council on compliance measures for users from the Nagoya Protocol on Access to Genetic Resources and the Fair and Equitable Sharing of Benefits Arising from their Utilisation in the Union. Adopted by the European Parliament 11 March 2014. Available at http://register.consilium.europa.eu/doc/srv?l=EN&f=PE%20131%202013%20INIT (accessed 23/1/2017).
15 See amendments to Federal Act on the Protection of Nature and Cultural Heritage adopted by the Swiss Parliament on 21 March 2014 in order to implement the Nagoya Protocol (unofficial translation). Available at: www.sib.admin.ch (accessed 23/1/2017).
16 Ibid.
17 United Nations Declaration on the Rights of Indigenous Peoples (UNDRIP), Article 3.
18 UNDRIP, Article 4.
19 UNDRIP, Article 5.
20 UNDRIP, Article 34.
21 UNDRIP, Article 31.
22 UNDRIP, Article 40.
23 Member states of ILO Convention 169: Argentina, Bolivia, Brazil, Central African Republic, Chile, Colombia, Costa Rica, Denmark, Dominica, Ecuador, Fiji, Guatemala, Honduras, Mexico, Nepal, Netherlands, Nicaragua, Norway, Paraguay, Peru, Spain and Venezuela.
24 ILO Convention 169, Article 6.
25 ILO Convention 169, Article 8.1.
26 ILO Convention 169, Article 8.2.
27 Member states of ILO Convention 107: Angola, Bangladesh, Belgium, Cuba, Dominican Republic, Egypt, El Salvador, Ghana, Guinea-Bissau, Haiti, India, Iraq, Malawi, Pakistan, Panama, Syrian Arab Republic, Tunisia.
28 See, 'Andean Community Decision 391: Establishing the Common Regime on Access to Genetic Resources' (1996) and 'Andean Community Decision 486: Establishing the Common Industrial Property Regime (2000)', 'African Model Legislation for the Protection of the Rights of Local Communities, Farmers and Breeders and for the Regulation of Access to Biological Resources' (OAU Algeria 2000), 'Swakopmund Protocol on the Protection of Traditional Knowledge and Expressions of Folklore Within the Framework of the African Regional Intellectual Property Organization' (ARIPO) Adopted by the Diplomatic Conference of ARIPO at Swakopmund (Namibia) on 9 August 2010, 'Guidelines for Developing Legislation for the Protection of Traditional Biological Knowledge, Innovations and Practices Based on the Traditional

Biological Knowledge, Innovations and Practices Model Law', Pacific Islands Forum Secretariat 2010.

29 See 'Statement by the Tulalip Tribes of Washington on Folklore, Indigenous Knowledge, and the Public Domain', 9 July 2003, WIPO Intergovernmental Committee on Intellectual Property and Genetic Resources, Traditional Knowledge and Folklore, Fifth Session, 5–17 July, 2003.

30 *Western Australia v Ward* [2002] HCA 28.

31 Cited in McRae *et al.* 2009, at 428.

32 Philippines Indigenous People's Rights Act of 1997, Article 32.

33 Ibid., Article 65.

34 Ibid., Article 26.

35 Act 20 Special System for the Collective Intellectual Property Rights of Indigenous Peoples, Official Gazette, No. 24,083, 27 June 2000, Article 15.

36 Law 27811, of 2002, *Introducing a Protection Regime for the Collective Knowledge of Indigenous Peoples Derived from Biological Resources*, Article 47.

37 Kenya Gazette Supplement No. 132 (National Assembly Bills No. 48) Nairobi, 11 August 2015.

38 Ibid.

39 WIPO/GRTKF/IC/1/3.

40 See WIPO/GRTKF/IC/4/15, for statements to the Fourth IGC meeting by the Indigenous Peoples' Biodiversity Network, the Saami Council and the Arctic Athabaskan Council, speaking also on behalf of the Assembly of First Nations, the Inuit Circumpolar Conference and the *Métis* National Council.

41 See, for example, Panel presentations by Alphonse Kambu (Kambu 2005), Ralph Regenvanu, Director, Vanuatu Cultural Centre, WIPO/GRTKF/IC/9/INF/7(b), Ms Jeannette Casing Serrano, Chairperson, National Commission on Indigenous Peoples, Philippines, WIPO/GRTKF/IC/9/INF/7(c) and Greg Young-Ing of the Creator's Rights Alliance, Canada, WIPO/GRTKF/IC/10/INF/5(c).

42 WIPO/GRTKF/IC/9/14 Prov. 2, para. 119.

43 WIPO/GRTKF/IC/9/14, para. 237.

44 WIPO/GRTKF/IC/10/INF/5(c) Competing Jurisdictions over Traditional Knowledge in the Northern Americas, para. 7.

45 WIPO/GRTKF/IC/8/15 Prov., para. 88 (USA) and WIPO/GRTKF/IC/18/11, para. 39 (China).

46 WIPO/GRTKF/IC/8/15 Prov., para. 91.

47 WIPO/GRTKF/IC/9/14 Prov. 2, para. 124.

48 WIPO/GRTKF/IC/9/14 Prov. 2, para. 221.

49 WIPO/GRTKF/IC/9/14 Prov. 2, para. 230.

50 WIPO/GRTKF/IC/9/14 Prov. 2, para. 221.

51 WIPO/GRTKF/IC/13/9 African Group Proposal on the Protection of Traditional Knowledge, Traditional Cultural Expressions and Genetic Resources, Geneva 2008.

52 WIPO/GRTKF/IC/9/14 Prov. 2, para. 103.

53 WIPO/GRTKF/IC/11/5(a), page. 59.

54 WIPO/GRTKF/IC/9/14 Prov. 2, para. 111.

55 WIPO/GRTKF/IC/14/12, para. 47.

56 See Peru Law 27811.

57 WIPO/GRTKF/IC/19/10 Annex, page 2, Article 3.2.

58 WIPO/GRTKF/IC/11/5(a), page 58.

59 WIPO/GRTKF/IC/12/5(b), page 80.

60 WIPO/GRTKF/IC/11/15 Prov., page 47.

61 WIPO/GRTKF/IC/9/14 Prov. 2, para. 210.

62 WIPO/GRTKF/IC/11/5(b), Annex, page 13.

63 WIPO/GRTKF/IC/8/15, para. 93.

64 WIPO/GRTKF/IC/8/15, para. 93.

65 WIPO/GRTKF/IC/2/8, 33.
66 Ibid.
67 WIPO/GRTKF/IC/4/3, 69, 19; WIPO/GRTKF/IC/3/10.
68 WIPO/GRTKF/IC/4/3, 38–9.
69 WIPO/GRTKF/IC/4/3, 47 (iii).
70 Ibid. See further the case study on the Australian case *M*, Payunka, Marika and Others v Indofurn Pty Ltd* (1994) 30 IPR 209 discussed in Janke, T. (2003).
71 Rodrigo De la Cruz (2006) 'Regional Study in the Andean Countries: Customary Law in the Protection of Traditional Knowledge', Final Report Revised for WIPO. Available at: www.wipo.int/export/sites/www/tk/en/resources/pdf/study_cruz.pdf (accessed 23/1/2017).
72 Ibid., 40.
73 WIPO/GRTKF/IC/13/11, para. 81.
74 Tobin, B. (2008) *The Role of Customary Law in Access and Benefit-Sharing and Traditional Knowledge Governance: Perspectives from Andean and Pacific Island Countries*, UNU-IAS, WIPO, Geneva.
75 WIPO pleaded a lack of funds as the basis for failing to publish the UNU-IAS/WIPO report in 2008. When it eventually did release the report it placed an unedited version online, something it could just as easily have done in 2008. Furthermore, WIPO dated the document as a 2013 report, even though it had not been updated from 2008. The obvious lack of coverage of key issues such as the Nagoya Protocol, which was adopted in 2010, undermined the credibility of the report among experts in the field. Pers. comm. Miranda Forsyth, Canberra, 2015.
76 See WIPO/GRTKF/IC/10/5, WIPO/GRTKF/IC/16/5, WIPO/GRTKF/IC/18/5.
77 Article 2, Draft provisions.
78 Ibid., Article 6.
79 Ibid., Article 5.
80 Ibid., Article 1(5) The application, interpretation and enforcement of protection against misappropriation of traditional knowledge, including determination of equitable benefit sharing and distribution of benefits, should be guided, as far as possible and appropriate, by respect for the customary practices, norms, laws and understandings of the holder of the knowledge, including the spiritual, sacred or ceremonial characteristics of the traditional origin of the knowledge.
81 Ibid.
82 Ibid.
83 WIPO/GRTKF/IC/18/5, Article 1.
84 Ibid. Article 3(1)(e).
85 WIPO/GRTK/IC/18/5, Article 1(3)(iii).
86 WIPO/GRTKF/IC/19/INF/8.
87 Ibid.
88 Ibid.
89 *R v Van der Peet* [1996] 2 SCR 507.
90 WIPO/GRTKF/IC/27/4, Draft Article 3.
91 WIPO/GRTKF/IC/28/5.
92 WIPO/GRTKF/IC/28/5 and WIPO/GRTKF/IC/28/4.
93 Pers. Comm. Alejandro Argumedo, Geneva, 1993.
94 WIPO/GRTKF/IC/27/4, Draft Article 7, WIPO/GRTKF/IC/27/5, Draft Article 6.
95 WIPO/GRTKF/IC/27/4, Draft Article 6.4.
96 WIPO/GRTKF/IC/27/5, Draft Article 5.5.
97 WIPO/GRTKF/IC/18/5, Article 5.1.
98 WIPO/GRTKF/IC/18/5, Article 3.
99 Nagoya Protocol, Article 12.

Bibliography

Alexander, M., Chamundeeswari, K., Kambu, A.,Ruiz, M.N. and Tobin, B. (2003) *The Role of Registers and Databases in the Protection of Traditional Knowledge: A Comparative Analysis*, United Nations University Institute of Advanced Studies, Tokyo.

Barcham, M. (2007) 'The limits of recognition', in B. Smith and F. Morphy (eds), *The Social Effects of Native Title*, Australian National University Press, Canberra.

Borrows, J. (2002) *Recovering Canada: The Resurgence of Indigenous Law*, University of Toronto Press.

Bowrey, K. (2006) 'Alternative Intellectual Property? Indigenous Protocols, Copyleft and New Juridifications of Customary Practices', *Macquarie Law Journal* 6, 65–95.

Coombe, R. (2005) 'Protecting Traditional Environmental Knowledge and New Social Movements in the Americas: Intellectual Property, Human Right, or Claims to an Alternative Form of Sustainable Development?' *Florida Journal of International Law* 17, 115.

Cuskelly, K. (2011) *Customs and Constitutions: State Recognition of Customary Law around the World*. IUCN, Bangkok, Thailand.

De la Cruz, R. (2006) 'Regional Study in the Andean Countries: Customary Law in the Protection of Traditional Knowledge', Final Report Revised for WIPO, WIPO.

Dodson, M. (2007) 'UNPFII Study on Customary Laws Pertaining to Indigenous Traditional Knowledge and to What Extent Such Customary Laws Should be Reflected in International and National Standards Addressing Traditional Knowledge', Document E/C.19/2007/10, ECOSOC, Geneva.

Dutfield, G. (2000) *Intellectual Property Rights, Trade and Biodiversity: Seeds and Plant Varieties*, Earthscan, London.

Fletcher, M.L.M. (2007) 'Rethinking Customary Law in Tribal Court Jurisprudence' *Michigan Journal of Race and Law*, 13, 57–97.

Forsyth, M. (forthcoming) 'Navigating the mythscape of biodiversity and associated traditional knowledge in Pacific Island countries', in B. Tobin (ed.), *Customary Law, Traditional Knowledge and Self-Determination*, Cambridge University Press.

Gibson, J. (2005) *Community Resources: Intellectual Property, International Trade and Protection of Traditional Knowledge*, Routledge, London.

Gilbert, J. (2006) *Indigenous Peoples' Land Rights under International Law: From Victims to Actors*, Transnational Publishers, Aldersey, New York.

Graber, C. (2012) 'Stimulating trade and development of indigenous cultural heritage by means of international law: issues of legitimacy and method', in C. Graber, K. Kuprecht and J. Lai, *International Trade in Indigenous Cultural Heritage*, Edward Elgar, Cheltenham, UK.

Halewood, M. (2005) 'Common Law Aboriginal Knowledge Protection Rights: Recognizing the Rights of Aboriginal Peoples in Canada to Prohibit the Use and Dissemination of Elements of Their Knowledge', PhD Diss., York University, York.

Halewood, M. (forthcoming) 'Canadian Common Law Aborignal Knowledge Protection Rights: An opportunity for expanding the recognition of customary law', in B. Tobin, (ed.), *Customary Law, Traditional Knowledge and Self-Determination*, Cambridge University Press.

Inter-American Commission on Human Rights (IACHR) (2010) 'Indigenous and Tribal People's Rights over their Ancestral Lands and Natural Resources: Norms and Jurisprudence of the Inter American Human Rights System', OEA/Ser.L/V/II. Doc. 56/09, Organization of American States, Washington DC.

Janke, T. (1998) *Our Culture: Our Future*, Report on Australian Indigenous Cultural and Intellectual Property Rights, Australian Institute of Aboriginal and Torres Strait Islander Studies and the Aboriginal and Torres Strait Islander Commission, Canberra, Australia.

Janke, T. (2003) *Minding Culture: Case Studies on Intellectual Property and Traditional Cultural Expressions*, WIPO, Geneva.

Kambu, A. (2005) 'A South Pacific Perspective on the Protection of Traditional Knowledge, Traditional Cultural Expressions and Related Genetic Resources: The Case of Bindeku/ Kamaneku Tribes of the Highlands Region of Papua New Guinea', WIPO/GRTKF/IC/8/ INF/6(A).

McRae, H., Nettheim, G., Anthony, T. *et al.* (2009) *Indigenous Legal Issues Commentary and Materials*, 4th edn, Thomson Reuters, Sydney.

Olsen,V. (2008) 'Marco legal para los derechos de los pueblos indígenas en Colombia', Human Rights Everywhere, Santander.

Posey, D. (ed.) (1999) *Cultural and Spiritual Values of Biodiversity: A Complementary Contribution to the Global Biodiversity Assessment*, Intermediate Technology Publications.

Raven, M. (2012) 'Protocols: Devices for Translating Moralities, Controlling Knowledge and Defining Actors in Indigenous Research, and Critical Ethical Reflection', *Australian Aboriginal Studies* 2010/2, 33–47.

Reynolds, L. (2006) 'Mayan Law Still Lacks Official Recognition', Latin America Press, Lima. www.globalissues.org/HumanRights/indigenous/#CustomaryLaw backward orrelevantjusticesystems

Swiderska, K., Shrumm, H., Hiemstra,W. *et al.* (eds) (2012) 'Biodiversity and Culture: Exploring Community Protocols, Rights and Consent', *Participatory Learning and Action*, 65, International Institute for Environment and Development, London.

Taubman, A. (2005) 'Saving the Village: Conserving Jurisprudential Diversity in the International Protection of Traditional Knowledge', in K. E. Maskus and J. H. Reichmann (eds.), *International Public Goods and Transfer of Technology Under a Globalized Intellectual Property Regime*, Cambridge University Press.

Tobin, B. (2008) 'The Role of Customary Law in Access and Benefit-Sharing and Traditional Knowledge Governance: Perspectives from Andean and Pacific Island Countries', UNU-IAS, WIPO, Geneva.

Tobin, B. (2014a) 'Biopiracy by Law: European Union Draft Law Threatens Indigenous Peoples' Rights over their Traditional Knowledge and Genetic Resources', *European Intellectual Property Review*, 36(2), 124–36.

Tobin, B. (2014b) *Indigenous Peoples, Customary Law and Human Rights: Why Living Law Matters,* Routledge, London.

Tobin, B. (2015) 'Traditional Knowledge Sovereignty: The fundamental role of customary law in protection of traditional knowledge', in M. Rimmer (ed.), *Indigenous Intellectual Property: A Handbook of Contemporary Research*, Edward Elgar, Cheltenham, UK.

Tobin, B. (2016) 'Conocimiento tradicional, derecho consuetudinario e implementación del Protocolo de Nagoya: desafíos para los países andinos', in S. Roca, *Propiedad intelectual, conocimientos tradicionales y recursos genéticos*, Universidad ESAN, Lima.

Tsosie, R. (2007) 'Cultural Challenges to Biotechnology: Native American Genetic Resources and the Concept of Harm', *Journal of Law, Medicine and Ethics* (Fall), 396–411.

Williams, R. (1990) *The American Indian in Western Legal Thought: Discourses of Conquest*, Oxford University Press.

WIPO (2000) 'Matters Concerning Intellectual Property and Genetic Resources, Traditional Knowledge and Folklore: Document prepared by the Secretariat', WIPO, Geneva, WO/GA/26/6.

WIPO (2001a) 'Matters Concerning Intellectual Property and Genetic Resources, Traditional Knowledge and Folklore: Document prepared by the Secretariat', WIPO/GRTKF/IC/1/3, WIPO, Geneva.

WIPO (2001b) 'Intellectual Property Needs and Expectations of Traditional Knowledge Holders: WIPO Report on Fact-finding Missions on Intellectual Property and Traditional Knowledge' (1998–1999) WIPO, Geneva.

WIPO (2006) 'Draft Customary Law Issues Paper 1: Customary Law and the Intellectual Property System in the Protection of Traditional Cultural Expressions and Traditional Knowledge: Issues Paper – version 3.0', WIPO, Geneva.

WIPO (2013) 'Customary Law, Traditional Knowledge and Intellectual Property: An Outline of the Issues', WIPO, Geneva.

WIPO (2016) 'Glossary of Key Terms Related to Intellectual Property and Genetic Resources, Traditional Knowledge and Traditional Cultural Expressions', WIPO, Geneva, WIPO/GRTKF/IC/30/INF/7.

Yazzie, The Hon. R. (1994) '"Life Comes From It": Navajo Justice Concepts', *New Mexico Law Review*, 24, 175–90.

Zion, J.W. (1988) 'Searching for Indian Common Law', in B.W. Morse and G.R.Woodman (eds), *Indigenous Law and the State*, Foris Publications, Bilbao.

Part 3

Stakeholder perspectives

11 An indigenous perspective on the WIPO IGC

Maui Solomon

Introduction: 'round pegs and square holes'

The key problem from the outset is that the IGC process has set about placing 'traditional knowledge round pegs into intellectual property rights square holes'. The two just do not fit neatly together. This is because, as was noted by Darrell Posey in 1999:[1]

> Intellectual Property Rights are inadequate and inappropriate for protection of traditional ecological knowledge and community management of resources because they:
>
> - Recognise individual, not collective rights;
> - Require a specific act of invention;
> - Simplify ownership regimes;
> - Stimulate commercialisation [which may not always be negative];
> - Recognise only market values;
> - Are subject to economic powers and manipulation;
> - Are difficult to monitor and enforce;
> - Are expensive, complicated and time consuming.

In contrast, the values, rights and philosophies of indigenous peoples place greater emphasis on responsibilities to mother nature and one another rather than on a strictly 'rights' dominated system. In these knowledge systems emphasis is given to collective rights (and concomitant obligations), inter-generational responsibilities, cultural identity, the sacred and environmental sustainability. For these reasons, Posey had for years advocated the development of a system of traditional knowledge and resource 'rights' that went 'beyond intellectual property rights' and published a book, with Graham Dutfield, in 1996 outlining how this could be done.[2] Sadly, this seminal work has sat on the shelf and gathered dust when it provided in the early years of TK and IP discourse a roadmap for how this complex task could be tackled. After more than twenty years it continues to provide, in the author's opinion, greater clarity and direction on this subject matter than the IGC process has managed to come up with to date.

Can the IP system be adapted to accommodate TK and the 'needs and expectations of indigenous peoples'?

The IGC has made it clear that its focus regarding TK 'is to remain tightly tied to WIPO's overall IP mandate'[3] and although acknowledging that indigenous peoples have much broader aspirations and expectations than the status quo, the FFM concludes that its worldwide fact-finding mission may have 'created unrealistic expectations' and that there was therefore a need to 'manage expectations among TK holders and others regarding the role of IP in TK protection'.[4] On the other hand the FFM report noted that while there were '*conflicts*' and '*conceptual difficulties*' expressed by those who were consulted, these were not '*insuperable obstacles*' and that the IP system had shown that it could adapt to accommodate such diverse innovations as software and layout-designs in the digital environment; ipso facto, '*it is not inconceivable that IP principles might provide effective protection for traditional knowledge.*'[5]

Over the past fifteen years or so, there have been mixed messages from the IGC to indigenous peoples. On one hand, there is a desire to better know what are the 'needs and expectations' of indigenous peoples for the 'effective protection of their traditional knowledge'. On the other hand, there is the underlying but overriding purpose of finding ways to make TK (somehow) conform to existing IP principles. These two messages are fundamentally at odds with each other. It is therefore not surprising that the process is into its sixteenth year with the majority of draft text still bracketed as signalling significant disagreement among member states.

The IGC meetings have been characterised by indigenous peoples and member states talking past one another. For example, at the conclusion of the eleventh meeting of the IGC (which I attended), a Joint Statement of the Indigenous Consultative Forum was made on the issue of Traditional Cultural Expressions ('TCEs') that stated (inter alia):[6]

- An <u>international regime must enable the effective implementation, application and enforcement of Indigenous customary laws, knowledge and cultural practices</u>. In circumstances where there is a conflict, Indigenous customary laws and cultural practices shall prevail over domestic law or an international regime. We recognise that customary law extends to the cultural practices and knowledge of Indigenous peoples so as not to exclude those who have experienced serious dislocation within their States.
- It is clear that there are pressing problems in the regulation of TCEs, and that the IPR system does not offer a solution to some of the problems. There have been cases of Indigenous people using the IPR system to protect their TCEs. However, the reality is that there are far more cases of non-Indigenous people using the IPR system to take ownership over TCEs and TK using copyright, trademark and patents. In some such cases this has created an untenable situation whereby Indigenous peoples cannot legally access their own knowledge.

- The establishment of an entirely new system, a sui generis system, based on both customary law and western law could be developed and adopted to ensure adequate protection for TCE and TK needs.
- An international instrument must ensure that the right to prior informed consent of Indigenous peoples is guaranteed and protected as a fundamental principle and must be maintained throughout any access and benefit sharing arrangements where there is potential change of permitted use or third party involvement.
- Indigenous peoples are recognized as custodians and owners of their knowledge and traditional cultural expressions and have the exclusive right to control and manage their TCEs. (underlining added)

It is clear from this statement (and there have been many similar statements presented to the IGC over the years by the Indigenous Forum), what the 'needs and expectations' of indigenous peoples are in relation to the recognition and protections of their TK and TCEs. Chief among these are the development of *sui generis* systems, recognition of customary law, prior informed consent of the knowledge holders, a binding international instrument or regime and, perhaps the most important of all, that indigenous peoples are recognised as the custodians and owners of their own TK and TCEs (see Chapter 10 by Tobin on customary law, this volume). The Forum's statement acknowledges that while the IPR system may afford some *limited* protection against misappropriation of TK and TCEs, in the majority of cases it is non-indigenous peoples who use the IPR tools to gain monopoly rights over TK and TCEs, or simply use images, expressions and TK in the public domain without registering any rights.

However, the language of developed countries at the IGC meetings (ironically, mainly those with populations of indigenous peoples who have been colonised by those states), has been to consistently advocate for the IP status quo, non-disclosure requirements for genetic resources and TK, a non-binding regime, full disclosure of TK, a rejection of a 'human rights' based approach to finding solutions and a general 'dumbing down' or ignoring of the submissions made by indigenous peoples at IGC meetings. Little wonder that there have been walk-outs by various indigenous peoples groups over the duration of the IGC process.[7]

Misappropriation of TK continues unabated

While the IGC continues to dither over the development of an appropriate system of 'protection', misappropriation of TK and TCEs continues to happen around the world. A recent example involved the use by an American company, Fine Art America, of famous ancestral images of Maori Chiefs to adorn shower curtains they are selling in the USA for about $100 each.[8] These images have been taken from famous paintings by the nineteenth-century New Zealand artist Goffried Lindauer and were sourced by Fine Art from the internet. Legally they were not in breach of any 'law' because copyright in the paintings had long ago expired *but* from a cultural perspective the use of these ancestor figures on shower curtains and related products is culturally offensive. Following a very public outcry from Maori (including descendants of these Chiefs), the shower curtain images were withdrawn from the website. However, a recent search of the website shows that the same images are still being used on cushion covers, phone case covers and a range of other products for sale over the internet. The sale of these other products depicting *tupuna* or ancestor images is culturally offensive in the sense that it demeans the values and identity that Maori associate with these *tapu* (sacred) ancestral images.

This case highlights (once again) the inadequacy of the existing IPR system to protect against misappropriation or culturally offensive use of TCEs where these images have found their way into the public domain and where copyright or trademark rights over these images are not being sought by the company concerned (see Chapter 8 by Okediji on the public domain and Chapter 9 by Perlman on TCEs, this volume). Indigenous peoples have been bringing these kinds of examples to the IGC for the past fifteen years and calling for *sui generis* mechanisms to cater to these situations, but such calls have largely fallen on deaf ears. I have myself made interventions at IGC meetings to this effect.[9] At the eleventh meeting of the IGC a question posed by the Secretariat was, 'To what extent do existing IPRs already afford protection? What gaps need to be filled?' My response on behalf of my Moriori tribe was:

> There is a need for *sui generis* systems to be developed to fill these gaps because the IPR system does not provide protection for the values underlying TCEs. IPR can only protect some economic aspects of TCEs but not the values and identity of the culture which, to indigenous peoples, is the more important (underlining added).

Unless WIPO is prepared to 'evolve' the IP system to accommodate development of *sui generis* mechanisms, that take account of cultural factors (in addition to commercial ones), that have been strongly advocated by indigenous peoples at the IGC meetings, the notion that has been promoted at the IGC that the IP system can 'protect TK or TCEs' is an unfortunate hoax on indigenous peoples.

This is not to say that there are not member states which do support the intergenerational and holistic nature of indigenous peoples' cultural perspectives. For example, these perspectives are often reflected in the interventions of some states, particularly from the so-called 'developing countries' such as India, Brazil, and the African states attending the IGC meetings.[10]

As a consequence of the narrow IP focus of the IGC, TK 'protection' is being viewed from a set of rules contrary to its customary context. It would be like an international sporting body advising hockey teams that their sport is now to be governed according to the rules of basketball. It just won't work. Traditional knowledge is inextricably bound to the relationship that indigenous peoples have with their natural world and the intricate web of customary laws and protocols that give meaning to those relationships. Stripped bare of those values, TK will become a tool in the hands of others to exploit to their own advantage. One suspects that this may be the real intention of many of the developed countries.

Customary law and 'dumbing down' the 'protections' for TK and TCE within the IGC process

Indigenous peoples have consistently advocated for the recognition of their customary laws in any system of protection for TK and within any system developed by the IGC. For example, the Indigenous Consultative Forum at the eleventh meeting of the IGC urged that:

> Any international regime must enable the effective implementation, application and enforcement of Indigenous customary laws, knowledge and cultural practices.[11]

Some countries, through various phases of the IGC meetings, have also lent their support to the recognition of customary law in the development of mechanisms for recognising TK, including the African Group, New Zealand, China, India, Brazil and Peru. But not all countries were supportive with opposition coming from United States, Canada, Australia, Japan and others. However, as time has gone on, those countries that were supportive of recognising customary law have altered their positions and no longer appear to be supportive.

Since the initial meetings of the IGC there has been a gradual 'dumbing down' of the proposed mechanisms for the recognition and protection of TK and TCEs. As noted by Brendan Tobin, the latest draft of the articles for TK and TCE make no reference at all to recognising customary law of indigenous peoples and instead have invented a new term (unknown to international law) called 'cultural norms'. As Tobin wryly observes:

> At worst, the use of the term may demonstrate an attempt to beguile participants in the IGC into believing that references to 'cultural norms' are legally equivalent to and interchangeable with specific recognition of Indigenous peoples and local communities' laws and protocols.[12]

In 2005 and 2006 the IGC came closest (in my opinion), to the development of policies and objectives that would have provided (even if still limited) some of the 'protections' that indigenous peoples have been calling for. These articles and provisions represented the summation by the Secretariat of the submissions and interventions made by states, indigenous peoples and NGOs to the IGC since its inception in 2001 and included the findings of the FFM between 1998 and 1999.

For example, in relation to TK, in IGC Doc 8/5, Policy Objective (v), the language used was aimed at acknowledging the rights of the holders over their own TK:

> Empower holders of traditional knowledge and acknowledge the distinctive nature of traditional knowledge systems

> [The protection of traditional knowledge should aim to:]
>
> (b) be undertaken in a manner that empowers traditional knowledge holders to protect their knowledge by fully acknowledging the **distinctive nature of traditional knowledge systems and the need to tailor solutions that meet the distinctive nature of such systems**, bearing in mind that such solutions should be balanced and equitable, should ensure that conventional intellectual property regimes operate in a manner **supportive of the protection of traditional knowledge against misappropriation**, and should **effectively empower traditional knowledge holders to exercise due rights and authority** over their own knowledge.[13] (bold added)

Similar language was used in relation to TCEs in document 8/4 Policy Objective (v) to – 'effectively empower indigenous peoples. . . . to exercise due rights and authority over their own TCEs'.

Importantly, for indigenous peoples, the wording of Documents 8/4 and 8/5 also contemplated the adoption of a legally binding international instrument governing the use of TK and TCEs that would have influence at the national and regional levels of law and policy making.[14] The documents also provided for enforcement provisions against misappropriation, prior informed consent by the holders of TK, respect for the values of TK and development of *sui generis* systems of protection where requested or desired by indigenous peoples or local communities. It did not take long for some states to criticise these provisions on the pretext that 'more discussion was needed'.[15]

Even at this relatively early stage of the IGC process there were signs of the 'dumbing down' of language that might suggest that control was being handed to indigenous peoples over their own TK and TCEs – a non-negotiable outcome for many countries. For example, as noted in my Review Report for the NZ Delegation in 2006:[16]

> Although these policies purport to express worthy and strong sentiments for empowering indigenous peoples (e.g. particularly the words bolded in the above quotes) they are somewhat 'watered down' in the subsequent Guiding and Substantive Principles sections of both documents 8/4 and 8/5.
>
> For example, respect for customary laws and protection of TK is made subject to 'national law and policy' (document 8/5, Principle (h)); enforcement of protection against misappropriation of TK is to be guided 'as far as possible and appropriate' by respect for customary law (document 8/5, Article 1, 5); TK related to biological and genetic resources are subject to national laws governing ownership and access to these resources (document 8/5, General Principle (f) and Substantive Principles, Article 12, 1).
>
> Moreover, although the above Policy Objectives speak of tailoring solutions to meet the distinctive nature of TK systems ensuring that a balanced approach is adopted, the General and Substantive Principles appear to place significantly

> greater weight on the application of national regulatory laws and IP tools, rather than enhancing and promoting the use of customary laws and protocols for the protection of TK.

The pretence of offering indigenous peoples even this compromised form of 'protection' as outlined in Docs 8/4 and 8/5, has long since been dispensed with by the IGC. At the thirtieth meeting of the IGC (dealing with GRs), the regressive language and tone used by some developed countries concerning TK was telling. For example, a developed country delegation queried the basis upon which TK is even being considered for disclosure in patent applications: 'What was the public policy basis to require disclosure requirements for TK and not other knowledge.'[17] The fact that this very basic question is still being asked by a member state (in this case one of the most powerful and influential at WIPO), after fifteen years and thirty meetings, highlights the appalling lack of progress that has been made by the IGC. It also illustrates the 'thumbing of noses' by some developed countries at the recognition of the rights of indigenous peoples to their TK and GR under international instruments, such as the United Nations Declaration on the Rights of Indigenous Peoples adopted in 2007 (UNDRIP) and the Nagoya Protocol on Access and Benefit Sharing adopted in 2010. For example, Article 31 of the UNDRIP provides:

> 1 Indigenous peoples have <u>the right to maintain, control, protect and develop their cultural heritage, traditional knowledge and traditional cultural expressions</u>, as well as the manifestations of their sciences, technologies and cultures, <u>including human and genetic resources, seeds, medicines, knowledge of the properties of fauna and flora</u>, oral traditions, literatures, designs, sports and traditional games and visual and performing arts. They also have the right to maintain, control, protect and develop their intellectual property over such cultural heritage, traditional knowledge, and traditional cultural expressions. (underlining added)
>
> 2 In conjunction with indigenous peoples, <u>States shall take effective measures to recognize and protect the exercise of these rights.</u>[18]

Likewise, the Nagoya Protocol addresses traditional knowledge associated with genetic resources with provisions on access, benefit-sharing and compliance. Parties to the Nagoya Protocol under Articles 7, 12 and 16 are to take measures, in accordance with domestic legislation, to ensure that rights and interests relating to prior informed consent and fair and equitable benefit-sharing relating to indigenous peoples and local communities are provided for, keeping in mind customary laws and procedures as well as customary use and exchange of genetic resources.

It is therefore puzzling, in light of the fact that the same country is a signatory to the UNDRIP, and a signatory to the Convention on Biological Diversity 1992 (which gave rise to the Nagoya Protocol) that their delegate at the June 2016 meeting of the IGC would be asking why TK should be treated differently from any other knowledge. It is also puzzling, given that indigenous peoples and many countries have been explaining 'why' for the past fifteen years at the IGC. One can only assume that this is yet another stalling tactic designed to create further delays in making any meaningful progress in the IGC process.

The frustration shared by some of the other countries at this moribund state of affairs was reflected by the delegation of India stating that the IGC process needs to 'stop moving in circles' and instead start moving in a 'straight line'.[19] The thirtieth meeting of the IGC held in February 2016 was attended by only four indigenous representatives. As stated by a keynote speaker at that meeting, Aroha Te Pareake Mead:

> this surely presents a low point in this IGC process, a warning signal that the IGC may have lost credibility with many indigenous peoples because of the impasse and blockage that has occurred here and the lack of certainty over the recognition of indigenous peoples as rights holders.[20]

The stonewalling on the part of (mainly) developed countries towards genuine recognition and protection of TK within an IP framework, is one of pure self-interest. Simply put, those countries with the largest investment in IP rights globally want to ensure the status quo is maintained in order to protect their commercial investments – even if this means continuing the unethical practices of biopiracy and misappropriation of indigenous knowledge and resources. As one author has characterised it:

> What is clearly missing is a lack of political will to recognise and protect Indigenous peoples' rights by developed nations intent on protecting the research and commercials sectors' interest in maintaining as much traditional knowledge as possible within a freely accessible public domain.[21]

In re-stating their opposition to mandatory disclosure of the origins of GR and TK in patent applications, an industrialised country stated at the thirtieth meeting of the IGC that:

> It remained deeply concerned that new patent disclosure requirements would actually add new uncertainties in the patent system, particularly where the sanctions for noncompliance included invalidation of a patent.[22]

In other words, 'we want the status quo to remain'. The objections by this delegate are, to say the least, disingenuous. The IP system has continued to evolve over a period of two hundred years to accommodate a vast range of previously unforeseen innovations and will continue to do so in the future. TK and TCE comprise ancient and innovative knowledge and technology that are worthy of protection in their own right from unscrupulous exploitation. But, maintaining this intransigent position over a long period of time will, I envisage, result in a very much watered down and ultimately meaningless set of so-called 'protections' of TK. The IGC appears currently on track for achieving that outcome.

Conclusions

The Fact Finding Mission undertaken in 1998–99 raised the hopes of indigenous people around the world that their voices would at last be heard and the concerns they had been expressing for decades about the rapacious exploitation of their knowledge and natural resources would be controlled under new and innovative mechanisms to be developed through the IGC process, their rights recognised and their TK protected. Sadly, after sixteen years of effort and cost in attending these meetings, indigenous peoples appear to be no further ahead in achieving these hoped-for aims and aspirations. Indeed, they appear to have lost hope and confidence in the process reflected in the declining attendance at the IGC meetings.

Furthermore, the failure by member states to adequately resource the Voluntary Fund to enable effective participation of indigenous peoples in the IGC meetings and the opposition of the USA and other developed nations to indigenous peoples' participation being funded from WIPO's core budget (while at the same time claiming to support their participation)[23] is indicative of the level of importance that these countries attach to indigenous peoples' participation within the IGC process. Developed countries with large indigenous populations, including the USA, Australia, New Zealand and Canada have fiduciary and moral obligations to ensure that their First Nations peoples are adequately represented in these meetings. This is especially so given the nature of the subject matter being considered. Their collective efforts to date have been minimal and largely dismissive.

It is easy for indigenous peoples to feel disempowered and marginalised as this has been their lot for centuries in every country in which they have been colonised. However, they have managed to survive into the twenty-first century in spite of the efforts by their colonisers to suppress, and in some cases exterminate, them. Survivability and durability is a key strength of indigenous peoples and they will endure whatever the IGC process eventually comes up with. It took twenty-five years to get some agreement on the UNDRIP so not all hope is lost for the IGC process to come up with something useful. However, there will need to be a dramatic change in the 'political will' of countries to find a way to close the wide gaps that currently exist in the draft text if any meaningful recognition and 'protection' is to be given to the rights of indigenous peoples and their TK. As noted in the FFM Report, and throughout years of the IGC process, it has been stated that the IP system is capable of adapting to meet new and changing circumstances. On that note I will conclude by quoting the words of the UN Special Rapporteur on the Rights of Indigenous Peoples delivered at the February 2016 meeting of the IGC:

> I would like to point out that the flexibility of the intellectual property system has allowed it to adjust to changing contexts and technologies, such as the advent of the internet and the demand that the intellectual property system address the situation of the visually impaired. I believe that the time is ripe for WIPO to once again exercise its flexibility by finally addressing the intellectual property needs and expectations of traditional knowledge holders through the adoption of a treaty that recognizes and protects the rights of indigenous peoples to their genetic resources and traditional knowledge.

Author's note

The author's involvement with the WIPO IGC process began in June 1998 in Wellington, New Zealand, when a group of WIPO officials, led by Wend Wendland,[24] met with a number of Maori tribal claimants I represented in the Wai 262 Claim[25] to discuss a number of issues regarding the intellectual property needs and expectations of indigenous peoples. The visit was part of the Fact Finding Mission[26] (FFM) undertaken by WIPO representatives to the South Pacific in 1998 as a precursor to establishing the WIPO IGC. Since then I have participated in five meetings of the IGC and two inter-sessional meetings, predominantly in my capacity as a delegate for Hokotehi Moriori Trust.[27] I have also undertaken a comprehensive review of the IGC documents 8/4 and 8/5 for the Ministry of Economic Development in New Zealand. I have also made presentations to the IGC on aspects of GR, TK and TCE's from an indigenous tribal perspective and have taken part in various caucus meetings of the Indigenous Consultative Forum at IGC meetings. This paper is thus written from the perspective of an indigenous advocate on behalf of Moriori and Maori interests in the WIPO IGC process. It does not purport to express the view of indigenous peoples generally, although, in my experience attending the IGC meetings, many indigenous and traditional peoples groups[28] align with the general views expressed herein.

Notes

1 Posey, D.A. 'Introduction: Culture and Nature – The Inextricable Link' (1999), page 12 in *Cultural and Spiritual Values of Biodiversity: A Complementary Contribution to the Global Biodiversity Assessment.* (Edited by Darrell Addison Posey) Intermediate Technology Publications.
2 Posey, D.A. and Dutfield, G. *Beyond Intellectual Property: Toward Traditional Resource Rights For Indigenous Peoples and Local Communities* (1995), page 95, International Development Research Centre.
3 Op. cit. fn 2, page 232.
4 Ibid., page 235.
5 Fact Finding Mission (1998, WIPO) page 235. See below note 25 for full details.
6 Joint Indigenous Statement presented at the Closing of the Eleventh Session of the IGC by the Indigenous Consultative Forum (also known as the Indigenous Caucus).
7 A general walk-out by the majority of indigenous peoples happened at the eleventh meeting of the IGC and again at the twelfth meeting of the IGC.
8 In June 2016 it came to public attention that an American company called Fine Art America was selling shower curtains depicting famous Maori ancestor figures. See www.radionz.co.nz/news/te-manu-korihi/306885/portraits-on-shower-curtains-'profoundly-hurtful', for the full story (accessed 23/1/2017).
9 Intervention on behalf of Hokotehi Moriori Trust at Meeting Eleven of IGC on GRTKF, Geneva 3–12 July 2007, Agenda Item 7 – on Question 7, (11/4(a) Add).
10 For example, in a written statement by India to the seventh meeting of the IGC in November 2004, India states that 'We believe that Traditional Knowledge, Genetic Resources and Folklore are closely interlinked. There is a need to deal with all aspects of IP relating to TK, GR and folklore holistically'.
11 See note 7 above.
12 Tobin, B. 'Now You See It Now You Don't: The rise and fall of customary law in the IGC', this volume.

13 WIPO/GRTKF/IC/8/5.
14 WIPO/GRTKF/IC/8/5 – Article 2, Legal Form of Protection, page 15.
15 WIPO/GRTKF/IC/9/14 Prov. 2, paragraph 24.
16 Solomon, M. *Peer Review of WIPO Documents: The Protection of Traditional Cultural Expressions and Expressions of Folklore: Revised Objectives and Principles: (WIPO/GRTKF/IC/8/4) and The Protection of Traditional Knowledge: Revised Objectives and Principles (WIPO/GRTKF/IC/8/5)* paragraphs 6.26 to 6.28, page 43.
17 Statement by US delegation, WIPO/GRTKF/IC/30, paragraph 191.
18 UNDRIP, www.un.org/esa/socdev/unpfii/documents/DRIPS_en.pdf (accessed 23/1/2017).
19 Ibid., paragraph 585.
20 www.wipo.int/edocs/mdocs/tk/en/wipo_grtkf_ic_30/wipo_grtkf_ic_30_presentation_mead.pdf (accessed 23/1/2017).
21 Tobin, B. 'Now You See It Now You Don't: The rise and fall of customary law in the IGC', this volume.
22 US delegation statement, WIPO/GRTKF/IC/30, paragraph 191.
23 WIPO/GRTKF/IC/28, paragraph 34.
24 Wend Wendland is currently the Director of the Traditional Knowledge Division of WIPO.
25 The Wai 262 claim is the 262nd claim to have been filed in the Waitangi Tribunal, a specialist body set up in New Zealand to inquire into claimed breaches of the Treaty of Waitangi signed between the Chiefs and tribes and Queen of England in 1840. The Wai 262, filed in 1991 on behalf of six Maori tribes, sought recognition and protection by Maori to their *tino rangatiratanga* (sovereignty and self-determination) over genetic and biological resources, traditional knowledge and expression of that knowledge. The claim was heard between 1997 and 2007 and reported on in 2011. The report 'Ko Tenei Aotearoa – This is New Zealand', makes ground breaking recommendations for how the New Zealand government can better recognise and protect Maori rights to their GR, TK and TCEs but to date there has been no official response by the Government to this report. The author was senior legal counsel for three of six of the tribes namely, Te Rarawa (Del Wihongi), Ngati Kuri (Saana Murray) and Ngati Wai (Hori Parata).
26 This refers to the Global Fact Finding Mission undertaken by WIPO in 1998 and 1999 on the 'Intellectual Property Right Needs and Expectations of Indigenous Peoples', www.wipo.int/edocs/pubdocs/en/tk/768/wipo_pub_768.pdf (accessed 23/1/2017).
27 Hokotehi Moriori Trust is the mandated legal body to represent Moriori people (the indigenous peoples of Rekohu (Chatham Islands, New Zealand)). Rekohu is located 800 kilometres to the east of New Zealand.
28 For the sake of brevity, I have used the term 'indigenous peoples' in this chapter to refer to indigenous peoples, and traditional and local communities.

12 Genetic resources, traditional knowledge and disclosure obligations

Some observations from the life science industry

Dominic Muyldermans[1]

The complexity of natural product research

The life science industry is engaged in natural product research. For such natural product research, a wide variety of genetic resources (GRs) or traditional knowledge (TK) associated with such GRs might be needed (as natural starting material or, for example, as a testing tool).[1] These are often accessed from a number of sources in different jurisdictions, during the whole research and development process, which often takes many years and with no certainty as to whether this will eventually result in a commercial product.[2] These GRs can be collected in situ, but are often accessed through the acquisition of companies and other commercial purchases or from universities and gene banks.[3] Accessed GRs are combined with other available GR or TK to make novel combinations for evaluation and to further develop the most promising lead(s) into a product. With the rapid development of new biotechnological techniques, an interesting trait or lead could come from any GR.[4]

It needs to be stressed that not all natural product research leads to a product which is patented or even patentable. This can be the case because the product falls outside of patentable subject matter, it does not comply with the patentability criteria, or the entity that developed the product decides not to go for patent protection and instead decides to protect it as a trade secret.

Compliance with Access and Benefit-Sharing (ABS) obligations

The life science industry is dedicated to complying with all relevant rules on Access and Benefit Sharing (ABS), which includes obtaining the prior informed consent (PIC) for access and use of a specific GR, if and as needed; and entering into mutually agreed terms (MAT) specifying the benefit-sharing obligations.

1 Dominic Muyldermans is an independent attorney and legal advisor to ABS-int and CropLife International. The author would like to thank Emil Pot and Axel Braun for their valuable contributions.

It therefore fully endorses the implementation of effective rules and processes to monitor and enforce compliance with all relevant ABS rules.

The protection of GRs, TK and folklore has been under discussion in the WIPO Intergovernmental Committee on Intellectual Property and Genetic Resources, Traditional Knowledge and Folklore (WIPO IGC) since 2000. The life science industry has been actively engaged in this WIPO process since the beginning.

At the core of the negotiations at the WIPO IGC is the proposal to require patent applicants to disclose the country of origin or the source from which GRs were obtained to combat 'misappropriation' of GRs. The objectives as discussed within the IGC are focusing on: (1) ensuring compliance with ABS rules; and (2) avoiding the erroneous granting of patents, that is, ensuring that patents are only granted on inventions related to biological material which strictly comply with the patentability criteria. With regard to the objective related to compliance with ABS rules, the IGC is linking intellectual property and ABS, which in our opinion is not only not needed and ineffective since it is disregarding the existing legal framework on ABS, but it also raises questions as to conformity with established principles and rules of international patent law. With regard to the objective of avoiding the granting of erroneous patents, reference needs to be made to the fact that: (i) in principle no patent can be granted on the GR as such; (ii) the disclosure obligation is already part of the patent prosecution before a patent is granted;[5] and (iii) prior art databases and other facilitating instruments exist and are being further developed to review novelty and inventiveness of patent subject matter.

Not needed for effective enforcement of ABS rules:

New and specific legal frameworks on ABS

Since the start of IGC negotiations, the existing international and national legal landscape has changed fundamentally since a specific legal framework on ABS has come into existence. The Nagoya Protocol on Access to Genetic Resources and the Fair and Equitable Sharing of Benefits Arising from their Utilization to the Convention on Biological Diversity (Nagoya Protocol) entered into force on 12 October 2014. In the meantime many regions and countries have started to implement the Nagoya Protocol and have enacted regional or national rules and regulations on ABS. A key feature of the Nagoya Protocol is compliance and these regional and national rules include specific systems to ensure and check on compliance with ABS obligations. For example, the EU Regulation 511/2014 on compliance measures for users from the Nagoya Protocol on Access to Genetic Resources and the Fair and Equitable Sharing of Benefits Arising from their Utilization in the Union (EU Regulation) was adopted on 16 April 2014 and establishes a due diligence system to ensure effective monitoring of relevant ABS rules in the countries of the EU where a GR is used in research and development. It needs to be pointed out that the EU decided not to specifically interlink patent offices and the checks on the user's compliance with ABS obligations in the EU Regulation.

No interference per se *between patent law and ABS*

Whether material is protected by a patent or not does not, in principle, have an effect on the ABS obligations of a user of GRs. Patents are private proprietary rights granted pursuant to relevant patent laws, while obligations for ABS derive from the public sovereign rights of states (de facto establishing a public proprietary right) over their GRs. In addition to the rights obtained pursuant to the grant of a patent, the patent owner might also have certain obligations under the relevant rules on ABS due to the use of a GR. There is, in principle, no conflict between the affirmation of the sovereign rights of a State over its GRs and the recognition of the private rights held by a non-state actor pursuant to a granted patent (even in cases of a patent being granted for an invention involving the use of a GR – this does not infer such a patent could claim the GR as such).

In addition, patent matters might be included in the agreements related to ABS, e.g. in the MAT. This might include defining ownership and licensing of a patent on the products resulting from the use of the GR, and any financial arrangements related thereto (e.g. royalty payments). This situation is, however, not different from patent matters in other agreements, like research and development or distribution agreements which also contain contractual provisions on ownership of intellectual property rights and licensing or transfer of such rights, and payment schemes related thereto. This does not establish a link between patent rights and ABS compliance *per se*.

It can be concluded that using the patent system to ensure compliance with ABS is neither needed, nor effective, since there are specific ABS rules that focus on misappropriation, effectively ensuring compliance with ABS obligations. One of the instruments established by the Nagoya Protocol, as a key tool to facilitate compliance by enhancing transparency, is the so-called clearing house mechanism which enables monitoring of compliance.

Questionable compatibility with established rules of international patent laws

In addition, we believe that a disclosure obligation going beyond the scope of patent law would raise questions on compliance with international laws (especially TRIPs), would create practical issues, and would not serve the legislative purpose of the CBD, i.e. to foster the protection of biological diversity and enable the use and equitable benefit sharing from the use of GRs.

A disclosure obligation would be incompatible with the following key TRIPS conditions or requirements:

1 Incompatibility with *numerus clausus* of patentability requirements:
Art 27.1 of TRIPS provides that 'patents shall be available for any inventions (. . .) that are new, involve an inventive step and are capable of industrial application'. Disclosure requirements as defined by Art. 29 of TRIPS are limited to those facts that are required to enable a person skilled in the art to practise the invention. In addition to this, the applicant for a patent on plant

material in many cases makes a mandatory deposit of material that is essential to enable the claimed invention and which makes such material publicly available and conserves the viable biological material in the depository. A disclosure of source or origin obligation beyond what is needed to comply with the enablement condition would amount to adding a new patentability requirement and would have the result that patents may no longer be available for inventions that comply with the listed patentability requirements in these TRIPS articles.

2 Incompatibility with the *prohibition of discrimination*:
Art. 27.1 of TRIPS states that patents shall be available and patent rights enjoyable without discrimination as to the field of technology. Any type of disclosure obligation as a requirement for patentability for technologies dependent upon the use of GRs going beyond what is necessary in the context of the patent system would amount to discrimination.

3 Incompatibility with the *reasonableness requirement*:
Art. 62.1 of TRIPS states that additional requirements need to comply with reasonable procedures and formalities. This reasonableness needs to be assessed in the context of the patent system, more specifically in relation to the issue of patentability. An unnecessarily broad disclosure requirement would therefore go beyond what can be reasonably required to obtain a patent, and therefore imposes unreasonable obligations.

It can be concluded that using the patent system to ensure compliance with ABS rules is not only ineffective (as pointed out above), but it also raises questions as to compatibility with established rules of patent law.

Avoiding the erroneous granting of patents

The life science industry continues to support the goal of preventing the wrongful or erroneous grant of patents on inventions which are based upon or developed by using GRs or TK, by ensuring that all patentability criteria are fulfilled. This so-called 'defensive protection of GRs and TK' is ensured by the patent system itself and further enabled by information systems, e.g. databases. A key factor to ensure consistency in the application of disclosure in patent law is safeguarding the necessary relationship between the invention itself and the GRs and TK.

1 *No patent on the GR and/or TK used as such*
It is important to stress that a patent will in principle never protect the used GR or TK as such, and only an invention which might have been made by utilizing the GR and/or TK, and which complies with the patentability criteria, is eligible for patent protection. This exclusion is already specifically foreseen in patent law (vlg art. 52(2) EPC, 35 U.S.C. 101) and is therefore already dealt with within the patent system, by the patent offices when assessing patent applications and through case law. This is particularly the case when inventions consist of methods of treatment of, or any diagnostic

methods practised on, the human or animal body which are generally exempt from patenting or denied of a remedy for patent infringement. No private proprietary rights are thus claimed on the GR and/or TK as such which at all times will remain available for further use.

2 *Disclosure obligation and the enablement condition*

One of the fundamental tenets of patent law is that an applicant is granted a patent right protecting his/her invention only if they provide a complete disclosure of the invention. Hence, the patent application must disclose the invention in a manner sufficiently clear and complete to be carried out by a person skilled in the art, and the requirement that patent claims be supported sufficiently by the technical disclosure and written description in the patent. Lack of written description and/or insufficient disclosure may lead to refusal to grant the patent, or to its revocation in post-grant proceedings. In cases where the GR is not readily available there is already an implicit obligation to disclose the source of the GR, since otherwise third parties might not be able to carry out the invention (art. 83 EPC, 35 U.S.C. 112). Another option is to deposit the GR material under the Budapest Treaty at an international depositary authority (IDA), in case the GR comprises a micro-organism or other biological material that is accepted by the IDA.

The GR, and its source, is thus already disclosed within the context of a patent application and the scope of this obligation of disclosure is well defined in patent law, both from a legal and a policy point of view. Therefore, including a disclosure obligation which goes beyond what is needed and applied in the patent system would eclipse the necessary relationship between the GR and the claimed invention, thereby undermining the legal basis for the proprietary rights granted to the patentee.

3 *TK and the disclosure obligation*

Disclosure is also a key requirement in a patent application resulting from the use of TK. Disclosure of TK in a patent application is required if the TK is documented prior art which is relevant to the assessment of whether the invention as claimed is novel and non-obvious; and if TK is directly used in developing the invention (with the TK holder as a potential co-inventor).

The disclosure of TK to ensure defensive protection of TK is thus already foreseen by patent law. As pointed out above, going beyond the scope of this disclosure obligation would create legal uncertainty and ultimately undermine the value of TK itself.

4 *Databases and other facilitating instruments*

Databases on GRs and TK can facilitate access to relevant information by the patent offices while substantially examining the patent application. Databases include a range of information and materials. Such databases can be supplemented by guidelines for the patent offices to avoid situations where patents are granted that do not comply with the patentability criteria.

There are many initiatives under consideration to document TK. While some databases have the intent of providing a confidential record for the indigenous community, other databases might be instruments to ensure defensive protection.[6]

> The proposal by Japan of a 'One-Click Database Search System' composed of a WIPO portal site in combination with existing or yet to be developed databases in WIPO member states containing information about GRs, could be the central system for collecting the relevant data to assess prior art and guarantee high quality patents and avoid the granting of erroneous patents (see Chapter 6 by Robinson and Chiarolla in this volume).

It can be concluded that a disclosure obligation going beyond current patent laws is not needed for the granting of erroneous patents, since this can be done by applying the existing rules, supplemented by enhancing instruments such as databases to enable and improve prior art and TK searches. It is of paramount importance that no measures are imposed that negatively impact or interfere with the IP system as such, since this would substantially reduce the probability of innovation through natural product research. This is especially the case taking into account that natural product research requires the investment of many resources over a long period of time (even up to 20 years) to potentially come to a successful product. By fundamentally reducing the incentives for such natural product research, the benefits arising from the products resulting from such research will therefore substantially decrease.

Conclusion

Concluding, it needs to be stressed that in the life sciences sector, securing access to GRs and TK is important for the development of new products. Disclosure requirements going beyond what is needed and foreseen in patent law could reduce investments in the use of GRs and/or TK for innovative products. If fewer products are developed, there will be less revenue and thus benefits to be shared. Increased costs and legal uncertainty could significantly decrease the (monetary) value of certain GRs and TK, including those needed to address the conservation of biological diversity.

Notes

1 It is e.g. noted in Newman, D.J. and Cragg, G.M. (2012) 'Natural products as sources of new drugs over the 30 years from 1981 to 2010', *Journal of Natural Products*, 75 (3), 311–335, that 'in the area of cancer, over the time frame from around the 1940s to date, of the 175 small molecules, 131 or 74.8% are other than synthetic, with 48.6% actually being either natural products or directly derived therefrom'. See also Kingston, D.G. (2010) 'Modern natural products drug discovery and its relevance to biodiversity conservation', *Journal of Natural Products*, 74(3), 496–511.

2 It is pointed out by Cragg, G.M. and Newman D.J. (2005) *Pure Applied Chemistry*, 77, 1923–1942 that 'out of 114,000 extracts derived from approximately 12,000 species collected as part of the NCI natural products discovery program, only the two compounds taxol and camptothecin led to drugs currently in clinical use, with the possibility of the two additional compounds homoharringtonine and maytansine also leading to drugs'.

3 For an investigation of current exchange of genetic resources and use practices, based upon a 2011 survey of government and university researchers in the US, see Welch,

E.W., Shin, E. and Long, J. (2013) 'Potential effects of the Nagoya Protocol on the exchange of non-plant genetic resources for scientific research: Actors, paths and consequences', *Ecological Economics*, 86, 136–147.

4 For a description of the analytical challenge in accessing the natural products and their biosynthetic machinery, including the presentation of a complementary platform see, Bumpus, S.B., Evans, B.S., Thomas, P.M., Nta, J. and Kelleher, N.L. (2009) 'A proteomics approach to discovery of natural products and their biosynthetic pathways', *Nature Biotechnology*, 27(10), 951–956.

5 Art. 83 of the European Patent Convention stipulates that 'The European patent application shall **disclose** the invention in a manner sufficiently clear and complete for it to be carried out by a person skilled in the art'. In 35 U.S.C. 112, the specification of US patents 'shall contain a written description of the invention, and of the manner and process of making and using it, in such full, clear, concise, and exact terms as to enable any person skilled in the art to which it pertains, or with which it is most nearly connected, to make and use the same, and shall set forth the best mode contemplated by the inventor or joint inventor of carrying out the invention'.

6 It is important to note that any documentation of TK has to be undertaken with the prior informed consent of the relevant indigenous people, as per the Nagoya Protocol. This is especially important since special care needs to be applied to not document and disclose secret or sacred TK.

References

Bumpus, S.B., Evans, B.S., Thomas, P.M., Nta J. and Kelleher, N.L. (2009) 'A proteomics approach to discovery of natural products and their biosynthetic pathways', *Nature Biotechnology*, 27(10), 951–956.

Cragg, G.M. and Newman, D.J. (2005) 'Biodiversity: A continuing source of novel drug leads', *Pure and Applied Chemistry*, 77, 1923–1942.

Kingston, D.G. (2010) 'Modern natural products drug discovery and its relevance to biodiversity conservation', *Journal of Natural Products*, 74(3), 496–511.

Newman, D.J. and Cragg, G.M. (2012) 'Natural Products as sources of new drugs over the 30 Years from 1981 to 2010', *Journal of Natural Products*, 75(3), 311–335.

Welch, E.W., Shin, E. and Long, G.J. (2013) 'Potential effects of the Nagoya Protocol on the exchange of non-plant genetic resources for scientific research: Actors, paths and consequences', *Ecological Economics*, 86, 136–147.

13 Indigenous culture and Indigenous publishing

Balancing protection, preservation and promotion

Jens Bammel and José Borghino

Introduction

The deliberations of the World Intellectual Property Organization (WIPO) are of paramount importance for the global publishing industry and, therefore, for the International Publishers Association (IPA).[1] This is obvious in the case of WIPO's Standing Committee on Copyright and Related Rights (SCCR), which addresses the international copyright framework – but it is no less true of the agenda of the Intergovernmental Committee on Intellectual Property and Genetic Resources, Traditional Knowledge and Folklore (IGC).[2]

The IGC debates can potentially impact creators and publishers across the three main sectors of the publishing industry in the following ways:

- Literary authors who identify as members of Indigenous peoples will often use traditional knowledge, stories, customs or other aspects of their culture in their writing. And it is this individual experience or worldview, which often makes their work stand out. Publishing in this context is a powerful and recognised way to document, preserve, develop and pass on Indigenous cultures. It is also an essential part of the author's freedom of expression.
- Children and other learners have a right to an education that respects and involves their heritage and, to best accomplish this, educational textbooks produced for Indigenous readers should use stories from local Indigenous cultures. Again, it is this unique cultural perspective that is most effective in connecting with students and can teach them about their own culture and its relation to the rest of the world. Local, innovative and viable educational publishers are crucial to this process.
- Scholarly articles in scientific journals often record findings of an ethnological, historical, medical or pharmaceutical nature. Freedom of expression and freedom of scientific research is fundamental to the publication process of modern scientific and academic discourse and needs to be balanced carefully with respect for traditional secrecy.

While the scholarly community has formulated ethical standards to help regulate the scholarly communication of sensitive issues,[3] self-regulation is comparatively under-developed and less coherent in literary or educational publishing. There are good reasons for this. In the literary sphere, the biggest problem for most Indigenous peoples is not that publishing has been intrusive or exploitative, but that too few publishers are prepared to publish work by Indigenous authors. Publishing in general is a positive force in the recording, preservation and transmission of information, but Indigenous languages, traditional knowledge and other such cultural expressions are in decline in some places precisely because there is not enough publishing there.

Similarly, in the educational sphere, when misguided national governments, for example, prescribe a single textbook in a particular subject for an entire developing country, they unwittingly do irreparable harm to the teaching of local Indigenous languages and cultures. Leaving aside the obvious threat of ideological control they pose, such one-book policies immediately decimate the number of educational publishers operating in the local ecosystem and thereby drastically dampen the level of investment in that country's educational resources. In particular, in countries with multiple ethnicities, this lack of publishing diversity irrevocably cripples instruction of local cultural traditions using local languages. Whether children should be taught in local languages, or in a principal national language, is rightly the subject of impassioned debate around the world. The lack of national policies that promote a diverse and dynamic local publishing industry means that students must rely on books procured in massive single-title tenders. The result is one book, one version of history, no differentiation, no choice, a distinct lack of quality control and, usually, no prospect of a more respectful treatment of local cultures in education, let alone further competitive investment in the local industry.

From this perspective, the most pressing question for Indigenous communities in the developing world is not how to increase protection but something entirely different: how to get more Indigenous authors and local publishers to produce more works that help preserve and pass on local cultures.

But this is not what is being asked by WIPO's IGC. Rather, their question can be summarised as: what international instrument can be devised to insulate Indigenous peoples from the unwanted distribution and exploitation of their cultural expressions and traditional knowledge?

Such a formulation of the question assumes the following:

1 abusive distribution and exploitation is rife;
2 an international instrument is an efficient way to resolve this problem;
3 the basic tenets of such an instrument are ripe for consensus; and
4 existing Intellectual Property (IP) mechanisms cannot be usefully applied for this purpose.

What can WIPO do?

The past ten years of debate at the IGC have overlooked the real problem because of a fundamental misunderstanding of how WIPO's work can support the developing world. The misguided logic first assumes that, because most of today's IP-protected works are from developed countries, then the current IP system serves only them and is irredeemably tipped in their favour. It follows then, from this premise, that in order for the IP system and WIPO to better serve developing countries the system must protect those resources that developing countries have in abundance: Indigenous knowledge, expressions of folklore and genetic resources. This, it is posited, must be done through a new, *sui generis* right.

Ironically, the fallacy in this logic is most obvious to publishers of Indigenous works. They can see clearly that, in terms of literary culture, many developing countries are content rich but distribution poor. There is no dearth of great authors; rather, there is a lack of a book culture to sustain them locally. It is tragic, for example, that many African writers can only make a living from their writing if they are 'discovered' by publishers in developed countries or if they become academics, often teaching African literature in Western universities. Nobel Prize laureate Wole Soyinka, Chinua Achebe, Ngũgĩ wa Thiong'o and Buchi Emecheta, for example, have all spent many years at US universities.

The situation is even more complicated in educational and scholarly publishing. Any country can artificially create its own Indigenous, local educational publishing industry, but for such an industry to truly thrive the state must first institute policies that actively support and nurture it. Indigenous academics from developing countries can contribute their knowledge to academic journals, but they need a supportive academic culture that fosters their ambition to publish in top international journals, not one that hamstrings such ambition.

The creation of an international instrument through the IGC has also encountered a number of practical obstacles. On the one hand, for example, there is no national law regulating the publication of literary works which has been successfully tested and could serve as a model for such an instrument. And on the other hand, alternative conflict resolution mechanisms, often based on local Indigenous traditions, have been shown to successfully address many concerns.

It would also be useful if the objective of the instrument were clarified. Is this instrument about compensation or control? If it is about compensation, then how will it work? What is a realistic level of compensation? And with a new instrument would come new risks, new threats of litigation if a publisher were to breach the new laws, and, therefore, new costs. The last thing developing countries need is to raise the costs of local and Indigenous publishing.

Many argue that the IGC exercise is not about preventing publishing, but about respect, profit sharing and, ultimately, control. However, the basic legal questions have been raked over repeatedly by WIPO Member States without any satisfactory solutions being found. Who would be the holder of this new *sui generis* right? How will the right be determined and its rights-holders identified? From whom could users of Indigenous works reliably obtain permissions? What is the role of government? Can an idea be protected in perpetuity, even if it has neither a requirement for manifestation (as in copyright) nor disclosure (as in patents)? What will be the impact on the public domain? Will there be large swaths of ideas – not just specific expressions – that will perpetually escape becoming free to be shared by all? Are we seeing the ill-advised revival of the *domaine public payant*?[4]

These dangers are all-too evident in the real world. Parallel to, but independent from, the IGC debates, South Africa has come up with an extreme solution in its Intellectual Property Laws Amendment Act of 2013, which gives perpetual and extended rights of copyright to Indigenous works which are not original – a heretofore fundamental requirement for copyright – and where there is no identifiable author. Not only is this legislation considered by experts in the field[5] to be unworkable – although the law was passed in 2013, at the time of writing, 28 months later, it has not yet been brought into effect – but it also lays the groundwork for abuse. Publishers in South Africa and internationally consider the additional legal (and therefore financial) risks introduced by this Act to constitute a severe disincentive for authors who write about Indigenous communities and their practices and for publishers who would make that work available.

Protection versus freedom of expression

While these issues have been disputed over the past decade, another equally weighty and difficult problem has escaped the IGC's attention altogether: the issue of the relationship between protection on the one hand and freedom of expression on the other.

Granting a new right to collective entities must necessarily impact on the rights of individuals. The literary author who must ask permission to include a story in her book; the schoolbook writer who cannot incorporate a particular perspective in his textbook because it is controversial among Indigenous peoples; the scientist or academic who is prevented from reporting a fact or an observation – all of these people's fundamental human rights of freedom of expression and freedom to publish will be impinged upon. Many individuals interacting with Indigenous people have had to face such questions: what should be passed on? And how? Of course, there are strong ethical, legal and practical reasons why some observations should or may go unpublished. The question remains, however, whether a new international legal instrument will add a useful layer of protection, or whether it will be a superfluous or perhaps even dangerous, additional complication.

This concern becomes all the more important as we are talking about one of the most important human rights. Freedom of expression is enshrined in Article 19 of the Universal Declaration of Human Rights (1948), where it states: 'Everyone has the right to freedom of opinion and expression; this right includes freedom to hold opinions without interference and to seek, receive and impart information and ideas through any media and regardless of frontiers.'[6] Similarly, in the UN's International Covenant on Economic, Social and Cultural Rights (1966), freedom of expression is closely associated with the freedom to enjoy access to information.[7]

Without such freedoms, the dizzying creation and accrual of beneficial and liberating scientific knowledge over the past four hundred years is unimaginable, as is the global shift from absolutism to democracy. To quote Benjamin Franklin:

'Without freedom of thought there can be no such thing as wisdom; and no such thing as public liberty, without freedom of speech.'

The steady evolution of today's copyright framework has managed to achieve a careful balance between access and control, mainly by distinguishing the idea from its manifestation. We are all free to utter an idea previously expressed by someone else – we must only avoid copying the concrete manifestation of that idea for the limited term of its protection. It seems a dangerously broad restriction now to prohibit, in perpetuity, the copying, distribution, adaptation, translation or otherwise communication of ideas and concepts enshrined in traditional knowledge or expressions of folklore, as some have argued. Such a prohibition may not seem too extreme when limited, for example, to truly essential religious secrets, and most national laws include restrictions in the area of confidentiality and state secrets. But at this stage of the negotiations at WIPO there is no consensus to distinguish between different levels of protection.

Human rights

The lack of discussion of the human rights aspect of this matter at WIPO has been as frustrating to publishers as the lack of progress on the instrument has been for other parties. While it is true that other UN agencies specialise in human rights, an international instrument must ensure compliance with the entire corpus of international law, especially human rights law. The current silence about the interconnections between IP and human rights is in stark contrast to the joint statement made in 1998 by the then United Nations High Commissioner for Human Rights, Mary Robinson, and then Director General of WIPO, Kamil Idris, when they declared: 'If the intellectual property system is to remain an engine for economic, cultural and social development, we must continue to explore [crosscutting issues related to human rights and development] and stimulate further discussion.'[8]

Conclusion

In fact, even without the introduction of a new, *sui generis* protection, the current IP system has much to offer Indigenous peoples, providing valuable tools that are already helping them protect their intellectual property and traditional knowledge. There are countless examples of First Nation people using the existing IP system to their advantage – from the Toi Iho 'Maori Made' trademark[9] of New Zealand's Indigenous people, to the recent efforts by Jamaica to allow the trademarking of locally grown Jamaican Ganja,[10] there are many ways IP can support Indigenous economies. Alongside the glacial diplomatic progress of the IGC debate, WIPO's expert staff are playing a quiet but important role in advising how to make good use of the existing legal framework and a number of national IP offices have followed suit – for example, IP Australia successfully introducing the Dream Shield initiative.[11]

So, while Geneva keeps talking but gets no nearer to clarifying how an international instrument would work in practice, the world is moving ahead – turning creativity and knowledge, culture and traditions into marketable artefacts. Cultural richness is becoming economic wealth. For example, the Indonesian Ministry of Industry and Yayasan Batik Indonesia (the Indonesia Batik Foundation) is implementing product certification called Batik Mark for Indonesia's original batik prints and designs.[12]

On the issue of protecting traditional knowledge and folklore, Ambassadors in Geneva seem to be engaged in a dance full of empty political gestures. By contrast, publishers are the quintessential pragmatists – they have no interest in abstract legal discourses for their own sake or in symbolic declarations with little or no practical outcome.

The IPA hopes that WIPO's Member States will wake up and see that the current Copyright and IP system is already capable of delivering real wealth by unlocking the greatest asset of the developing world: its citizens' creativity, ingenuity, their innovative spirit and the cultural heritage they inhabit and carry with them, and which they are willing to share with the world.

Thankfully, the Indigenous publishers of the developing world are ready and waiting to play their role in recording, preserving and bringing to light this hitherto largely undiscovered treasure, in tandem with local Indigenous communities.

Notes

1 The IPA is a federation of national, regional and specialist publishers' associations. Its membership comprises 70 organisations from 60 countries in Africa, Asia, Australasia, Europe and the Americas. Based in Geneva, Switzerland, IPA represents the interests of the publishing industry in international fora and wherever publishers' interests are at stake. Founded in 1896, the IPA's initial aim was to ensure that countries throughout the world showed respect for copyright, and properly implemented the (then) new international copyright treaty, the Berne Convention. The IPA also promotes and defends freedom to publish, a fundamental aspect of the human right to freedom of expression.

2 The question of genetic resources does not affect publishing and is not discussed in this article.

3 See for example: www.aaanet.org/cmtes/ethics/Ethics-Resources.cfm (accessed on 23/1/2017).

4 See www.wipo.int/meetings/en/details.jsp?meeting_id=20207 (accessed on 23/1/2017), note on the meanings of the term 'public domain' in the intellectual property system with special reference to the protection of traditional knowledge and traditional cultural expressions/expressions of folklore, paragraphs 54 to 58.

5 For example, Professor O.H. Dean, 'South Africa: TK Legislation In The New Tradition', 2011, republished by *Intellectual Property Watch* at www.ip-watch.org/2011/10/06/south-africa-tk-legislation-in-the-new-tradition/?utm_source=daily&utm_medium=email&utm_campaign=alerts (accessed on 23/1/2017).

6 www.un.org/en/documents/udhr/index.shtml#atop (accessed on 23/1/2017).

7 www.ohchr.org/EN/ProfessionalInterest/Pages/CESCR.aspx, Article 15.1 (accessed on 23/1/2017).

8 In the joint Foreword to *Intellectual Property and Human Rights: A Panel Discussion to Commemorate the 50th Anniversary of the Universal Declaration of Human Rights*, WIPO, Geneva, 1998, www.wipo.int/mdocsarchives/WIPO_UNHCR_IP_PNL_98/WIPO_UNHCR_IP_PNL_98_INF%205_E.pdf (accessed on 23/1/2017).

9 www.toiiho.co.nz/ (accessed on 23/1/2017).

10 http://mythoughtsontechnologyandjamaica.blogspot.ch/2013/11/jamaica-and-switzerland-sign-historic.html (accessed on 23/1/2017).
11 www.ipaustralia.gov.au/tools-resources/publications-reports/dream-shield (accessed on 23/1/2017)
12 www.tnol.asia/arts-culture/18998-batik-mark-protecting-the-originality-of-indonesias-batik.html (accessed on 27 October 2015).

14 Disclosure requirements

Switzerland's perspective

Georges Bauer, Cyrill Michael Berger and Martin Girsberger

Introduction

Worldwide, in particular in tropical and sub-tropical regions, an enormous diversity of plants, animals and micro-organisms exist. The genetic resources (GRs)[1] of these organisms are a potential basis for modern biotechnology and can be useful in fields such as plant breeding, pharmacology, foodstuff production and cosmetics. In recent years, scientists have made considerable advances in these fields, including through genetic engineering. These advances have increased the possibilities for using genetic resources, therefore interest in and demand for these resources has also increased. Traditional knowledge (TK) of indigenous peoples and local communities related to genetic resources may also be of interest in this regard.

To resolve the numerous issues arising in this context, two main paths are being explored. The first of these paths is the so-called 'provider measures' which are to be taken by the State which either provide the genetic resources for use in research and development activities of science and industry, or in the territory of which are located the indigenous and local communities providing traditional knowledge for these purposes. In contrast, the second path is the 'user measures' which are measures to be taken by the State in which the genetic resources are used or where intellectual property rights (IPRs) are granted. Various measures can be taken within the context of these two paths. Among the provider measures, for example, the creation of databases is being considered. Databases would aid IP offices in their novelty searches before they issue a patent. On the other hand, the user measures being discussed are measures which increase transparency in the context of access to genetic resources and traditional knowledge and the sharing of the benefits arising out of their utilization (hereinafter 'transparency measures'). A variety of transparency measures is thinkable, including measures related to patents. One such widely discussed measure is the requirement for patent applicants to disclose certain information with regard to genetic resources and traditional knowledge in their patent applications. Both the disclosure requirement and databases are being discussed in the Intergovernmental Committee on Intellectual Property and Genetic Resources, Traditional Knowledge and Folklore (IGC) of the World Intellectual Property Organization (WIPO).

This chapter describes the proposals Switzerland submitted to the IGC on the requirement for patent applicants to disclose the source of genetic resources and traditional knowledge in patent applications. Additionally, it describes the parallel requirement Switzerland included in its Federal Act on Patents for Inventions (Patents Act).[2]

Reasons for the involvement of Switzerland

Switzerland's involvement in the IGC process and its recognition of the importance of the issues being discussed in that forum are based on its interest of having a fair and balanced intellectual property system. On the one hand, Switzerland supports the effective protection of biotechnological innovations through IPRs, in particular patents. On the other hand, a fair and balanced approach necessitates effective, efficient, practical and timely solutions to the issues arising in the context of access to genetic resources and traditional knowledge and the fair and equitable sharing of the benefits arising out of their utilization. While Switzerland is not a 'demandeur' when it comes to these issues, we believe that their resolution will result in a better and more robust, fair, predictable and transparent patent protection system.

In order to achieve this goal, Switzerland has taken a parallel approach in which a proposed amendment to the Patent Cooperation Treaty (PCT) was put forward, while at the same time revising the Patents Act. Switzerland has therefore concentrated its efforts towards the introduction of a new measure, namely the declaration of the source of genetic resources and traditional knowledge in patent applications.

The proposed disclosure of the source would allow parties to a contract on access and benefit-sharing (ABS) to verify whether the other contracting party is complying with its obligations. This would not only assist in and simplify the enforcement of these obligations, but would also permit verifying whether prior informed consent (PIC) has been obtained and whether provisions have been made for fair and equitable benefit-sharing.

The proposals submitted by Switzerland to WIPO present a practical and results-oriented way forward that could be introduced in a timely manner. Disclosing the source can be seen as the 'entry point' of the access and benefit-sharing system in the patent regime and would strengthen mutual supportiveness between the two regimes. The proposed way forward would enable countries to fulfil their international obligations, in particular those arising out of the Convention on Biological Diversity (CBD) and its Nagoya Protocol on Access to Genetic Resources and the Fair and Equitable Sharing of Benefits Arising from their Utilization (Nagoya Protocol).

Switzerland's proposal at the international level

In order to establish an international structure for the disclosure of the source of genetic resources and traditional knowledge in patent applications, Switzerland has sought to make changes at the international level that are very similar to what has been accomplished domestically through the amendment of the Patents Act. In brief, Switzerland submitted a proposal to amend the PCT to explicitly enable national patent legislation to require the declaration of the source of genetic resources and traditional knowledge in patent applications, if the invention is directly based on such resources or knowledge.[3] The proposal would afford the patent applicants the possibility of satisfying the new requirement at the time of filing an international patent application or later during the international phase. The proposed amendments to the PCT present a simple and practical solution to this issue. These amendments could be introduced in a timely manner and would not require extensive changes to the relevant international provisions. In addition, these changes could play a crucial role in maintaining today's high quality of patents.

Policy objectives

Switzerland proposes a disclosure of source requirement that accomplishes four policy objectives. They concern transparency, traceability, technical prior art and mutual trust (or 'the four T's' for short).

1 *Transparency*: With a requirement in national and international patent applications to disclose the source, the patent system would increase transparency in access and benefit-sharing with regard to genetic resources and traditional knowledge. The intention is to prevent 'bad' patents from being granted and to make sure that the sharing of benefits arising from the use of genetic resources and associated traditional knowledge takes place, thereby allowing the providers of genetic resources and traditional knowledge, especially developing countries and indigenous peoples and local communities, to better take advantage of the patent system.

2 *Traceability*: Traceability creates positive effects for both providers and users of genetic resources and traditional knowledge. On the one hand, disclosing the source in patent applications would allow the providers of genetic resources and traditional knowledge to keep track of the use of their resources or knowledge in research and development resulting in patentable inventions. On the other hand, achieving the objective of traceability also works in favour of the users of a genetic resource or traditional knowledge as disclosing the source creates legal certainty. The provenance of the genetic resource can be traced back and thus creates reassurance as to the legitimacy of its acquisition and/or utilization.

3 *Technical prior art*: Disclosing the source of genetic resources and traditional knowledge in patent applications would assist patent examiners and judges in the establishment of prior art with regard to inventions that somehow relate to these resources or this knowledge. In particular, it may facilitate the establishment of prior public use as well as the finding of lack of novelty or inventive step, as was the case with regard to the so-called 'neem tree oil' patent

(European Patent 0 436 257). This applies in particular to prior art regarding traditional knowledge, as disclosing the source would simplify searching the databases on traditional knowledge that are increasingly being established at the local, regional and national levels. In other words, the accessibility of prior art also helps to prevent the granting of 'bad' patents.

4 *Mutual trust*: The disclosure of the source would increase mutual trust among the various stakeholders involved in access and benefit-sharing, including among developing and developed countries, indigenous and local communities, private companies and research institutions. All of these stakeholders may be providers and/or users of genetic resources and traditional knowledge. Accordingly, disclosing the source would build mutual trust in the North–South relationship. Moreover, it would strengthen the mutual supportiveness between the ABS system and the patent system.

Amendment of the PCT and PLT

Switzerland submitted its proposals regarding the declaration of the source of genetic resources and traditional knowledge in patent applications to the WIPO Working Group on Reform of the Patent Cooperation Treaty in 2003. According to the proposal, the relevant national law may require the patent applicant to furnish: (1) a declaration as to the source of a specific genetic resource to which the inventor has had access, if the invention is directly based on such a resource. (2) The national provisions may also provide that a declaration as to the source of traditional knowledge related to genetic resources must be presented, if the inventor knows that the invention is directly based on such knowledge. Alternatively, (3) the national law may also require a declaration that the source referred to in (1) or (2) is unknown to the inventor or applicant, if this is in fact the case.

A possible amendment could read as follows:

> The patent applicant shall furnish:
>
> (i) a declaration as to the source of a specific genetic resource to which the inventor has had access, if the invention is directly based on such a resource;
> (ii) a declaration as to the source of traditional knowledge related to genetic resources, if the inventor knows that the invention is directly based on such knowledge;
> (iii) a declaration that the source referred to in (i) or (ii) is unknown to the inventor or applicant, if this is the case.

The trigger of the disclosure requirement – the concept of 'directly based on'

The source of genetic resources and traditional knowledge must be disclosed in the patent application, if the invention is 'directly based on' these resources or this knowledge. The intent behind 'directly based on' is to ensure that there is a minimal connection between the invention and the genetic resource or traditional knowledge. Furthermore, the concept of 'directly based on' should also prevent imposing an onerous requirement on the applicant.

The scope of the obligation for the patent applicant differs slightly between genetic resources and traditional knowledge. With regard to genetic resources, the idea is that the invention must be 'directly based on', 'a specific genetic resource to which the inventor has had access' before the disclosure requirement is applicable. The intention of this concept is to clarify that: (1) the invention must make immediate use of the genetic resource, that is, depend on the specific properties of this resource; and (2) that the inventor must have had physical access to this resource, specifically, its possession or at least contact which is sufficient enough to identify the properties of the genetic resource that are relevant for the invention. For example, the source of a plant would have to be declared in the patent application if the respective invention relates to a chemical compound which the inventor extracted from this specific plant.

With regard to traditional knowledge, the thinking behind the proposal is that 'the inventor knows' that the invention is 'directly based on such knowledge'. Unlike a resource, knowledge is not tangible in nature, thus physical access is impossible. Instead, the inventor must know that the invention is directly based on such knowledge. In other words, the inventor must consciously derive the invention from this knowledge. This is to avoid cases where, for example, the inventor is using a chemical compound derived from a plant to develop a new pharmaceutical, without knowing that an indigenous community has knowledge concerning the pharmaceutical use of this plant.

The object of disclosure – the concept of 'source'

According to the CBD and its Nagoya Protocol, as well as the International Treaty on Plant Genetic Resources for Food and Agriculture (the International Treaty) of the Food and Agriculture Organization (FAO), several entities may be involved in access and benefit-sharing. To reflect this multitude, the proposal of Switzerland applies the concept of 'source'.

The term 'source' should be understood in the broadest sense possible. Based on the international instruments above, the source should be declared as the entity that is competent: (1) to grant access to genetic resources and traditional knowledge, and/or (2) to participate in the sharing of the benefits arising out of their utilization.

Depending on the genetic resource or traditional knowledge in question, 'primary' and 'secondary' sources can be distinguished. Primary sources are the Contracting Party providing genetic resources (CBD Arts. 15, 16 and 19; Nagoya Protocol Arts. 5.1, 6.1, 11 and 15), indigenous and local communities (CBD Art. 8(j); Nagoya Protocol Arts. 5.2, 6.2, 7, 11, 12 and 16), and the Multilateral System established by the International Treaty (Arts. 10–13). Secondary sources are *ex situ* collections such as gene banks and botanical gardens as well as databases on genetic resources and traditional knowledge, and scientific literature.

The proposal requires the patent applicant to disclose only those sources about which the applicant, or inventor, has information at hand. These sources must be disclosed according to the following cascade.

If information is available on:

- *The primary source*, this primary source must be disclosed. For example, if the patent applicant knows that the source of a genetic resource is the Contracting Party providing this resource, this Contracting Party must be disclosed as the source.
- *The primary and one or several secondary sources*, the primary source must be disclosed, whereas the disclosure of the secondary source is optional. For example, if the patent applicant received the genetic resource from a botanical garden, but also knows the Contracting Party providing the genetic resource, this Contracting Party must be disclosed, whereas the disclosure of the botanical garden is optional.
- *A secondary source*, but not about a primary source, this secondary source must be disclosed. For example, if the patent applicant received the genetic resource from a botanical garden, but does not know the Contracting Party providing the genetic resource, the botanical garden must be disclosed as the source.
- *Several secondary sources*, but not about the primary source, the secondary source with the closest relationship to the primary source must be disclosed; the disclosure of the other secondary sources is optional. For example, if the genetic resource was provided from one botanical garden to several others, the first botanical garden in this chain must be disclosed, whereas the disclosure of the other botanical gardens is optional.

Only if the patent applicant, or the inventor, has no information at hand about the primary or the secondary source, may he or she disclose that the source is unknown. Considering the broad understanding of the term 'source', cases where neither the primary or a secondary source is known are likely to be rare.

As the disclosure of source requirement aids in traceability, once the source of the resource or traditional knowledge is declared, the disclosed source will make it possible to trace back from where the resource or knowledge came from. It would then be up to the providing country to determine whether their laws on PIC and mutually agreed terms have been appropriately complied with.

Sanctions

In order for the disclosure requirement to be meaningful, non-compliance by the patent applicant should carry adequate sanctions. The proposal by Switzerland distinguishes between pre-grant and post-grant sanctions:

- **Pre-grant sanctions**: If the patent office during patent examination finds a failure to comply, that is, that the patent application in question does not contain a disclosure of source when in fact it should contain such a disclosure, the patent office is to invite the patent applicant to comply with the disclosure requirement within an adequate time limit from the date of the invitation (the current PCT Regulations foresee a minimal time limit of two months). If the patent applicant does not comply with this invitation within the set time limit, the patent office may refuse the application or consider it withdrawn on the grounds of this non-compliance. If the applicant submits the required declaration of source, the patent office must accept this declaration and may not require any further document or evidence relating to the source declared unless it may reasonably doubt the veracity of the declaration concerned.

- **Post-grant sanctions**: If after the granting of the patent it is discovered that the applicant failed to disclose the source or submitted false information, this may not constitute grounds for either the revocation or invalidation of the granted patent. This is because in the view of Switzerland, revocation of a patent would destroy the very basis of benefit-sharing and cause unnecessary legal uncertainty. However, other sanctions provided under national law, including criminal sanctions such as fines, may be imposed.

Switzerland's National Patent Act

On the domestic front, Switzerland introduced a disclosure requirement, similar to the PCT proposal, to its national patent law in 2008. The objective of the disclosure requirement is to increase transparency in access and benefit-sharing, an objective that is accomplished through three main features:

1. *Scope*
   Patent applicants are required to disclose the source of any genetic resources and traditional knowledge when they submit a national or an international patent application. Furthermore, if the source is unknown, this must be confirmed by the applicant in writing. In addition, another provision stipulates that this obligation also applies to international applications.[4]
2. *Trigger*
   The trigger mechanism dictates when disclosure becomes necessary. According to the Patent Law, disclosure is necessary, 'provided the invention is directly based on this resource/knowledge'.[5] This is the same trigger mechanism as is proposed to be used for the amendment of the PCT.
3. *Sanctions*
   Without this feature, the incentive to abide by a disclosure requirement disappears. Like the PCT proposals, the Swiss Patents Act distinguishes between pre-grant and post-grant sanctions:

   - During the pre-grant phase, if the patent application does not meet the requirements of the Patents Act, the Swiss Federal Institute of Intellectual Property shall set a timeline for the applicant to remedy the deficiencies.[6] If these deficiencies are not remedied within the given timeline, the application would be rejected.[7]

- In the post-grant phase, the sanction for intentionally providing false information concerning the source is a fine of up to 100,000 Swiss francs. In addition, a judge may order the publication of the judgment.[8] It should be noted, however, that the Patents Act Law does not provide for the revocation of the patent.

Databases on traditional knowledge and the disclosure requirement

There have been several cases in the past in which patents were granted for inventions that were based on or utilized traditional knowledge and did not meet the criteria for novelty and/or inventive step. The granting of these 'bad' patents can usually be explained by a lack of accessibility to prior art by the patent authorities. The very nature of traditional knowledge can act as a barrier because traditional knowledge is often transmitted orally and is seldom written. In addition, this knowledge is usually passed down in unfamiliar languages.

Therefore, while the international community has been debating how to resolve these issues for the past decade and a half, several governments, indigenous communities and non-governmental organizations have been proactive and have begun to establish databases. Some delegations have even encouraged this as the best manner of protection, under the perception that it would be the least burdensome method. However, all of these databases are likely to be established with differing forms and formats, and have differing structures. Not having an established harmonized system that patent authorities can access will hinder their attempts to search for prior art. In addition, the costs of establishing a database may act as a barrier for some governments and local communities. On several occasions, indigenous representatives have also pointed to the fact that databases are too burdensome a tool for indigenous peoples and local communities, since very often they are unfamiliar with the specific procedures to be followed in order to register their respective knowledge and are not convinced about the benefits and utility of such databases.

While databases may serve a good purpose and could be a useful tool, patent examiners cannot rely solely upon them. Disclosing the source of traditional knowledge and genetic resources in patent applications can assist patent examiners in their search for prior art, particularly with regard to traditional knowledge, because the disclosure would simplify the search of an already existing database.

Notes

1 'Genetic resources' are defined in Article 2 of the CBD as meaning genetic material – that is, any material of plant, animal, microbial or other origin containing functional units of heredity – of actual or potential value (Biodiversity Convention, 1992).

2 Federal Act on Patents for Inventions (Patents Act, PatA), CC 232.14, available at: www.admin.ch/opc/en/classified-compilation/19540108/index.html (accessed 10/1/2017).

3 See WIPO/GRTKF/IC/11/10, Declaration of the Source of Genetic Resources and Traditional Knowledge in Patent Applications: Proposals by Switzerland, 6 June 2007 www.wipo.int/meetings/en/doc_details.jsp?doc_id=152238 (accessed 10/1/2017).

4 See art. 49a and art. 138 Patents Act.

5 Federal Act on Patents for Inventions (Patents Act, PatA), CC 232.14, available at: www.admin.ch/opc/en/classified-compilation/19540108/index.html(accessed 10/1/2017). Art. 49a (1) paras. a and b Patents Act.

6 Article 59 (2) Patents Act.

7 Article 59a (3), para. b Patents Act.

8 Article 81a Patents Act.

15 India's position in the Intergovernmental Committee for the Protection of Traditional Knowledge

*Biswajit Dhar**

Introduction

For more than a decade and a half, India has consistently been emphasising the need to protect genetic resources (GR)[1] and traditional knowledge (TK), the latter defined in a broad sense so as to include folklore or the more current expression, 'tradition cultural expressions' (TCE). India has accorded considerable importance to the protection of GRs, TK/TCE, which has been reflected through its twin actions. First, India has taken several domestic legal/administrative measures to protect GRs, TK and TCE, and second, India has consistently raised the issue of safeguarding the interests of its local communities who have been the custodians of biological resources and have held traditional knowledge, in relevant international forums, in particular, the World Trade Organization (WTO) and the World Intellectual Property Organization (WIPO).

Three main legislative measures have been adopted thus far. These include the adoption of two laws, namely the Biological Diversity Act (BD Act) and the Protection of Plant Varieties and Farmers' Rights Act (PPVFRA). The former was enacted for meeting India's commitments as a party to the Convention of Biological Diversity (CBD), while the latter provides a legal regime for protecting the rights of the producers of new varieties of plants, which is in keeping with the country's commitments to the Agreement on Trade Related Aspects of Intellectual Property Rights (TRIPS) under the WTO. A third legislative measure that India has taken is the revision of the Patents Act to prevent misappropriation of TK and illegal accession of genetic resources.

The BD Act gave legal effect to the three pillars of the CBD, namely, 'conservation of biological resources, sustainable use of its components and fair and equitable sharing of benefits arising out of the utilization of genetic resources'.[2] According to this Act, commercial exploitation of genetic resources and associated TK can be made after an access permit is obtained from the competent national authorities and benefit-sharing arrangements are worked out in consultation with these authorities.

* Professor, Centre for Economic Studies and Planning, School of Social Sciences, Jawaharlal Nehru University, New Delhi.

The PPVFRA, as its title suggests, provides a mechanism for protecting the innovations made by traditional farmers. Like the formal plant breeders, traditional farmers can also protect the plant varieties that they have developed.

The Patents Act was amended to include a number of provisions to check illegal accession of biological resources and to prevent misappropriation of TK from India. The first of these is the so-called 'disclosure' requirement, wherein the patent applicants are required to 'disclose the source and geographical origin of the biological material . . . when used in an invention'.[3] A patent can be revoked if the applicant makes a wrong mention of 'the source or geographical origin of biological material used for the invention'.[4] Further, the Patents Act does not allow patenting of 'an invention which, in effect, is traditional knowledge, or which is an aggregation or duplication of known properties of traditionally known component or components'.[5] And, finally, patents can be revoked if the claimed invention 'was anticipated having regard to the knowledge, oral or otherwise, available within any local or indigenous community in India or elsewhere'.[6]

India's interventions in the WTO and the WIPO have been broadly based on its above-mentioned domestic legislative framework. Together with other like-minded countries, India has been arguing for the adoption of an international regime that takes cognisance of the need to prevent illegal bio-prospecting and to protect the interests of the holders of TK. These efforts began in the WTO with India and several developing countries arguing that the TRIPS Agreement include provisions for regulating use of GRs and to protect TK, if inventions for which patents are claimed use the resources and/or the knowledge. Concurrently, these issues were taken up in the WIPO by the Intergovernmental Committee on Intellectual Property and Genetic Resources, Traditional Knowledge and Folklore (henceforth, the IGC).

This chapter intends to discuss in some detail the position India has taken in the IGC negotiations. Our view is that the context and the content of the IGC negotiations can be better understood if the essentials of the negotiations in the WTO are discussed at the outset.

GR and TK protection in the WTO

India was among the first countries to argue in the WTO that the global intellectual property regime formalised after the adoption of the TRIPS Agreement needed amendments in order to prevent misappropriation of TK and unauthorised accessing of biodiversity. In the second WTO Ministerial Conference in 1998, the then Minister of Commerce, Mr Ramakrishna Hegde, presented the following arguments in support of India's case:

> A situation, where indigenous biotechnology, developed over the ages in countries such as India, is being used without any flow back of benefits from patentees to original developers calls for amendments in the TRIPS Agreement. The imbalances in the TRIPS Agreement and its tilt against the holders of indigenous know-how, mainly based in developing countries, misaligns it

> with another major international agreement, namely, the Convention on Biodiversity'.[7]

In its first formal proposal for the amendment of the TRIPS Agreement, India put forth a mechanism through which the inventor using biological material and associated TK could access such material legally. The proposal was for such an inventor to enter into a 'material transfer agreement . . . to use the biological material and a transfer of information agreement'. India proposed that such 'an obligation could be incorporated through inclusion of provisions in Article 29 of the TRIPS Agreement requiring a clear mention of the biological source material and the country of origin'.[8]

A joint submission by twelve countries in 2002, including India, elaborated the so-called 'disclosure proposal'.[9] These countries opined that the 'TRIPS Agreement and the Convention on Biological Diversity should be mutually supportive and promote the sustainable use of resources' and that therefore, '[M]odifications in the TRIPS Agreement are necessary to ensure that it will not run counter to the objectives of the CBD'.[10] They proposed that the TRIPS Agreement should be amended thus: 'Members shall require that an applicant for a patent relating to biological materials or to traditional knowledge shall provide, as a condition to acquiring patent rights: (i) disclosure of the source and country of origin of the biological resource and of the traditional knowledge used in the invention; (ii) evidence of prior informed consent through approval of authorities under the relevant national regimes; (iii) evidence of fair and equitable benefit sharing under the relevant national regimes.'[11]

In a subsequent submission, the proponents of the 'disclosure proposal' explained that their key demand would improve the effectiveness of the patent system in the following ways: (a) reduce instances of 'bad' patents; (b) enable the patent offices to ascertain more effectively the 'inventive step' claimed in patent applications; (c) enhance the ability of countries to track 'bad' patents in the instances where they are granted and challenge the same; (d) improve compliance with their national laws on prior informed consent and fair and equitable benefit sharing prior to accessing a biological resource/associated traditional knowledge. These countries argued that '[P]lacing the onus on a patent applicant to disclose the basis of its claims is a step that can pre-empt any misuse of patent laws and thereby prevent misappropriation of knowledge and resources'.[12]

The proponents of the disclosure proposal further pointed out that several countries, including India, were compiling databases of traditional knowledge at the national level, which could assist in helping patent examiners to reject patent requests relating to unauthorised use of the knowledge of traditional communities. They were, however, of the view that given the vast spread of TK among the local communities, and especially the preponderance of such knowledge as a part of the oral traditions, comprehensive documentation of TK was not possible. Under such circumstances, the ideal approach was to insist that patent applicants disclose the source and nature of the TK and location of the biological material, for the users of the knowledge and material were in the best position to share this information.

As mentioned above, India's interventions in the WTO on ways of safeguarding the interests of holders of TK and to ensure sustainable use of the biological resources in keeping with the objectives of the CBD, was an important plank of the position that it took in the IGC process. The successful implementation of the 'disclosure provisions' in its national patent law, which has been facilitated by the enactment of the Biological Diversity Act, helped India in clearly articulating its position in the IGC.

India's statements in IGC

India's support for the adoption of a *sui generis* system for the protection of GR, TK and TCE has been built around at least three planks. The first of these was based on the need to protect the interests of the local communities holding and practising TK/TCE and who are the GRs, without which the TK/TCE would usually not manifest themselves.[13] The other two were responses to the arguments made by the Group B countries who have been arguing that databases, such as the Traditional Knowledge Digital Library (TKDL) developed by the Indian Government, and contractual arrangements between the providers and users of the GR and TK/TCE, would be sufficient for safeguarding the interests of the custodians and holders of the resources and knowledge, respectively. Thus, Group B countries have been reluctant to bring a very large body of knowledge held by the traditional communities within the framework of intellectual property protection, and more importantly, by recognising the special characteristics of the local knowledge systems and their holders.

Why should TK/TCE and GRs be protected?

India has pointed out that over the millennia, many forms of TK/TCE have evolved, which are continuously being codified and are widely held and used by the holders, who are also the practitioners of such knowledge. This necessitated the establishment of a legal framework for safeguarding the interests of the practitioners, not only from misappropriation through the grant of 'bad' patents, but to also ensure that they are granted exclusive rights over their knowledge, in the exercise of which, they can secure adequate incentives.

In the absence of any protection for these forms of knowledge internationally, piracy and misappropriation were increasing.[14] India has had to get patents granted on the wound-healing properties of the turmeric plant[15] and the fungicidal properties of the neem plant (*Azadirachta indica*),[16] to name a few, even though these properties had been known to Indians for ages and had also been codified in various ancient texts of Indian systems of medicines. The system of Yoga was an ancient Indian system of living in which the physical postures were but a small part of the overall concept of being. But today India is also watching with consternation the efforts to copyright some yogic postures[17] and even to attach a trademark to Yoga.[18]

The above-mentioned developments provide sufficient justification for India's position that it was essential to have norms and standards at the earliest for protecting TK and to ensure that the rightful holders of this knowledge can receive their deserved incentives. One of the sectors that could gain considerably from the efforts to protect the TK holders would be traditional health practices. The continued importance of traditional healers in large swathes of the developing world in this direction would act as a powerful complement to the initiatives that various intergovernmental institutions, especially the World Health Organization (WHO) have taken for the promotion of traditional medicines over the past two decades.[19,20,21]

In the absence of such a system of protection, the efficiency of the intellectual property system in general, and the patent system in particular, is being undermined. The reality of post-grant opposition to patents and other forms of IPRs was that they were not only cumbersome but also expensive to follow across international borders. The large scale on which misappropriation and piracy of TK, TCE and GR took place made it that much more difficult for a country such as India to fight each and every such misappropriation.

Limitations of databases and contractual arrangements

As a first step towards addressing the problem of misappropriation of TK, the Government of India has developed the TKDL, an extensive database of TK that has been used by the practitioners of traditional knowledge. TKDL has proved its utility by challenging the grant of 'bad' patents, thereby preventing 'misappropriation' of TK. Although the TKDL is rich in content,[22] India pointed out in its submission that the database did not represent more than a 'minuscule part of the codified and oral knowledge'[23] existing in the country. Further, TK is a living knowledge system and, consequently, the TKDL can never be a comprehensive database of TK. India therefore argued that because of its inherent limitation, a database like the TKDL cannot provide an effective solution to the problem of misappropriation of TK. In other words, TKDL and other such databases cannot be a substitute for an effective *sui generis* system of protection of TK.

India also challenged the view that private contractual agreements between TK holders belonging to the local communities and prospective users of the knowledge, namely the large corporate interests, can help in protecting the interests of the former. This approach, in India's view, would leave the TK holders worse-off, given the asymmetry of information and power relationship between the two parties.[24] The benefits for the TK holders from such contractual agreements would remain minimal no matter how carefully such contracts are drafted, and even after giving due consideration to the huge imbalance between the negotiating capacities of the providers and users.

Rationale for the adoption of a *sui generis* system

India has argued that solutions to the protection of TK through a *sui generis* system had to be sought in terms of positive protection and defensive protection.[25] In its view, the IGC needs to develop systems to formally recognise TK and accord them the same level of protection – both positive and defensive – as were available to innovations based on modern science.

The importance of a *sui generis* system for protecting TK stemmed from the fact that the extant forms of intellectual property can provide protection to TK only to a limited extent.[26] Conventional forms of IPRs are not effective for protecting TK on at least two counts: (a) most of the knowledge is not held by individuals, but by a collective; and (b) the knowledge is usually developed over a period of time, usually over generations, and therefore does not meet the criteria of novelty or inventiveness as understood in the case of patents. Therefore, India's view has been that since the extant intellectual property system did not adequately protect the interests of the holders of TK and folklore as also being the custodians of GR, modern cultural and manufacturing industries were able to commercially exploit the products using GR as also embodying TK, without the consent of the communities and without sharing the benefits with the providers of the knowledge and GR.[27]

India further alluded to the fact that both TK and TCE were dynamic in nature[28] – a large segment of this knowledge was no longer confined to particular communities but was widely held and used. Thus, TK and TCE belonged to the knowledge system where holders were communities, regions or countries. Included in this knowledge system was codified knowledge covering, among others, traditional medicine, agriculture and environment. The codification was done in the ancient scriptures and the knowledge was passed on from generation to generation on the basis of those scriptures or through recognised courses of study. Therefore, rights on TK had to be granted in perpetuity. Providing rights for a limited period to reward the creativity of individuals would thus not be applicable in the case of TK.

Notwithstanding the above-mentioned limitations of the intellectual property system, several intellectual property laws existing in India include provisions aimed at guarding against misappropriation and misuse of TK. However, most of these, except the one for protecting geographical indications, provide only defensive protection against misappropriation and misuse of TK. India's efforts to protect TK therefore suggest that only a *sui generis* system of protection can provide effective legal protection to TK. Further, given the transboundary nature of the infringements, this system of protection had to be introduced on a global basis to be effective.[29]

Essentials of a *sui generis* system

According to India, the *sui generis* system for the protection of TK (and also TCE) and ensuring exploitation of biological resources in a sustainable manner must have the following overarching objectives. These include: (i) empowerment of the holders of TK; (ii) ensuring prior informed consent of the holders before potential users are granted the permitted use of the TK, TCE and GRs; (iii) promoting fair and equitable sharing and distribution of monetary and non-monetary benefits arising from the use of TK, TCE and GRs; (iv) facilitating the continuing customary use, development, exchange and transmission of TK, TCE and GRs by and between custodians/holders of the knowledge and resources; and (v) encouraging, rewarding and protecting tradition-based creativity and innovation.[30]

India agreed with the position held by several WIPO members that for realising the objectives of protection of TK, TCE and GRs, the IGC would need to utilise fully the options that can possibly be used under the IP system. The IGC also needed to address the issue of misappropriation, which should not only prevent future misappropriation but also help to evolve ways and means to provide remedial solutions for the past wrongs. This clearly established, in India's view, the need for an agreed international framework for faster invalidation procedures of large-scale misappropriation, which happens due to lack of access to TK-related prior art in foreign patent offices.[31]

Such a framework, while taking a holistic view of the entire IP system, should also provide for disclosure, prior informed consent and a benefit-sharing mechanism. India argued that disclosure of TK was essential for a full description of the invention.[32] The most transparent and credible way of realising this objective was the inclusion of a requirement in patent applications for mandatory disclosure of the TK, and also the associated GRs, used in an invention. India emphasised that it was essential that the norm-setting exercise being undertaken by the IGC include a mandatory requirement in patent applications to disclose the source and/or the origin of the TK and associated GRs.[33]

India stated that the 'disclosure' requirement would effectively make sure that misappropriation did not happen and that positive benefits can accrue to communities from where TK/GRs were used. In this context, India referred to the provisions to prevent misappropriation of TK/TCE that exist in its patent law. These include: (i) mandatory disclosure of the source and geographical origin of biological material used in inventions while applying for patents in India; (ii) provisions for opposition to patents and revocation of patent if disclosure was not made or was wrongfully made; and (iii) provisions for opposition to patents and revocation of patents if the invention was anticipated by the knowledge, oral or otherwise, available within any local or indigenous community in India or elsewhere.[34]

India had successfully demonstrated that, through routine monitoring of patenting activity, it was possible to revoke the patents that have misappropriated TK. However, considering that thousands of 'bad' patents have been granted, challenging the grant of such patents individually is not an effective remedy for the developing countries since they lack the necessary resources. It was therefore necessary to opt for the proactive solution, best provided by introducing the disclosure requirements for patent applicants.

The 'positive' protection to owners of TK, according to India, requires a grassroots-based approach, wherein control over the knowledge should remain in the hands of the communities and tribes who are the holders and practitioners of such knowledge.[35] Positive protection should be provided to TK for encouraging commercial and industrial use of TK along with adequate benefit sharing for TK holders. Relevant provisions for providing information on source or origin, prior informed consent, and evidence of benefit sharing while keeping IP protection should be included. India's suggestion has been that specific laws governing education practice and the search of codified TK systems should be duly recognised when sharing benefits arising from commercial use of such knowledge.[36] The structure of 'positive' protection that India has suggested uses some of the elements of the existing forms of IP protection, which have been re-cast after spelling out the realities of the traditional knowledge systems, some of which are described below.

As regards eligibility of protection, India's view was that codified and regulated TK like the traditional systems of healthcare, namely health care such as Ayurveda, Siddha and Unani, should be included to be accorded protection as a priority, through legal or other measures.

According to India, the *sui generis* system of protection should first assert collective rights of the communities over the knowledge and genetic resources. These rights should include rights to collective ownership over TK, collective management of TK; the right to prevent misappropriation, to insist on prior informed consent, to benefit sharing, to deny access and to recognise the identity of TK. India pointed out that efficacy of protection of TK and TCE may not be evaluated in monetary terms alone, since the knowledge holders also wanted their moral rights to be acknowledged and not be infringed. Hence, it was both the economic and moral rights that were to be protected as a means of rewarding the holders of TK.[37]

The implementation of the regime providing both 'positive' and 'defensive' protection, according to India, needs to be carried out through an identified 'national authority'. This authority should perform several roles, including identification of the beneficiaries and ensuring the flow of benefits to them by being able to negotiate on their behalf with potential users of TK/TCE and GRs.[38]

Protection of both TK and TCE should be based on common principles for there is a fairly strong interface between the two forms of knowledge. This was in keeping with India's position that a holistic view on the protection of TK and TCE must be taken. The principles of protection should reflect the combination of positive and defensive rights, including effective safeguards against illicit use. India argued that protection should be comprehensive and include promotion, preservation and conservation, and reiterated the need to prevent misappropriation and to grant positive rights to both undisclosed as well as codified and non-codified expressions of folklore already in the public domain.

India argued that even when a specific form of TK had lost its community identity and was widely used, the benefits arising from its use must flow to the communities that were holding/using it. In such a situation, an agency/authority needs to be designated under a national law which, after collecting the benefits from the users on behalf of the knowledge holders, will transmit the benefits to the communities.[39] This mechanism of benefit sharing is similar to that laid down under the Convention on Biological Diversity, which gives the states the sovereign rights over the biological resources, but in effect the state acts like a trustee whose sole responsibility is to ensure that custodians of the TK and GRs get adequate benefits arising from the commercial exploitation of the knowledge and the resources. The manner in which this is to be ensured is left to national legislations. India therefore insisted that the definition of the TK/TCE holder needed to be broadened to include the country or region as a whole as the sole holder of TK in the case of codified but non-customary knowledge systems of health care such as Ayurveda, Siddha, Unani and Yoga in India.[40]

Finally, stating its position regarding the exceptions and limitations arising from the implementation of protection for TK, particularly for tradition-based products, India cautioned that these exceptions should not adversely affect the scientific, commercial and industrial use within the country and out of the country, provided that benefit sharing was ensured.

Conclusions

As a *demandeur* of protection for traditional knowledge, traditional cultural expressions and genetic resources, India has provided an elaborate framework in the IGC negotiations, which, in its view, could be adopted as a legally binding treaty under the World Intellectual Property Organization. This framework has its genesis in India's experience of having implemented its domestic legislations, especially the Patents Act and the Biological Diversity Act, for protecting the interests of the holders of traditional knowledge and the custodians of genetic resources. These laws include provisions to prevent misappropriation of TK and unauthorised accessing of biological resources. Moreover, the Biological Diversity Act has a mechanism for incentivising the TK holders and custodians of GR, namely, through the sharing of benefits arising from the commercial exploitation of the knowledge and resources. At the same time, India has been able to develop a large database on traditional knowledge, the Traditional Knowledge Digital Library, which has been used to challenge the grant of 'bad' patents, mostly through the European Patent Office. Based on these experiences, India has been considering the adoption of a domestic law that can offer 'positive' protection. The key elements of such a law have also been put forth, along with the instruments for implementing it.

However, India is deeply aware that its efforts at protecting genetic resources and traditional knowledge at the domestic level will go unrewarded if a complementary legal regime at the international level is not established. This arises from the fact that while Indian authorities are able to arrest the grant of 'bad' patents in their own jurisdiction, the misappropriation of traditional knowledge and use of biological resources without seeking permission of the national competent authority set up under the BD Act is continuing unabated in foreign jurisdictions. For India, therefore, adoption of a set of legally binding instruments at the end of the process that was initiated through the WIPO IGC would provide a lasting solution to the problems that it has flagged during the negotiations since WIPO member states would have then endorsed a framework for preventing unsustainable use of biological resources and for protecting the interests of the traditional knowledge holders through the recognition of their rights over their knowledge.

Notes

1 The terms 'genetic resources' and 'biological resources' are generally used interchangeably, although conceptually these are not identical. The Convention on Biological Diversity (CBD), the international treaty that addresses issues related to genetic resources/biological resources, provides the definitions of these terms that clearly distinguish one from the other. To begin with, CBD defines genetic material 'as any material of plant, animal, microbial or other origin containing functional units of heredity'. This definition suggests that all forms of life on earth make up available genetic material. 'Genetic resources', according to the CBD, means 'genetic material of actual or potential value'. This definition, in other words, emphasises the intrinsic value of the genetic material available on the earth. And, finally, 'biological resources' are defined to include 'genetic resources, organisms or parts thereof, populations, or any other biotic component of ecosystems with actual or potential use or value for humanity'. The important term used here is 'ecosystems', which suggests that all the genetic material constantly interacts with one another under different environmental conditions.
2 Article 1.
3 Section 10(4)(d)(ii)(D).
4 Section 64(1)(p).
5 Section 2(p).
6 Section 64(1)(q).
7 WTO (1998), 'India: Statement Circulated by Mr. Ramakrishna Hegde, Minister of Commerce', WT/MIN(98)/ST/36, 18 May, paragraph 15.
8 WTO (2000), 'Communication from India', IP/C/W/195, 12 July, paragraph 16.
9 WTO (2002), 'The Relationship between the TRIPS Agreement and the Convention on Biological Diversity and the Protection of Traditional Knowledge', IP/C/W/356, 24 June.
10 WTO (2002), 'The Relationship between the TRIPS Agreement and the Convention on Biological Diversity and the Protection of Traditional Knowledge', IP/C/W/356, 24 June, p. 1.
11 WTO (2002), 'The Relationship between the TRIPS Agreement and the Convention on Biological Diversity and the Protection of Traditional Knowledge', IP/C/W/356, 24 June, p. 5.
12 WTO (2003), 'The Relationship between the TRIPS Agreement and the Convention on Biological Diversity and the Protection of Traditional Knowledge: Submission by Bolivia, Brazil, Cuba, Dominican Republic, Ecuador, India, Peru, Thailand, Venezuela', IP/C/W/403, 24 June, pp. 2–3.
13 Several of the TCE do not directly depend on the GRs to manifest themselves, as for example forms of music. However, almost the entire domain of TK, a large part of which was represented by traditional medicines, depends on GRs.
14 Some evidence of the extent of misappropriation of TK became known after the European Patent Office (EPO) started using the TKDL for examining patent applications. The complete list of the patents revoked by the EPO and those under examination is available at the following website: www.tkdl.res.in/tkdl/langdefault/common/Outcome.asp?homepage=sub (accessed 10 January 2017).
15 Patent # US5401504A on 'Use of turmeric in wound healing' was granted in 1995, but was subsequently withdrawn after the Indian organisation, Council of Scientific and Industrial Research had challenged its grant before the US Patent and Trade Mark Office. For details, see, Jayaraman, K.S. (1997), 'US patent office withdraws patent on Indian herb', *Nature* 389, 6, 4 September.
16 In 2005, the European Patent Office had revoked the patent that it had granted to W.R. Grace, of Columbia, Maryland, and the US Secretary of Agriculture in 1994. The patent specification included a method for controlling fungi on plants using a

hydrophobic oil extracted from the seeds of the neem tree. For details, see, Sheridan, Cormac (2005), 'EPO neem patent revocation revives biopiracy debate', *Nature Biotechnology* 23, 511–512.

17 The most recent case in this regard was the ruling by U.S. Court of Appeals for the 9th Circuit that 'a sequence of yoga poses and breathing exercises was not entitled to copyright protection'. See, United States Court of Appeals for the Ninth Circuit (2015), Bikram's Yoga College of India v. Evolution Yoga, Llc, Argued and Submitted, May 8, 2015, Pasadena, California (accessed 10 January 2017 from: http://cdn.ca9.uscourts.gov/datastore/opinions/2015/10/08/13-55763.pdf).

18 'Branding the Missionaries of Position', Business Bloomberg, February 25, 2011 (accessed 10 January 2017 from: www.bloomberg.com/bw/magazine/content/11_10/b4218078711181.htm).

19 After a series of initiatives that were taken by its regional offices since the later 1990s, the WHO developed the Strategy for Traditional Medicine for 2002–2005; WHO (2002), WHO Traditional Medicine Strategy 2002–2005, Geneva.

20 The importance of traditional medicines was re-emphasised through the 'Beijing Declaration', adopted at the conclusion of the WHO Congress on Traditional Medicine in 2008. 'Beijing Declaration: Adopted by the WHO Congress on Traditional Medicine, Beijing, China, 8 November 2008' (accessed 10 January 2017 from: www.who.int/medicines/areas/traditional/TRM_BeijingDeclarationEN.pdf?ua=1).

21 Following the Resolution on Traditional Medicine adopted at the Sixty-Second World Health Assembly (WHA 62.13), the WHO traditional medicine strategy for 2014–2023 was adopted in 2013. See, WHO (2013), WHO Traditional Medicine Strategy: 2014–2023, Geneva.

22 According to the agencies responsible for the TKDL, the database currently contains more than 290,000 formulations from the texts of traditional medicine systems of India which are Ayurveda, Unani and Siddha. Details are available at: www.tkdl.res.in/tkdl/langdefault/common/SourceInfo.asp?GL=Eng (accessed 10 January 2017).

23 WIPO (2012), Intergovernmental Committee on Intellectual Property and Genetic Resources, Traditional Knowledge and Folklore, Twentieth Session (Geneva, February 14 to 22, 2012): Report, WIPO/GRTKF/IC/20/10, paragraph 96, p. 28.

24 WIPO (2004), Intergovernmental Committee on Intellectual Property and Genetic Resources, Traditional Knowledge and Folklore, Seventh Session (Geneva, November 1 to 5, 2004): Report, WIPO/GRTKF/IC/7/15, paragraph 175, pp. 90–91.

25 WIPO (2004), Intergovernmental Committee on Intellectual Property and Genetic Resources, Traditional Knowledge and Folklore, Seventh Session (Geneva, November 1 to 5, 2004): Report, WIPO/GRTKF/IC/7/15, paragraph 175, pp. 90–91.

26 WIPO (2008), Intergovernmental Committee on Intellectual Property and Genetic Resources, Traditional Knowledge and Folklore, Eleventh Session (Geneva, July 3 to 12, 2007): Report, WIPO/GRTKF/IC/11/15, paragraph 430, p. 154.

27 WIPO (2002), Intergovernmental Committee on Intellectual Property and Genetic Resources, Traditional Knowledge and Folklore, Fourth Session (Geneva, December 9 to 17, 2002): Report, WIPO/GRTKF/IC/4/15, paragraph 74, p. 26.

28 WIPO (2008), Intergovernmental Committee on Intellectual Property and Genetic Resources, Traditional Knowledge and Folklore, Eleventh Session (Geneva, July 3 to 12, 2007): Report, WIPO/GRTKF/IC/11/15, paragraph 405, p. 150.

29 WIPO (2008), Intergovernmental Committee on Intellectual Property and Genetic Resources, Traditional Knowledge and Folklore, Eleventh Session (Geneva, July 3 to 12, 2007): Report, WIPO/GRTKF/IC/11/15, paragraph 350, p. 138.

30 WIPO (2008), Intergovernmental Committee on Intellectual Property and Genetic Resources, Traditional Knowledge and Folklore, Eleventh Session (Geneva, July 3 to 12, 2007): Report, WIPO/GRTKF/IC/11/15, paragraph 430, p. 154.

31 WIPO (2004), Intergovernmental Committee on Intellectual Property and Genetic Resources, Traditional Knowledge and Folklore, Sixth Session (Geneva, March 15 to 19, 2004): Report, WIPO/GRTKF/IC/6/14, paragraph 81, p. 37.

32 WIPO (2008), Intergovernmental Committee on Intellectual Property and Genetic Resources, Traditional Knowledge and Folklore, Eleventh Session (Geneva, July 3 to 12, 2007): Report, WIPO/GRTKF/IC/11/15, paragraph 106, p. 57.
33 WIPO (2012), Intergovernmental Committee on Intellectual Property and Genetic Resources, Traditional Knowledge and Folklore, Twentieth Session (Geneva, February 14 to 22, 2012): Report, WIPO/GRTKF/IC/20/10, paragraph 96, p. 28.
34 WIPO (2002), Intergovernmental Committee on Intellectual Property and Genetic Resources, Traditional Knowledge and Folklore, Fourth Session (Geneva, December 9 to 17, 2002): Report, WIPO/GRTKF/IC/4/15, paragraph 140, p. 47.
35 WIPO (2002), Intergovernmental Committee on Intellectual Property and Genetic Resources, Traditional Knowledge and Folklore, Fourth Session (Geneva, December 9 to 17, 2002): Report, WIPO/GRTKF/IC/4/15, paragraph 140, p. 47.
36 WIPO (2007), Intergovernmental Committee on Intellectual Property and Genetic Resources, Traditional Knowledge and Folklore, Tenth Session (Geneva, November 30 to December 8, 2006): Report, WIPO/GRTKF/IC/10/7 Prov. 2, paragraph 169, p. 84.
37 WIPO (2008), Intergovernmental Committee on Intellectual Property and Genetic Resources, Traditional Knowledge and Folklore, Eleventh Session (Geneva, July 3 to 12, 2007): Report, WIPO/GRTKF/IC/11/15, paragraph 350, p. 138.
38 WIPO (2005), Intergovernmental Committee on Intellectual Property and Genetic Resources, Traditional Knowledge and Folklore, Eighth Session (Geneva, June 6 to 10, 2005): Report, WIPO/GRTKF/IC/8/15 Prov. 2, paragraph 94, p. 48.
39 WIPO (2010), Intergovernmental Committee on Intellectual Property and Genetic Resources, Traditional Knowledge and Folklore, Sixteenth Session (Geneva, May 3 to 7, 2010): Report, WIPO/GRTKF/IC/16/8, paragraph 189, p. 56.
40 WIPO (2007), Intergovernmental Committee on Intellectual Property and Genetic Resources, Traditional Knowledge and Folklore, Tenth Session (Geneva, November 30 to December 8, 2006): Report, WIPO/GRTKF/IC/10/7 Prov. 2, paragraph 169, p. 84.

16 The WIPO IGC

A U.S. perspective

Dominic Keating[1]

Introduction

WIPO IGC negotiations are at a substantive impasse. The Consolidated Texts are heavily bracketed, demonstrating a lack of agreement on almost all provisions. Demandeur countries, including the Africa Group, the Group of Latin American and Caribbean Countries and much of the Asia Pacific Group, seek to establish *sui generis* systems of protection that would provide exclusive rights to owners of traditional knowledge and folklore. They also seek to establish a new patent disclosure requirement that would facilitate a transfer of wealth between industrialized and non-industrialized institutions. Non-demandeur countries, including the United States, Japan, Canada and Korea, are concerned with these approaches. Among other things, they believe these approaches might have a stifling effect on innovation and creativity.

This chapter will elaborate a U.S. perspective on the IGC negotiations by discussing the fundamental role of innovation and creativity in the U.S. economy, the importance of intellectual property in supporting innovation and creativity, key proposals contained in the Consolidated Text, where these proposals originated, and how these proposals might interfere with intellectual property, innovation and creativity. This chapter will then propose four steps that the IGC can take to resolve the impasse.

The role of innovation and creativity in the U.S. economy

Innovation and creativity drive U.S. economic growth and competitiveness (Economics and Statistics Administration [ESA] and U.S. Patent and Trademark

1 Dominic Keating serves as Director, IP Attaché Program at USPTO and Adjunct Associate Professor at American University, Washington College of Law. He earned a Master of Science degree in biotechnology from Johns Hopkins University. The views and comments expressed herein are solely the opinion of the author, do not reflect the performance of duties in the author's official capacity, and are not endorsed by, nor should be construed as, any viewpoint official or unofficial of the United States Patent and Trademark Office or American University, Washington College of Law. The author confirms to the best of his knowledge that no information contained herein is privileged, confidential or classified.

Office [USPTO], 2012). Two-thirds of U.S. economic growth since the 1940s is due to increases in productivity due to innovation (Blank, 2012). According to the Bureau of Economic Analysis (BEA), an estimated 4.3 percent of U.S. gross domestic product, worth $698.7 billion, was attributed to arts and cultural production industries (2015). Also, innovation has been clearly linked to gains in the standard of living (Cohen, 2011).

According to a report by the Economics and Statistics Administration (ESA) and the U.S. Patent and Trademark Office (USPTO), the total number of U.S. jobs supported directly and indirectly by IP-intensive industries is 40.0 million. This is 27.7 percent of all jobs in the U.S. economy (2012). These industries also account for more than $5 trillion, or more than 35 percent of U.S. Gross Domestic Product. Among these industries, the core copyright industries account for nearly 5.5 million jobs in the United States (Siwek, 2014). Direct and indirect jobs in the biopharmaceutical sector account for approximately 3.4 million jobs in the United States. IP-related jobs are well-paid, with annual total compensation that is substantially higher than the average for all U.S. workers (ESA-USPTO report, 2012).

Creative works such as books, movies and television programs disseminate essential information and provide beneficial entertainment. Innovations in the life sciences are critical to the health and livelihoods of Americans and foreigners alike. According to a report by the U.S. Department of Agriculture, in 2014, more than 168.9 million acres of biotech crops were planted in the United States and at least 420 million had been planted worldwide. These crops have helped to improve food security and sustainability, as well as to improve the incomes of farmers. Immunizations prevent 33,000 deaths in the United States and at least 2.5 million deaths worldwide on an annual basis. More than 11,000 potential new medicines are now in development (Long and Works, 2013).

The importance of intellectual property

Innovation is incentivized by a vibrant intellectual property system. Patents protect innovation by providing the inventor with a twenty-year, limited exclusive right to a patentable invention. Trademarks help consumers to identify the source of high quality products and services. They also help innovative companies to gain market share by distinguishing the goods or services of one entity from those of another. Copyrights protect creative works such as books, movies and software code. All of these forms of protection incentivize the development of innovative and creative products, create jobs and drive economic growth.

Proposals contained in the Consolidated Texts

The WIPO IGC is negotiating separate, 'Consolidated Texts' for intellectual property and genetic resources, intellectual property and traditional knowledge and intellectual property and folklore. The Consolidated Texts are parallel in structure. They each contain a list of terms or definitions, a preamble and a set of policy objectives or guiding principles. This is followed by articles that attempt to define the subject matter of the instrument, scope of the instrument, exceptions and limitations, relationship of the proposed instrument with other international agreements, sanctions and remedies and transboundary cooperation. Articles containing a disclosure requirement are in the genetic resources and traditional knowledge texts only.

Articles entitled 'scope of the instrument' define the demandeurs' proposed mechanism for obtaining protection in the traditional cultural expressions text and one of the demandeurs' proposed mechanisms for obtaining protection in the traditional knowledge text. In both of these texts, the focus is on a *sui generis* system of protecting traditional cultural expressions or traditional knowledge. Articles entitled 'disclosure requirement' define the demandeurs' proposed mechanism for obtaining protection in the genetic resources text and one of the demandeurs' proposed mechanisms for protection in the traditional knowledge text.

The mechanisms for protection in the Consolidated Text on traditional knowledge focus on a *sui generis* system that would allow traditional knowledge beneficiaries to exclude others from accessing or disclosing protected traditional knowledge, and a new patent disclosure requirement. The mechanism for protection in the Consolidated Text on traditional cultural expressions focuses on a *sui generis* system that would allow beneficiaries to prevent others from unauthorized access to, disclosure, fixation, use or modification of the protected traditional cultural expression. This mechanism, like the *sui generis* system discussed in the traditional knowledge text, would have a potentially unlimited duration. The mechanism for protection in the Consolidated Text on genetic resources focuses on a patent disclosure requirement.

Proposals for preventing the erroneous grants of patents and voluntary codes of conduct for industry are also included in the 'scope of protection' in the genetic resources and traditional knowledge texts (World Intellectual Property Organization [WIPO], June 2, 2014). Much of this language was incorporated into the text from proposals for joint recommendations that were co-sponsored by the delegations of Canada, Japan, Norway, the Republic of Korea and the United States (WIPO, May 9, 2014a and WIPO, May 9, 2014b). Among other things, these proposals would encourage the use of databases of genetic resources and associated traditional knowledge to prevent the erroneous grants of patents and would establish a WIPO portal site for searching genetic resources and associated traditional knowledge.

Origins of key proposals in the Consolidated Texts

The WIPO General Assembly established the IGC in 2000, as a forum to discuss access to genetic resources and benefit sharing, the protection of traditional knowledge and the protection of expressions of folklore (WIPO, August 25, 2000). In doing so, the General Assembly cited regional consultations on folklore, the outcome of consultations among member states on genetic resources, and a commitment of the Director General at the Adoption of the Patent Law Treaty (WIPO, October 3, 2000).

The IGC texts on traditional knowledge and traditional cultural expressions root back to two documents prepared by the WIPO International Bureau in 2002 that set forth 'Elements of Sui Generis Systems. . .'. Included in these elements are policy objectives, subject matter of protection, criteria for protection, ownership and rights. In 2004, the WIPO Secretariat introduced core objectives documents for TK and TCEs that included elaborate, instrument-like annexes. The WIPO General Assembly authorized the IGC to begin text-based negotiations in 2009. This decision came at a time of disparate views and expectations by Member States, frustrations amongst some WIPO Members with the lack of progress in the Committee and strong, political pressure for an outcome. Since that time, the IGC has met fourteen times to conduct text-based negotiations.

The elements contained in the WIPO documents were derived in part from 'Model Provisions for National Laws on the Protection of Expressions of Folklore against Illicit Exploitation and Other Prejudicial Actions' that were developed by the United Nations Educational, Scientific and Cultural Organization (UNESCO) and WIPO in 1982. The model provisions include sections on principles, subject matter of protection, rights, sanctions and exceptions (UNESCO and WIPO, 1985). A Group of Experts met in 1982 to discuss the possibility of developing a treaty based on the draft provisions. The Group ultimately decided against a treaty-based approach due to uncertainties related to the proposed treaty. Specifically, they were unable to identify specific expressions of folklore to be protected, they were concerned about folklore that crossed national boundaries and they were not able to determine a viable dispute settlement system (von Lewinski, 2008). Four Regional Consultations on the Protection of Expressions of Folklore, organized by WIPO and UNESCO in 1999, were unable to resolve these intransigent issues (WIPO, October 20, 2002; September 30, 2002; May 20, 2002).

The scope of the Consolidated Text on genetic resources is based on proposals for patent disclosure requirements that were introduced in the Convention on Biological Diversity (Convention on Biological Diversity, 1996), the WTO Committee on Trade and the Environment (1997), the WTO TRIPS Council (2004) and the WIPO Standing Committee on Patents (SCP) (1999), starting from 1994. These proposals were not accepted in CTE, the TRIPS Council and the SCP due to concerns that they would create uncertainties in patent rights, reducing patent incentives for innovation, among other concerns. Also, several disclosure requirements have been implemented in national laws, but it appears that they have not been successful in benefiting indigenous and local communities or reducing the erroneous grants of patents. To the contrary, industry regularly raises concerns over these national laws.

U.S. concerns with proposals in the Consolidated Texts

The incorporation of treaty-like articles into the WIPO meeting documents, starting with the second year of the IGC, has built expectations of a treaty-based outcome. These articles quickly became the focus of the IGC, and they remain the focus of the IGC today. The IGC may not have fully considered, however, whether the treaty-like articles evolving from the WIPO-UNESCO process would be an appropriate basis for a workable outcome. Treaty-focused expectations have also foreclosed other, more practical, outcomes from the negotiations. For example, in the Sixth Session of the IGC, the Brazilian delegation attempted to block progress on model contractual clauses for access to genetic resources and benefit sharing, perhaps out of concerns that it would distract the IGC from its focus on a treaty-based outcome.

The most significant concerns with all three IGC Consolidated Texts relate to *sui generis* systems and disclosure requirements, due to their potential negative impact on intellectual property, innovation and creativity.

Sui generis *systems*

Sui generis systems of protection are unique systems. Like other systems of protection, they are informed by prior, related national experiences. The negotiation of the Paris Convention for the Protection of Industrial Property of 1883 and the TRIPS Agreement of 1995, among other international legal instruments, benefited from a long history of patent protection under national laws. This history included the Venetian Patent Statute of 1474 and the English Statute of Monopolies of 1624. National experience is especially important for informing negotiations of international instruments regarding *sui generis* protection, due to the unique nature of this form of protection.

The International Convention for the Protection of New Varieties of Plants (UPOV) is a successful *sui generis* system of protecting intellectual property. It builds upon decades of national experience in protecting plants, including the U.S. Plant Patent Act of 1930 and the patenting of cultivated sports, mutants, hybrids and newly found seedlings under a 1954 amendment to the U.S. Patent Act of 1952 (35 U.S.C. § 161). UPOV went into force in 1962. It has been revised twice and it remains in force today, with a membership of 72 Parties.

The Washington Treaty on Integrated Circuits, in contrast, is a *sui generis* system that was negotiated with little guidance from national experience and never went into force. The Semiconductor Chip Protection Act (SCPA) was passed in the United States in 1984, in order to prevent the unauthorized copying of American chip designs (Radomsky, 2000). Japan and the European Communities followed suit in 1985 and 1986, respectively, with similar legislation (Radomsky, 2000). At the time, there was no intellectual property protection for chip design. Chip designs, which cost millions of dollars to develop, could be rapidly copied by foreign competitors. The Washington Treaty was adopted in 1989, just five years after passage of the SCPA.

With limited national experience to draw from, including only one appellate court decision under the SCPA, the Washington Treaty was doomed to failure. Among other reasons, industry could not rely upon this protection because it contained a reverse engineering exception and a requirement of substantial similarity for infringement. Industry shifted focus to enforcing or cross-licensing process patents for the manufacture of chips and product patents for devices that contain the integrated circuit. The Washington Treaty was not viable because it was not based upon a sufficient body of national experience.

Similarly, the IGC is negotiating *sui generis* systems of exclusive rights without one or more successful national models for guidance. Also, the draft text is not designed to address specific problems at the national level. For these reasons, the negotiators lack a foundation for a successful outcome.

Furthermore, the *sui generis* systems that are being negotiated would provide exclusive rights to the owner without a corresponding benefit to the public. The patent system, in contrast, provides a limited exclusive right to an inventor in exchange for a public disclosure of the invention. At 18 months from the filing date, the information contained in a patent specification is made accessible to the world through publication. Twenty years from the filing date, the invention becomes part of the public domain. All are free to make, use, sell and import the invention at that point in time.

Equally troublesome is the often ubiquitous subject matter of the *sui generis* system: knowledge and culture. WIPO Members must inquire whether it is practical to remove knowledge and culture from the public domain. If this knowledge or culture is not currently in the public domain, WIPO Members must ask whether it would be practical or desirable to protect it against public use.

The text defining *sui generis* protection includes an option for an indefinite term of protection. Such a term of exclusivity would create a monopoly. Products and services that are provided in commerce are continuously adapted and improved, in order to compete with other products and services. On the other hand, products and services provided under a monopoly tend to become outdated and no longer able to meet the needs of the consumer. A perpetual monopoly might allow an exclusive licensee to charge a premium to consumers for an unlimited term, while failing to adapt a product or service to the needs of the consumer. This would be contrary to public policy objectives. In contrast, patents and copyrights provide limited exclusive rights that expire, allowing the protected subject matter to ultimately fall into the public domain.

Disclosure requirement

Proposals for new disclosure requirements in patent laws generally require applicants for patents directed to genetic resources and associated traditional knowledge to disclose in their patent applications the source and origin of genetic or biological material used to develop a claimed invention, evidence of prior informed consent and mutually agreed terms for using the genetic resource, and evidence of equitable benefit sharing related to the invention. Sanctions may include the rejection of a non-compliant patent application or the revocation of a non-compliant patent (WTO, 2003). These requirements could have a devastating impact on research and development in the field of biotechnology and pharmaceuticals due to uncertainties that they would introduce into patent protection.

Biotech and pharmaceutical inventions tend to become more valuable over time, after regulatory authorities have approved them for marketing and medical professionals have recognized their value. Competitors are therefore motivated to challenge patents on these inventions after regulatory approval. According to a study by Tufts Center for the Study of Drug Development (2014), the development of a new drug often takes more than ten years and costs approximately $2.6 billion. A patent term, however, expires twenty years from the filing date. This leaves many biotech and pharmaceutical companies with only a ten-year term of effective patent protection for their products, between the grant of marketing approval and the expiration of the patent, in which to recuperate their substantial investment.

A new disclosure requirement would allow competitors to challenge a pharmaceutical or biotech patent during this critical period of time on the basis of the disclosure requirement. For example, a competitor could potentially challenge a patent based on a lack of prior informed consent from the correct tribe or the correct tribal leader, allegations that terms were not 'mutually agreed' or allegations that benefits shared were not 'equitable'. According to the Biotechnology Industry Association, 'a disclosure requirement that could block the grant of a patent for an otherwise eligible invention, or could be used to invalidate that patent, (would) create unacceptable risks for (its) Members' (2004). The uncertainties created by disclosure requirements would increase risk in a resource-intensive drug development process, thereby reducing necessary investment and associated research and development.

Some national laws, such as the Indian National Biodiversity Act of 2002, would allow a national authority to demand royalties in an amount that the authority may deem fit in exchange for allowing a patent application that is in condition for allowance to be granted, allowing an inventor to commercialize his invention or export a product containing a genetic resource from the country. These laws would take advantage of the information that is provided by a patent applicant through a new disclosure requirement to impose financial penalties on an inventor. They could also be used to transfer wealth from the patent owner, to the country of origin of genetic materials used in making an invention, via an imposed royalty.

Recommendations

After more than five years of formal negotiations, the IGC Consolidated Texts leave little or no room for agreement between WIPO Members. They contain hundreds of brackets, representing the divergent positions of Member States. Also, signs of frustration with the text-based negotiations are evident. At the 2014 General Assembly meeting, the Africa Group blocked consensus on a 2015 work program for the IGC. This meant that the IGC did not meet in 2015.

Based on the foregoing considerations, four recommendations may help WIPO Members to move beyond the impasse in the IGC:

(1) Move beyond the Consolidated Text
(2) Negotiate a declaration to reconcile objectives and principles
(3) Develop soft law instrument(s) to further satisfy the mandate
(4) Conclude IGC negotiations and revert to IGC or ad hoc Meetings of Experts.

Move beyond the Consolidated Text

IGC negotiations are based on the Consolidated Text for historical reasons, despite the fact that they are making little or no progress on the basis of this text. With high expectations for a treaty negotiation and little progress on the Consolidated Text, demandeurs have been introducing proposals from the IGC into the Design Law Treaty negotiations in the WIPO Standing Committee on the Law of Trademarks, Industrial Designs and Geographical Indications; the Standing Committee on Patents; and the Committee on Development and Intellectual Property to create pressure for an IGC-related treaty negotiation. A number of industrialized countries have opposed these proposals based on their substance. Rather than catalyzing a treaty negotiation, therefore, these proposals are likely to create further gridlock in WIPO and reduce its credibility. A more effective approach would be to examine specific problems that WIPO Members face, examine national solutions to those problems and determine whether an international solution is warranted.

A U.S.–India collaboration to prevent erroneous grants of patents will help to illustrate this approach. In 1995, two expatriate Indian inventors obtained U.S. patent protection (U.S. Patent No. 5,401,504) on a method of using turmeric in healing wounds (Das and Cohly, 1995). The broadest claim was directed to 'a method of promoting healing of a wound by administering turmeric to a patient afflicted with wound'. The specification of the patent disclosed that turmeric had been used as a dietary agent and a traditional medicine for the treatment of sprains and inflammatory conditions in India. This patent was assigned to the University of Mississippi Medical Center.

The Indian Council of Scientific and Industrial Research (CSIR) filed a re-examination proceeding, administratively challenging the claims of the patent. CSIR provided numerous documents, translated from Hindi and other South Asian languages, demonstrating that the invention had been previously used in India. USPTO used this documentation to reject all of the patent claims, invalidating the patent.

In 1997, USPTO granted a patent to Texas-based RiceTec Corp. for its hybridized basmati rice (U.S. Patent No. 5,663,484; Sarreal *et al.*, 1997). The company used traditional cross-breeding methods to produce a photoperiod-resistant rice plant that has similarities to Indian basmati rice. Unlike traditional basmati varieties, it is suited for growth in North America. The Indian CSIR filed a request for reexamination with the USPTO in April 2000 on three of the patent's twenty claims. CSIR argued that these claims were so broad that they would include 90 percent of all basmati grown anywhere in the world. RiceTec subsequently amended its patent claims to remove these three claims, among other claim amendments.

Informed by these experiences, USPTO and CSIR entered into a Traditional Knowledge Digital Library (TKDL) access agreement in 2009. Under the terms of the agreement, USPTO Patent Examiners have access to more than 200,000 traditional medicine formulations on Ayurveda, Unani and Siddha comprising 30 million pages, in a text-searchable format. These steps help to reduce the erroneous granting of patents and they help to improve patent quality. The USPTO took additional steps to expand its access to commercial prior art databases related to traditional knowledge and genetic resources. The USPTO currently has access to more than 70 databases that are relevant to traditional knowledge and traditional medicines (2015).

In this example, the Indian and U.S. governments identified an international problem related to the erroneous grants of patents on inventions related to traditional knowledge and genetic resources. They worked together to find a solution that will help to mitigate the problem, on a bilateral basis. In a similar manner, WIPO Members could share their concerns in the IGC related to the erroneous grants of patents. They could then determine if a multilateral solution is warranted. If it is warranted, they could develop a solution to this specific problem. The same approach could be used to address a full spectrum of problems faced by WIPO Members.

Negotiate a declaration to reconcile objectives and principles

A common set of objectives and principles is foundational to progress in the IGC. Although WIPO Members have discussed objectives and principles in the IGC, they have not reached a consensus. WIPO Members will continue to negotiate at cross purposes to each other if they do not reconcile objectives and principles.

A declaration may be the most effective means of refining objectives and principles in the IGC. Declarations may be used by WIPO Members to declare common aspirations. They may also be used to interpret legal instruments. A declaration in the IGC might include, for example, aspirations to preserve, promote and respect traditional knowledge and traditional cultural expressions, while preserving incentives to innovate and create.

Months before the Doha Ministerial Conference launched a round of multilateral trade negotiations in the WTO, a group of African countries asked the TRIPS Council to examine the relationship between the TRIPS agreement and access to medicines. These countries viewed TRIPS and health as a threshold issue to be addressed in the round. Positions were divided and difficult to reconcile amongst WTO Members. In order to reconcile positions on this issue and allow the round to progress, WTO Members negotiated the Doha Declaration on TRIPS and Public Health. This subsequently facilitated the amendment of TRIPS Article 31 and helped to build confidence in the WTO process.

Develop soft law instrument(s) to further satisfy the mandate

WIPO Members should fulfill the mandate in a balanced and step wise approach through the development of soft law, such as a resolution, joint recommendation, tool-kit or action plan. This soft law should be consistent with common objectives and principles, as well as the aspirations contained in the declaration. It should also be rooted in successful, national experiences.

Soft law is not legally binding. However, it can be a valuable tool for solving international problems. It can also be a platform for future work. For example, a joint recommendation on the protection of well-known marks was negotiated in the SCT and adopted by the Assembly of the Paris Union for the Protection of Industrial Property and the WIPO General Assembly in 1999. The purpose of this joint recommendation was to accelerate the development and implementation of harmonized principles and rules.

The United States and co-sponsors have tabled a proposed joint recommendation on patent examination and a proposed joint recommendation for a WIPO-hosted traditional knowledge database portal.[1] Both of these joint recommendations would help to reduce the erroneous grants of patents. Although these proposals have not been negotiated by the Committee, they are illustrative of a soft law recommendation that could be negotiated and adopted by the IGC.

Conclude IGC negotiations and revert to IGC or Ad Hoc Meetings of Experts

In order to avoid unrealistic expectations and reduce frustrations and gridlock, WIPO Members should notionally establish a time period for concluding IGC negotiations. Within this time period, WIPO Members should set reasonable goals to negotiate one or more feasible international legal instruments under the mandate. After the expiration of the time period, and with the blessing of the IGC, the General Assembly should terminate the negotiating mandate and require the IGC to return to its pre-negotiation work program. Alternatively, the IGC could become an Ad Hoc Committee of Experts. The focus of work, after the conclusion of the negotiating mandate, would be on examining actual problems and sharing relevant national experiences to address these problems, as well as harmonizing objectives and principles.

Note

1 World Intellectual Property Organization (2014, May 9), *Joint Recommendation on Genetic Resources and Associated Traditional Knowledge* (Document Code WIPO/GRTKF/IC/28/7), retrieved from www.wipo.int/meetings/en/doc_details.jsp?doc_id=274082; and World Intellectual Property Organization (2014, May 9), *Joint Recommendation on the Use of Databases for the Defensive Protection of Genetic Resources and Traditional Knowledge associated with Genetic Resources* (Document Code WIPO/GRTKF/IC/28/8), retrieved from www.wipo.int/meetings/en/doc_details.jsp?doc_id=274080 (accessed 22/3/2017).

References

Biotechnology Industry Organization. (2004, November 5). BIO's statement to IGC on genetic resources and patent disclosure requirements. Retrieved from www.bio.org/advocacy/letters/bios-statement-igc-genetic-resources-and-patent-disclosure-requirements (accessed 22/3/2017).

Blank, R. (2012, March 9). 'Acting Deputy Commerce Secretary Rebecca Blank Highlights Administration's Gender Equality Efforts on Trip to Switzerland.' [Web blog post]. Retrieved from https://obamawhitehouse.archives.gov/blog/2012/03/09/acting-deputy-commerce-secretary-rebecca-blank-highlights-administration-s-gender-eq

Bureau of Economic Analysis, U.S. Department of Commerce. (2015, January 12). 'Spending on Arts and Cultural Production Continues to Increase.' [News release] Retrieved from www.bea.gov/newsreleases/general/acpsa/acpsa0115.pdf (accessed 22/3/2017).

Cohen, A.J. (2011). *Innovation and Economic Growth*. Retrieved from www.goldmansachs.com/our-thinking/archive/archive-pdfs/gsr.pdf (accessed 22/3/2017).

Convention on Biological Diversity. (1996, September 22). *The Impact of Intellectual Property Rights Systems on the Conservation and Sustainable Use of Biological Diversity and on the Equitable Sharing of Benefits from its Use*. (Document No. UNEP/CBD/COP/3/22). Retrieved from www.iisd.ca/biodiv/cop3/3_22_vfinal.htm (accessed 22/3/2017).

Das, S.K. and Cohly, H.H.P. (1995). *U.S. Patent No. 5,401,504*. Washington, DC: U.S. Patent and Trademark Office.

Economics and Statistics Administration (ESA) and the U.S. Patent and Trademark Office (USPTO) of the U.S. Department of Commerce. (2012). *Intellectual Property and the U.S. Economy: Industries in Focus*. Retrieved from www.uspto.gov/sites/default/files/news/publications/IP_Report_March_2012.pdf (accessed 22/3/2017).

Long, G. and Works, J. (2013, January). *Innovation in the Biopharmaceutical Pipeline: A Multidimensional View*. Retrieved from http://phrma.org/sites/default/files/pdf/2013innovationinthebiopharmaceuticalpipeline-analysisgroupfinal.pdf (accessed 22/3/2017).

Patents for Plants, 35 U.S. Code § 161.

Radomsky, Leon. (2000). 'Sixteen Years after the Passage of the U.S. Semiconductor Chip Protection Act: Is International Protection Working?' *Berkeley Technology Law Journal*, 15 (3), 1049–1094.

Sarreal, E.S., Mann, J.A., Stroike, J.E. and Andrews, R.D. (1997). *U.S. Patent No. 5,663,484*. Washington, DC: U.S. Patent and Trademark Office.

Siwek, S. (2014, December). *Copyright Industries in the U.S. Economy: The 2014 Report*. Retrieved from www.iipawebsite.com/pdf/2014CpyrtRptFull.PDF (accessed 22/3/2017).

Tufts Center for the Study of Drug Development. (2014, November 18). 'Cost to Develop and Win Marketing Approval for a New Drug Is $2.6 Billion.' [News release] Retrieved from http://csdd.tufts.edu/news/complete_story/pr_tufts_csdd_2014_cost_study (accessed 22/3/2017).

United Nations Educational, Scientific and Cultural Organization, World Intellectual Property Organization. (1985). *Model Provisions for National Laws on the Protection of Expressions of Folklore Against Illicit Exploitation and other Forms of Prejudicial Action, 1982*. Retrieved from www.wipo.int/wipolex/en/details.jsp?id=6714 (accessed 22/3/2017).

U.S. Patent and Trademark Office. (2015, March). 'Traditional Knowledge and Medicine Dictionaries/Databases.' Retrieved from www.uspto.gov/patent/laws-and-regulations/comments-public/traditional-knowledge-and-medicine-dictionariesdatabases (accessed 22/3/2017).

United States Department of Agriculture, Economic Research Service, Economic Research Report Number 162 (2014, February). Genetically Engineered Crops in the United States. Retrieved from www.ers.usda.gov/webdocs/publications/45179/43668_err162.pdf?v=41690 (accessed 22/3/2017).

von Lewinski, Silke. (2008). *International Copyright Law and Policy*. (p. 536). New York. NY: Oxford University Press.

World Intellectual Property Organization. (2014, June 2). *Consolidated Document Relating to Intellectual Property and Genetic Resources*. (Document Code WIPO/GRTKF/

IC/28/4). Retrieved from www.wipo.int/meetings/en/doc_details.jsp?doc_id=276319 (accessed 22/3/2017).

World Intellectual Property Organization. (2014a, May 9). *Joint Recommendation on Genetic Resources and Associated Traditional Knowledge*. (Document Code WIPO/GRTKF/IC/28/7). Retrieved from www.wipo.int/meetings/en/doc_details.jsp?doc_id=274082 (accessed 22/3/2017).

World Intellectual Property Organization. (2014b, May 9). *Joint Recommendation on the Use of Databases for the Defensive Protection of Genetic Resources and Traditional Knowledge associated with Genetic Resources*. (Document Code WIPO/GRTKF/IC/28/8). Retrieved from www.wipo.int/meetings/en/doc_details.jsp?doc_id=274080 (accessed 22/3/2017).

World Intellectual Property Organization. (2002, October 20). *Intergovernmental Committee on Intellectual Property and Genetic Resources, Traditional Knowledge and Folklore*. (Document Code WIPO/GRTKF/IC/4/3). Retrieved from www.wipo.int/edocs/mdocs/tk/en/wipo_grtkf_ic_4/wipo_grtkf_ic_4_3.pdf (accessed 22/3/2017).

World Intellectual Property Organization. (2002, September 30). *Intergovernmental Committee on Intellectual Property and Genetic Resources, Traditional Knowledge and Folklore*. (Document Code WIPO/GRTKF/IC/4/8). Retrieved from www.wipo.int/edocs/mdocs/tk/en/wipo_grtkf_ic_4/wipo_grtkf_ic_4_8.pdf (accessed 22/3/2017).

World Intellectual Property Organization. (2002, May 20). *Intergovernmental Committee on Intellectual Property and Genetic Resources, Traditional Knowledge and Folklore*. (Document Code WIPO/GRTKF/IC/3/9). Retrieved from www.wipo.int/edocs/mdocs/tk/en/wipo_grtkf_ic_3/wipo_grtkf_ic_3_9.pdf (accessed 22/3/2017).

World Intellectual Property Organization. (2000, October 3). *Report* (adopted by the Assembly). (Document Code WO/GA/26/10). Retrieved from www.wipo.int/meetings/en/doc_details.jsp?doc_id=1520 (accessed 22/3/2017).

World Intellectual Property Organization. (2000, August 25). *Matters Concerning Intellectual Property and Genetic Resources, Traditional Knowledge and Folklore*. (Document Code WO/GA/26/6). Retrieved from www.wipo.int/edocs/mdocs/govbody/en/wo_ga_26/wo_ga_26_6.pdf (accessed 22/3/2017).

World Intellectual Property Organization. (1999, September 8). *Protection of Biological and Genetic Resources*. (Document Code SCP/3/10). Retrieved from www.wipo.int/meetings/en/doc_details.jsp?doc_id=1143 (accessed 22/3/2017).

World Trade Organization. (1997, September 29). *Committee on Trade and Environment – Cluster on Market Access – Item 8: The Relationship of the TRIPS Agreement to the Development, Access and Transfer of Environmentally-Sound Technologies and Products (EST&PS) – Input from India*. (Document Symbol WT/CTE/W/66). Retrieved from https://docsonline.wto.org (accessed 22/3/2017).

World Trade Organization. (2004, September 21). *Elements of the Obligation to Disclose the Source and Country of Origin of Biological Resource and/or Traditional Knowledge Used in an Invention*. (Document Symbol IP/C/W/429). Retrieved from https://docsonline.wto.org (accessed 22/3/2017).

World Trade Organization. (2003, June 24). *The Relationship between the TRIPS Agreement and the Convention on Biological Diversity and the Protection of Traditional Knowledge*. (Document Symbol IP/C/W/403). Retrieved from https://docsonline.wto.org/dol2fe/Pages/SS/DirectDoc.aspx?filename=t%3A%2Fip%2Fc%2Fw403.doc& (accessed 22/3/2017).

17 Access and benefit sharing, intellectual property rights and the protection of traditional knowledge in Costa Rica

Participation in and possible impacts of the IGC's work

Jorge Cabrera Medaglia

Introduction

Costa Rica is considered a megadiverse country and a pioneer in the regulation and implementation of legislation regarding access to genetic resources and benefit sharing. This chapter analyzes the current situation and challenges to improve the implementation of the ABS legal framework, including that which is related to traditional knowledge, the country's participation in the IGC and the impact of the IGC's work and results achieved to date to facilitate and support the ongoing national processes on these issues.

Costa Rica's legal framework for ABS and for the protection of traditional knowledge

This section looks at the relationship between genetic resources and intellectual property rights in the Biodiversity Law and related provisions.

General considerations on the Biodiversity Law

The most important legal framework for ABS and the protection of traditional knowledge is the Biodiversity Law (BL) No. 7788 of 1998.

The general goal of the BL is to promote the conservation and sustainable use of biodiversity and to ensure the fair and equitable sharing of benefits derived from it (Article 1). The whole BL responds to this goal as put forth by the CBD. For example, it establishes the environmental function of the land (Article 8), general principles of the law (Article 9), objectives (Article 10), criteria for applying the law (Article 11); National Systems for Protected Areas' administrative structure (including the administration of the national wild protected areas, Articles 22 to 43), the guarantee of environmental safety (biosafety and exotic organisms, Articles 44 to 48), the conservation and the sustainable use of the ecosystems and species

(Articles 49 to 61), regulations on access to genetic resources (Articles 62 to 76), intellectual property rights (Articles 77 to 85), education and public awareness and research and transfer of technology (Articles 86 to 91), environmental impact assessment (Articles 92 to 97), incentives (Articles 98 to 104) and procedures and sanctions (Articles 105 to 113). All of these elements are in accordance with the three objectives of the CBD.

The BL is applied to the components of biodiversity that are under the sovereignty of the State, as well as to the processes and activities carried out under its jurisdiction or control, independently from those effects manifested inside or outside of the national jurisdiction. This Law shall specifically regulate the use, management, associated knowledge and sharing of the benefits and derived costs of the utilization of the biodiversity components (Article 3). Also, Article 6 (public domain) establishes that biochemical and genetic properties of the components of the wild or domesticated biodiversity belong to the public domain. The State will authorize the exploration, research, bioprospecting, use and utilization of biodiversity components which constitute part of the public domain, as well as the use of all genetic and biochemical resources, through access standards established in Chapter V of this Law. Likewise, in accordance with Articles 62 and 69, all research or bioprospecting programs on genetic or biochemical material derived from biodiversity that aim to be carried out in the Costa Rican territory require an access permit, unless they are covered for some of the exceptions provided for under this Law. These exceptions (Art. 4), basically refer to access to human genetic resources; the exchange of genetic and biochemical resources and the associated traditional knowledge resulting from the traditional practices of indigenous peoples and local communities, when they are non-profit; and the Public Universities which had the term of one year (up to May 7 1999)[1] to establish their own controls and regulations for their research (implying access) and where they are non-profit.

In this regard the access regulations are applied to genetic resources on public or private land, terrestrial or marine environments, under *ex situ* or *in situ* conditions and in indigenous territories. Likewise, relevant access provisions of the Law are applied to indigenous territories, but additionally the indigenous or customary rules of these peoples should be taken into account, as well as the community *sui generis* intellectual rights. Similarly, the law recognizes for communities and indigenous peoples the right to oppose access to their resources and associated knowledge for cultural, spiritual, economic reasons, or for other reasons.

The Biodiversity Law establishes the National Commission for Management of Biodiversity (CONAGEBIO) as a decentralized organ of the Ministry of Environment and Energy (Article 14). Among its responsibilities are the following:

- develop policies and responsibilities set up in Chapters IV, V (Access to genetic and biochemical elements and protection of associated knowledge) and VI of the Law and coordinate them with several responsible organizations on the subject (subsection 2)
- formulate and coordinate policies for access to biodiversity elements and associated knowledge

- ensuring adequate scientific-technical transfer and fair sharing of benefits, which under Chapter V are denominated as General Access Norms (subsection 3).

The Commission shall implement its agreements and resolutions and institute its procedures through the Executive Director of the Technical Office (Article 16). This entity is formed by government entities such as the Ministry of Environment (President), Ministry of Foreign Trade, Ministry of Health, Ministry of Agriculture, Costa Rican Institute of Fishing and Aquaculture, National Commission of Rectors; Indigenous Table, Farmers Table, private sector associations (Union Costarricense de Cámaras y Asociaciones del Sector Empresarial Privado), Costa Rican Federation for Conservation of the Environment, and the National System of Areas of Conservation Director (Art. 15). This Commission shall formulate policies on access and distribution of benefits and can revoke Technical Office resolutions on this subject (Article 14).

In accordance with Article 62, CONAGEBIO shall propose access policies on genetic and biochemical elements of biodiversity *ex situ* and *in situ* and act as a binding consultation organ for procedures of request of intellectual rights protection on biodiversity.

The Commission has a Technical Office for support, which will be integrated by an Executive Director and the staff indicated in the Regulation of Law. The Technical Office will have as a main attribution to process, reject and control the access applications to resources of biodiversity (Art. 17 a); coordinate access with Conservation Areas, the private sector, indigenous peoples and local communities (Art. 17.b); organize and maintain an updated register of access applications to biodiversity elements, *ex situ* collections and those from natural and juridical persons dedicated to genetic manipulation (subsection c); and gather information to update regulations relative to compliance of agreements and guidelines on biodiversity (subsection d).

The access procedure is regulated in detail in the Law. The competent organization to grant the access is in the first instance the Technical Office.

In Chapter V, section I, the requirements and procedures for access to genetic and biochemical elements and protection of associated knowledge are defined. The Law regulates basic requirements for access that include prior informed consent, distribution of benefits, protection of associated knowledge, and the way in which activities will contribute to conservation (Art. 63). The Law establishes the procedures to be followed (Art. 64), cultural objection rights (Art. 66), registers of access applications and protection of confidential information (Art. 67). In Section II, the Law regulates in more detail the topic of permits for research and bioprospecting (Art. 69), their term, limitations and characteristics (Arts. 70 and 71), requirements for access applications (Art. 72), authorization of the Technical Office with respect to those agreements subscribed among individuals that provide for access to genetic and biochemical elements (Art. 74) and the possibility of framework contracts with universities and other research centers duly registered (Art. 74). It is established that up to 10 percent of the research budgets and

50 percent of the royalties should be provided in favor of the Area of Conservation, private ownership or indigenous land, as well as the administrative costs (Art. 76).

TK protection in the national legal system[2]

The Costa Rican protection system for traditional knowledge is based on several premises:

a) The legal system of access assures prior informed consent and the sharing of benefits related to TK. The Technical Office and eventually the National Commission for Biodiversity Management itself has the authority to control compliance with these requirements (Arts. 63, 65, 66, 72, *inter alia*).
b) From this point of view, it is a combination between access mechanisms, contracts or licences and a *sui generis* scheme based on registers.
c) The existence and validity of different forms of knowledge and innovation and the need to protect them using appropriate mechanisms (Art. 77), such as patents, trade secrets, copyright, plant breeder rights, community *sui generis* intellectual rights, *inter alia* (Art. 78).
d) Legislation is aimed at protecting knowledge through a registration system, an approach that has been supported by comparative law. Thus, it shall proceed to inventory community *sui generis* intellectual rights that the communities request to protect (Art. 84).
e) For the definition of the extent, nature and requirements of these rights, a participative process of consultation should begin with indigenous peoples and farmer communities (Art. 83 of Law of Biodiversity). This process will determine how community *sui generis* intellectual rights will be used; who will exercise ownership and identify the beneficiaries (Art. 85).
f) Before assigning rights and obligations, whether private or collective, the following specific issues need to be considered:

- the material and associated knowledge to be protected;
- the procedures for protection/registration;
- any granted rights and against whom these will be enforced;
- relevant enforcement mechanisms.

– **Material and associated knowledge to be protected**: are the knowledge, innovations and practices of local communities and indigenous peoples, although these terms are not defined. This knowledge is associated with genetic and biochemical elements (Art. 82), including knowledge of medicine, agriculture, and other kinds of biological and genetic resources, but not including folklore. The possibilities of a regulation for sectors (medicine, etc.) carried out in a progressive manner should be analyzed. However, some issues are yet to be defined, such as the criteria for determining what is to be protected. Otherwise, there is a risk of taking more from the public domain than is justified, creating speculative demands and preventing a clear understanding of the rights and the obligations of third parties and the actions that infringe or contravene the agreed rights. Evidently some characteristics of these rights, for example, their evolution and adaptability must be recognized in order to take necessary precautions in the registry system.

- **The registration procedure**: in Costa Rica, this procedure is voluntary and declaratory, free of charge, unofficial, informal and for an indefinite term. Other elements (e.g. the possibility of opposition, existence of expert examiner, reciprocity in the recognition of the rights of TK protected in other countries; grounds for revocation and annulment; rights to appeal), must be included in the procedures. The relationship between this protection and others that can be claimed through the traditional IPR instruments must likewise be acknowledged (e.g. collective marks, geographical indications, etc.).
- Any **granted rights**: Costa Rican Law has little to say on this issue, but some rights can be extracted from several provisions (e.g. the need for prior informed consent and for mutually agreed terms for benefits in cases of access and use of TK; the right to cultural objection to oppose its use for religious or spiritual reasons). In general, the rights derived from the existence of community intellectual *sui generis* rights must be defined. This issue is as critical as that of the scope (limitations) of the said rights. Another issue requiring regulation has to do with the collective or private nature of such rights. In that sense, especially in local communities, the assumption that all knowledge is generated collectively is difficult to uphold.
- **Relevant enforcement of rights**: if no substantial mechanisms are established for enforcement of the right, the protective regulations would not suffice. In Costa Rica this issue is addressed only through a fine for illegal access. The mandatory consultation with Technical Office (TO) of the CONAGEBIO, regarding IPR applications, that uses resources or knowledge and the binding nature of the decisions of the TO could also be conceptualized as an enforcement mechanism to guarantee substantive rights on TK and associated genetic resources.

In summary the success of the expected system will depend greatly on the results of the participatory process envisaged in the BL. Unfortunately, this process has not been developed yet, except for pilot cases as explained later.

Progress in implementing this system has been slow. After some unfinished efforts and partial consultations, in 2015, the CONAGEBIO developed the project named 'Pilot Plan concerning the Sui Generis Intellectual Community Rights in two Indigenous territories', whose general goal was to develop a pilot plan in two indigenous territories that can define the rules and requirements for a national participatory process for the establishment of the Sui Generis System. A draft 'Regulation proposal regarding the sui generis communal intellectual rights related to the protection of traditional knowledge and the intangible elements associated' was developed in the past and was also consulted with the two indigenous peoples in order to receive their feedback and views. This process was developed in coordination with the Costa Rican Indigenous National Table (Mesa Nacional Indígena). The results from this Pilot Plan will serve as the foundation to establish in the future, at a national level, the participatory process with indigenous and rural/local communities to determine the nature, scope and requirements of the Sui Generis Intellectual Community Rights (as foreseen by the BL) and to consequently develop all the legal instruments to be enacted to regulate the system of protection.

As part of the Pilot Plan's methodology, 12 round-tables/meetings were carried out in two indigenous territories of the indigenous people of the 'Bribri': Talamanca-Bribri and Cabagra. One of the main outcomes of this Pilot was the development of a detailed 'Guide' to orient the consultation process. Finally, a Global Environment Facility (GEF) project is under preparation to support the national participatory process for the development of the *sui generis* intellectual community rights, including the establishment of TK registry and for the development of Bio-cultural community Protocols (in line with the Nagoya Protocol).

Relationship between IPRs, genetic resources and the protection of TK: the case of the CAFTA-DR implications on the BL

Costa Rica was the last country in the region to ratify the Free Trade Agreement between Central America, the Dominican Republic and the United States[3] (US, Dominican Republic and Central America Free Trade Agreement, known as CAFTA-DR).

The Patent Law, No. 6867, from April 5 1983, and its amendments,[4] establish as patentable all creations derived from human intellect and which can be applied in industry. It can be a product, a machine, a tool or a manufacturing process (Art. 1). However, the law specifies that the following will not be considered an invention:

a) Discoveries, scientific theories, mathematical methods and isolated computing programs.
b) Purely aesthetic creations, literary or artistic works.
c) Publicity or business plans, principles or economic methods and those related to purely mental, intellectual or gaming activities.
d) The juxtaposition of known inventions or mixture of known products, their variations in structure or use, dimensions or materials, except when it relates to a combination or fusion that cannot function separately or when qualities or characteristic functions are modified to obtain an industrial result that is not obvious to a technician in that material.

It specifies that plant varieties will have legal protection through a special law. It also indicates that the following cannot be patented:

a) Inventions whose commercial exploitation must be necessarily and objectively impeded to protect the public order, morality, health or the life of persons or animals, in order to preserve plants or avoid serious damage to the environment.

b) Diagnostic, therapeutic and surgical methods for the treatment of people or animals.
c) Plants and animals.
d) Procedures that are essentially biological and used for the production of plants or animals.

The amendments introduced by Law No. 8632 to the Trademark Law and other Laws (one of the laws enacted to fulfill the CAFTA-DR obligations) reformed paragraphs c and d of Article 4 of the Patent Law (exclusions to patentability), in the following way: 'c) Plants and animals, except microorganisms, as long as they are not microorganisms as found in nature'; d) The essentially biological processes for the production of plants or animals that are not the biological or microbiological processes.'

During the process of drafting the BL the issue of intellectual property rights and their relationship with biodiversity inevitably arose. Thus, the BL establishes that intellectual property rights shall be congruent with the objectives of the Patent Law by virtue of the principle of integration (Article 79). The Law originally excluded the following: DNA sequences from patent processes; plants and animals; unmodified microorganisms; essential biological processes for plant and animal production; the processes of nature or natural cycles; inventions essentially derived from the knowledge of biological traditional practices or in the public domain; inventions that are produced monopolistically that may affect the processes or basic agricultural products used for food and health purposes (Article 78).

However, this article was modified by an amendment of an IPR Law[5] which was enacted to comply with the IPR commitments of the CAFTA-DR. The amendment indicates that: a) DNA and RNA sequences are excluded from patent protection to the extent they don't fulfill the patent requirements; b) microorganisms as they are found in nature are not patentable; c) it was clarified that non-biological and microbiological processes can be protected; d) it was added to the exclusion those inventions whose commercial exploitation shall be impeded to protect the public order, the morality, the health or life of human beings and animals and plants and to prevent serious damage to the environment.

Authorities should consult the TO before granting protection of intellectual or industrial property-related innovations that involve biodiversity elements. The submission of a certificate of origin and prior informed consent is required in order to gain these IP protections. A well-grounded opposition by the TO shall prevent protection from being granted (Article 80).[6] The BL provides that beneficiaries enjoying protection of intellectual or industrial property rights are subject to a compulsory licence in case of duly justified emergency. In the event of a justified emergency, this licence will allow the use of such rights for the benefit of the community. This provision is aimed at solving an emergency, without involving compensation or royalty payment (Article 81).

The Biodiversity Law recognizes the existence of the certificate of origin in the case where national genetic resources are accessed and requires the presentation of this certificate before the competent office in order to issue IP rights. Similarly, a consultation is required with the Technical Office of CONABEGIO in the cases of innovations based on biodiversity elements of Costa Rica. It should be noted that the presentation of the certificate guarantees that the access procedure was followed. This includes the negotiation of prior informed consent, the establishment of mutually agreed terms and the sharing of benefits from the utilization of genetic and biochemical resources.

However, neither the Plant Variety Protection Law No. 8631 – nor its regulations – expressly requires the National Seed Office (the competent authority to grant plant breeders' rights) to consult with the TO before a plant breeder's rights are issued. Protection exceptions are made for wild plants which are not modified (Article 2). Likewise, within the concept of 'notoriously known variety' ('variedad notoriamente conocida' in Spanish) will be included all the varieties that are protected by community *sui generis* intellectual rights, whether those rights have been registered or not, in accordance with what is established in Articles 82 and 84 of the Biodiversity Law No. 7788, to the extent that the variety is adequately described and it is possible to verify its existence (Article 4).

To the present date, no patent applications have been identified that have made use of national genetic resources.[7] There was only one patent application which was consulted to the TO. A patent application (file No. 7415) named 'Substituted piridinonas as modulatories of p38 map quinase'. The summary is below:

> Formula 1 components and its pharmaceutically acceptable salts are described in which R1, R2, R3, R4 and R5 are described in this application. These components are useful for the treatment of illnesses and ailments caused by or exacerbated by p38 map unregulated quinase activity and/or FNT. Also described are pharmaceutical compositions that contain compounds, methods to prepare compounds and treatment methods that use the compounds.[8]

The TO issued a letter of response which commented, among other points, that in the forms used by the Registry there is no indication of the responsibility of the applicant to disclose the origin of the materials used in the invention. However, the TO concluded that the strains of *N. meningitidis* were originated in the United States, Germany and Holland. It also considered that the claims do not include any information that refers to the patenting of microorganisms, DNA sequences, nucleotoids, and/or amino acids as are found in nature. Therefore, this patent application does not infringe upon what is stipulated in BL.

The challenge for the implementation of the Article 80 consultation relates to the fact that the Registry is made up of specialized intellectual property lawyers that are not familiar with the Biodiversity Law. Another challenge is that this consultation is carried out at the substantive review of the application which delays the completion of the patent processing.

Furthermore, the issue of the relationship between ABS and IPR was raised in the process of enactment of the laws required to implement the CAFTA-DR.[9]

The Constitutional Court (Resolution No. 8–13832 of 2008) declared unconstitutional the last Project of the named 'Implementation Agenda' of the CAFTA-DR. The implementation Agenda included all the amendments or new laws needed to be in full compliance with the CAFTA-DR obligations.

The last draft laws (several amendments to the BL, Project No. 16955) to be approved by the Parliament were consulted by the Congressmen to the Constitutional Court. The most relevant modification of the Project was in connection to the content of Article 78 paragraph 6 of the BL (exclusions for the IPR protection).

The proposed amendment limited the exclusion provided in Article 76 only to patents and added the qualification 'as far as the invention does not meet the patentability requirements of the Law No. 6867 of 1983' (patent law).

The Constitutional Court resolution declared that the proposed modification affected the rights and interest of the indigenous peoples and therefore a consultation process should have been carried out in accordance with the ILO 169 Convention (especially Article 6).

Finally, the Draft Law was modified and the changes proposed in Article 78 were withdrawn. However, later on by the decree No. 34959 MINAET-COMEX of 2008, the proposed amendment was again introduced this time as an 'executive decree' (not a Law) regulating Article 78 paragraph 6.

A subsequent constitutional issue was raised against decree 34959 that was subsequently annulled by the Constitutional Court due to the lack of appropriate consultations with the relevant stakeholders (Resolution No. 18147–12 of 2012) reaffirming the previously held view of the Court on the relationships between IPR exclusions and the rights of the indigenous peoples. After the annulment of the decree the existing framework remains with the same wording of Article 78 of the BL with no particular consequences arising from the Court ruling.

Participation and relevance of the WIPO Committee for the national processes on TK, genetic resources and IPR

Negotiations in the IGC have eventually resulted in draft provisions on three international instruments for the protection of traditional knowledge, traditional cultural expressions (TCE), and intellectual property issues related to genetic resources. Many technical documents have also been produced in the IGC process over the last fifteen years on different issues of relevance for national processes to protect TK and to understand and define the relationship between IPR and genetic resources.

Costa Rica's participation in the Committee is scarcely documented. Negotiations are carried out by Geneva-based negotiators with apparently limited coordination with CONAGEBIO and national intellectual property authorities. In the case of CONAGEBIO,[10] it has been consulted on limited occasions particularly as far as the substantive issues under negotiation such as on traditional knowledge and genetic resources are concerned. Recently CONAGEBIO was consulted in preparations for IGC 29 held in February 2016 particularly with respect to the latest draft on genetic resources. CONAGEBIO has consistently reaffirmed the importance of the introduction of disclosure requirements in patent applications and has highlighted – as explained earlier – the need of full implementation of the principles embedded in Article 80 of the Biodiversity Law; in other words, that the outcome of the IGC should be supportive of the implementation of national laws on these matters.

Finally, the Committee's work in general can be useful for the development and implementation of the TK, IPR and genetic resources provisions of national legal regimes in the following aspects:

- WIPO has prepared an extensive number of documents on positive and defensive measures for the protection of TK. The country could also benefit from the extensive information and resources developed at the IGC in the areas of both TK and genetic resources. The technical inputs of the IGC could help in the implementation of outstanding and challenging issues facing Costa Rica. For instance, they could be relevant as technical inputs for national processes particularly in the case of the tool kit to document and catalog TK; the glossary of terms; the studies on legal options for protection of TK (including case studies and practical experiences); documents on customary law; technical studies on disclosure requirements; information on widely disseminated TK and the issue of public domain.
- IGC outcomes could provide additional conceptual and political support of the disclosure requirements established in the BL (especially Article 80). An inclusion of appropriate disclosure provisions in the text of the instruments under negotiation compatible with what is already provided for in the national legislation may contribute to solve some of the political difficulties encountered so far for its full implementation. The draft provisions on IPR and Genetic Resources[11] on subject matter of instrument (Article 1); disclosure requirements (Article 2) and sanctions and remedies (Article 45) – all bracketed – can provide the additional orientation needed to facilitate the technical implementation of the disclosure requirements established in the BL. Lack of further details on the functioning of the general provisions on disclosure of the BL has also prevented the full operationalization of the existing provisions, due to the concerns of the IP offices and other stakeholders on the insufficient development/lack of clarity of the disclosure requirements in national legislation. Furthermore, a clear formulation of this requirement in the international instrument to be adopted in light of the current mandate of the IGC, may reinforce the legal foundations for this mechanism, dissipating some remaining internal opposition to the requirements of the BL due to the potential conflict with IPR rules. In light of the final outcome of the Nagoya Protocol text on check-points of Article 17 (IPR offices are not mentioned as a check-point) the inclusion of disclosure requirements in the IGC text will provide additional support to the rationale and legal foundations of this mechanism.

- IGC draft provisions on TK[12] could also support conceptually and provide orientation for the development of a *sui generis* intellectual community rights system in the country since much of the content of the present draft is relevant for the development of domestic regimes. Some of the Draft Articles under discussion at the IGC are relevant and match the issues to be agreed in the participatory process envisaged in Article 82 (determination of the nature, scope, requirements) and of Article 85 (how the intellectual community *sui generis* right will be used and who are the beneficiaries) of the Biodiversity Law. The Draft Articles already have addressed these issues (again under brackets), for instance subject matter (Article 1); beneficiaries of protection (Article 2); criteria/scope for protection (Article 3). Other draft provisions are also relevant to guide the participatory process such as: sanctions and remedies (Article 4); exceptions and limitations (Article 6); term of protection (Article 7); transitional measures (Article 9); and transboundary cooperation (Article 12).

Conclusions

Costa Rica has developed, not free of difficulties and controversies, a sophisticated regime regarding the protection of traditional knowledge and access to genetic resources and benefit sharing, addressing particularly the relationship between genetic resources and intellectual property rights.

The country's participation in the IGC has not been adequately documented, nor is it possible to clearly determine what has been the precise stand regarding the various issues under negotiation.

However, the work done and the potential outcomes of the IGC can support and provide important inputs towards the improvement of the existing legal regime, to facilitate the finalization of the ongoing process and to overcome some of the technical and political difficulties encountered so far to fully implement critical provisions of the Biodiversity Law on these issues.

Notes

1 Only the University of Costa Rica developed their own Regulation of Access of 1999.

2 For a detailed study of the legal system, see 'El sistema sui generis para la protección de los conocimientos tradicionales: un estudio de la Ley de Biodiversidad de Costa Rica y normativa relacionada', by Jorge Cabrera Medaglia. Document submitted to Informal Panel of WIPO on national experiences about *sui generis* systems of protection. Geneva, 9 July 2003.

3 The CAFTA-DR was approved October 7, 2007 through referendum. The Law is No. 8622 published in La Gaceta on September 21, 2007.

4 One thesis holds that reforms to the Patent Law of 2000 tacitly derogated the exclusions of the Law of Biodiversity since they were promulgated later on; it excludes some, but not all, of the aspects provided in Article 78 from the patent process.

5 Law No. 8686 published in La Gaceta on November 26, 2008.

6 Decree No. 34958-MINAET-COMEX, regulation to Article 80 of the BL limits the opposition on the grounds of lack of compliance with the patent requirements.

Likewise, in the case of non compliance a fine is established. The processing of the applications seems to be allowed even if no certificate of compliance was produced. A constitutional action was brought against this regulation and the decree was also annulled by the Constitutional Court.

7 Personal communications of Karen Quesada from the Industrial Property Registry and of Maribelle Alvarez of the CONAGEBIO Technical Office.

8 Through the resolution from 8:07am on July 30, 2009, the application was rejected due to the lack of an invention unit, lack of clarity and because the claims are related to treatment methods which are excluded from patentability according to Article 1 of Law No. 6867.

9 A similar case occurred with the implementing legislation of the Agreement on Commercial Promotion between Peru and the United States (Law No. 29316 published on January 14 in the Official Peruvian Diary). Law No. 29316 has amended the substantive requirement of having an access contract for genetic resources with the Peruvian State or a license contract with the indigenous communities for the use of their traditional knowledge, as a condition for being able to obtain a patent that uses GR or TK. In the case that there is no contract, a sanction exists to penalize the applicant, but this is not a cause for annulment or invalidity of the patent. Decision 486 was modified in two fundamental moments of the IP process: the patent application and the declaration of annulment of patent that has already been granted.

10 Communications with Alejandra Martínez y Maribel Alvares staff of CONAGEBIO.

11 WIPO/GRTKF/IC/30/4. Consolidated document relating to Intellectual Property and Genetic Resources, Traditional Knowledge and Folklore.

12 WIPO/GRTKF/IC/31/4. The Protection of Traditional Knowledge: Draft Articles.

Reference

Cabrera Medaglia, Jorge, El sistema sui generis para la protección de los conocimientos tradicionales: un estudio de la Ley de Biodiversidad de Costa Rica y normativa relacionada. Documento presentado al Panel Informal de la OMPI sobre experiencias nacionales sobre sistemas sui generis de protección. Geneva, 9 July 2003.

18 New Zealand, *mātauranga Māori* and the IGC

Jessica C. Lai

Introduction

Since the turn of the millennium, New Zealand has actively sought innovative means to protect *mātauranga Māori* (the Māori knowledge system) in its intellectual property laws. Concurrently, the Waitangi Tribunal heard the Wai 262 Claim on *mātauranga Māori* and the Crown's Treaty of Waitangi obligations, and the Crown has continued to negotiate in the World Intellectual Property Organization (WIPO) Intergovernmental Committee on Intellectual Property and Genetic Resources, Traditional Knowledge and Folklore (IGC). This chapter looks at this complex constellation and argues that New Zealand both can and needs to address the Wai 262 Report and its domestic laws, before purporting to negotiate treaties on behalf of Māori, and to thereby act as a world leader in this arena.

New Zealand's position within the WIPO IGC is unique for many interrelated reasons. First, unlike many indigenous peoples, the Māori are not a disenfranchised people without political power. They play an important and present role in the everyday life, politics and economy of New Zealand. Second, New Zealand is typically considered to be relatively avant-garde when it comes to protecting traditional cultural expressions (TCEs), traditional knowledge (TK) and related genetic resources (GRs) within the realm of intellectual property (IP). Finally, and arguably most importantly, the Treaty of Waitangi (that regulates the relationship between Māori tribes and the Crown) is part of New Zealand's constitutional law. For reasons that will become clear, these factors shape New Zealand's international relationships, including within the IGC.

This chapter gives a brief overview of New Zealand's legal environment, including the role of the Treaty of Waitangi and its importance with respect to TCEs, TK and GRs. It then examines New Zealand's approach to international negotiations specifically with respect to Māori interests.

New Zealand's legal environment

Signed on behalf of the English Crown and Māori, the Treaty of Waitangi is the founding document of modern New Zealand and is a core part of its constitutional law (see Orange 1987). Though the Treaty has no direct legal effect unless implemented into domestic law, Māori view the Treaty as the 'main reference point for the recognition of respective roles, responsibilities and authority over cultural heritage, values and traditions' (Smith 2003, 41). The Treaty of Waitangi is interpreted chiefly through the Treaty of Waitangi Act 1975, which – according to its preamble – is an 'Act to provide for the observance, and confirmation' of the Treaty Principles, through the therein established Tribunal, which makes 'recommendations on claims relating to the practical application' of the said Principles.[1] The reference to the 'Principles' of the Treaty, rather than the Treaty provisions themselves, is partly due to the differences between the English and Māori texts, both of which are official versions (Treaty of Waitangi Act 1975, s 2 definition of 'Treaty'), and because the Treaty is not capable of literal application in modern contexts (Frankel 2011, 122). The Tribunal assesses issues that arise because of the two different texts and related Crown duties and omissions thereof (see also Chapter 11 by Solomon in this volume for a relevant Indigenous perspective).

According to the English text, Article 2 of the Treaty guarantees Māori 'the full exclusive and undisturbed possession of their Lands and Estates Forests Fisheries and other properties which they may collectively or individually possess so long as it is their wish and desire to retain the same in their possession'. The modern English translation of the Māori text, in contrast, guarantees 'unqualified exercise of their chieftainship [*tino rangātiratanga*] over their lands, villages and all their treasures [*taonga*]'.[2] The Waitangi Tribunal has made it clear that *taonga* includes all dimensions of a tribal group's estate, material and non-material heirlooms and *wahi tapu* (sacred places), ancestral lore and *whakapapa* (genealogies) (Waitangi Tribunal; Wai 38, 210). Though the Treaty does not mention IP or cultural heritage, it is clear that these can be considered *taonga* (Zografos 2005, 930; Gray 2005, 73).

In 1991, six *iwi* (tribes) – Ngāti Kuri, Ngāti Wai, Te Rarawa, Ngāti Porou, Ngāti Kahungunu and Ngāti Koata – filed the Wai 262 Claim. It is often referred to as the Māori claim over fauna and flora or IP, but it is far more encompassing than this. The Tribunal described it as being 'about mātauranga Māori – the unique Māori way of viewing the world, incorporating both Māori culture and Māori traditional knowledge. It is no stretch to describe this claim as being about the survival of Māori culture and its ongoing place in this country' (Wai 262, 1). Thus, the Claim extended to the 'knowledge system itself, and its internal mechanism for transmission, dissemination, tuition, and development' (Solomon 2005a, 358). The Claim and the Report challenged the right of the Crown to make decisions and set policies that affect Māori interests in their culture, including IP rights (IPRs), without consultation or shared decision-making (for discussion, see Zografos 2005, 935–936; 2010, 65–70).

Released in 2011, the Wai 262 Report recommended that no one should be able to use Māori cultural heritage in public in an offensive or derogatory manner (Wai 262, 93). This was recommended for *taonga* works or *mātauranga Māori* (Māori knowledge), both of which have *kaitiaki* (guardians), and also *taonga*-derived works, which do not have *kaitiaki*. Even if use is not offensive or derogatory, if it is of a *taonga* work or *mātauranga Māori*, the relevant *kaitiaki* would be able to challenge future commercial use, so long as there is no IPR that pre-dates the implementation of the recommendations (Wai 262, 89). In such situations, *kaitiaki* would have a right to consultation and possibly a right to consent to use.

Many of these recommendations would broaden existing New Zealand law that interfaces IP regimes with the Māori knowledge system (Frankel 2012). Under the current Trade Mark Act 2002, it is an absolute ground of refusal (and, thus, the Commissioner must refuse) to register a mark 'the use or registration of which would, in the opinion of the Commissioner, be likely to offend a significant section of the community, including Māori' (s 17(1)(c)). Through this, the Māori are specifically designated as a significant section of the community. The 2002 Act also established the Māori Trade Marks Advisory Committee,[3] which is forwarded any sign which 'is, or appears to be, derivative of a Māori sign, including text and imagery', to advise the Commissioner on whether it 'is, or is likely to be, offensive to Māori' (s 178; see Frankel 2005; Lai 2014, 99–110). Similarly, the Geographical Indications (Wines and Spirits) Registration Act 2006 was amended (before it came into force) to mirror the Trade Marks Act 2002, such that no geographical indicator can be registered if it would be likely to offend a significant section of the community, including the Māori (New Zealand 2015, 2).

The Wai 262 Report also recommended that a Māori Advisory Committee be set up to advise the Commissioner of Patents on TK, related GRs and Māori values. At the time the Report was released, such a Committee was already recommended in a Bill, which was realised in 2013 (Patents Act 2013, ss 225–228). Under the Patents Act 2013, the Māori Patents Advisory Committee has the task of advising the Commissioner of Patents in his/her decision of whether the 'commercial exploitation' of an alleged invention would be contrary to morality or *ordre public*. Specifically, it would advise on whether an alleged invention is derived from Māori TK or 'indigenous plants or animals' and, if so, whether 'the commercial exploitation of that invention is likely to be contrary to Māori values' (ss 275–278; see Lai 2014, 135–142).

The Tribunal additionally made the following recommendations that are not in the 2013 Act:

1 The Committee should also be able to advise the Commissioner on the requirements of patentability, e.g. on whether there is novelty and an inventive step;
2 The Committee must be able to advise the Commissioner on any existing and conflicting *kaitiaki* interests, even if the requirements to patentability are satisfied;
3 The Committee should maintain an advisory role, but should be able to advise the Commissioner as it sees fit (not just when requested) and should be able to prepare guidelines and protocols;
4 The Commissioner should be required to approach the Committee when an application raises 'Māori issues', rather than it being optional;
5 The Commissioner or his or her representative should sit jointly with the chairperson of the Committee to increase the expertise of the former; and
6 It should be mandatory that patent applicants disclose whether any *mātauranga Māori* or *taonga* species contributed towards the invention (Wai 262, 201 and 203–205).

The Wai 262 Report is not binding, but has symbolic and political force. Shortly after the Report was released, Attorney-General Christopher Finlayson (2011) and Prime Minister John Key (Chapman and McCammon 2011) stated that, though there is some emerging international consensus to assist the Government, any response should not be rushed as there were other areas of the Report that were novel and there was a need to consider and balance relevant public and private interests specific to New Zealand. At the time of writing, the Government had not officially responded to the Report. The New Zealand Government is unlikely to want to make any major changes before the status of the free-trade agreement (FTA) with the United States, Canada, Australia, Malaysia, Singapore, Japan, Mexico, Peru, Vietnam, Brunei and Chile (the Trans-Pacific Partnership Agreement (TPP), discussed further below) is settled[4] and before it knows the outcome of the IGC negotiations.

New Zealand in the IGC

New Zealand is known for its conscientiousness when it comes to its international obligations. That is, it does not sign international agreements that it knows it does not or cannot implement. This being the case, New Zealand is mindful of its Treaty obligations – including with respect to TK, GRs and TCEs – when it negotiates at the international level. The Government has acknowledged this on several occasions in different fora. For instance, in 2002, it noted in the World Trade Organization (WTO) forum that New Zealand is 'to some extent, in a unique position' regarding TK, because the Treaty is a part of New Zealand's constitutional law (MFAT 2002). Similarly, the New Zealand Government has indicated that though it partakes in international negotiations relating to TK, GRs and TCEs, ultimately New Zealand's key concerns are linked to the Treaty. In 2002, the New Zealand Delegation to the IGC stated that '[t]he Treaty must be foremost in our minds when considering the protection of traditional cultural expressions. In considering solutions we must maintain our ability to meet our Treaty obligations' (New Zealand 2002, para. 41; see also Solomon 2005b).

One sees this practice not only in the way New Zealand behaves in negotiations specifically relating to TK, GRs and TCEs, but also in its general FTAs (see also Chapter 19 by Roffe on preferential trade agreements, in this volume). New Zealand's FTAs frequently include clauses that allow the Government to give preferential treatment to the Māori in accordance with the Treaty of Waitangi, and to protect TK, GRs and TCEs.[5] We also see this in the IP chapter of the TPP, dated 26 January 2016. The TPP has a specific Article on 'Cooperation in the Areas of Traditional Knowledge' associated with GRs, with respect to understanding the issues involved and the relationship between TK associated with GRs and examination and prior art (Art. 18.16). More specifically, Annex 18-A to the IP Chapter, requiring New Zealand to accede and ratify the International Union for the Protection of New Varieties of Plants (UPOV) 1991 (para. 1), stipulates that this 'shall [not] preclude the adoption by New Zealand of measures it deems necessary to protect indigenous plant species in fulfilment of its obligations under the Treaty of Waitangi' (para. 2). The interpretation of the Treaty is not subject to the TPP's dispute settlement provisions (para. 4), nor is the consistency of any measures implemented by New Zealand to fulfil its Treaty's obligations with respect to its obligations to ratify UPOV 1991 (para. 3). Regarding paragraph 2, a dispute settlement panel can only be established to assess whether any measure implemented under it is inconsistent with a Party's rights under the TPP (para. 4).

Given its willingness to introduce novel laws and the central nature of the Treaty when it does, the outcome of the IGC deliberations are of limited importance to New Zealand in terms of potential changes to substantive national laws. Indeed, New Zealand has noted its concern that the IGC instruments not 'limit the ability of New Zealand to tailor its protection system to suit the particular domestic circumstances' (New Zealand 2007, 16). As an illustration, New Zealand is particularly conscious of its multi-culturalism, the importance of cultural interflow, freedom of expression and the development of the New Zealand identity (Wai 262, 74–77; Lai 2013). Moreover,

> it is also important to retain flexibility for countries to develop solutions and mechanisms appropriate to their own unique characteristics and circumstances. While the development of sui generis systems at the international level is an objective that many states support, this should not preclude the development of country or region-specific alternative approaches to protecting the knowledge and practices of indigenous communities. This is particularly important given the 'culturally distinctive' nature of TK and TCEs, and the possibility of other domestic legal sources of rights in relation to TK and TCEs that may need to be taken into account (e.g. indigenous and human rights, Treaty of Waitangi).
>
> (New Zealand 2007, 16)

As a result, New Zealand has favoured a 'menu of options approach' to the different instruments – as opposed to a 'one-size-fits-all' solution – including the use of non-IP or 'practical tools', such as codes, guidelines, checklists and model clauses (WIPO IGC 2004, paras 41 and 88).

Regardless of New Zealand's concerns regarding the retention of flexibility, its delegations have taken an active role in the IGC negotiations (e.g. WIPO General Assembly 2015). This is both with regard to the drafting of the various texts (e.g. WIPO IGC 2012) and also in ensuring indigenous involvement (e.g. WIPO IGC 2004, para. 17; WIPO Program and Budget Committee 2014).[6] New Zealand participates in part because it is in its interest to shape international law to its liking, rather than have it shaped by others. Moreover, the development of protection at the international level is of importance to New Zealand (as many misuses and misrepresentations of Māori TK, GRs and TCEs occur extra-territorially) and is something that New Zealand cannot bring about by itself. In written comments to the IGC, New Zealand noted that misuse and misrepresentation often occur because third parties are not aware of customary laws and protocols, which are often in common among different indigenous and local communities around the world, and – in this sense – the development of an international code of ethics, guidelines and/or best practice mechanisms could be greatly beneficial (New Zealand 2007, 15–16) (see Chapter 10 by Tobin on customary law and Chapter 6 by Robinson and Chiarolla on codes of conduct in this volume).

Furthermore, the IGC is seen as a means to develop extra-territorial protection via the introduction of concepts like reciprocity, national treatment and most-favoured nation into the realm of TK, GRs and TCEs (New Zealand 2007, 17) (see Chapter 21 by Frankel which discusses national treatment, in this volume). Highlighting the central nature of the Treaty, in the same written comments, New Zealand stated that certain protection of TK, GRs and TCEs stemming from Treaty obligations might not be reciprocated (New Zealand 2007, 17). However, this could be difficult given that, at the time of writing, national treatment was the means of international protection used in the drafts for TCEs and TK (WIPO IGC 2014a, Art. 11; 2014b, Art. 11). A further beneficial outcome of the IGC proceedings could be the introduction of disclosure requirements in patent law regarding TK and GRs (WIPO IGC 2014c, Art. 3), which is seen as a means of bridging the divide between the Agreement on Trade-related Aspects of Intellectual Property Rights (TRIPS)[7] and the Convention on Biological Diversity (CBD)[8] (Frankel 2011, 131; Heath and Weidlich 2003, 80–81), and which New Zealand has stated belongs within the jurisdiction of WIPO (ICTSD 2009, 3) (see also Chapter 5 by Bagley on disclosure, in this volume). While such a requirement was not included in the Patents Act 2013, the Wai 262 Report recommended that patent applicants should have to disclose: (1) 'the source and country of origin of any genetic or biological resource that contributed in any material way to the invention'; and (2) 'mātauranga Māori that was used in the course of research, including traditional knowledge that is not integral to the invention but that prompted the inventor to take the course of research that led to the relevant patent application' (Wai 262, 203–205).

New Zealand's contributions to the substantive negotiations have waned in recent years, as has the progress of the IGC. The result has somewhat spiralled as New Zealand has refused to sign the Nagoya Protocol to the CBD[9] until it has resolved the Wai 262 Claim (MFAT 2011), which is in turn dependent on the outcome of the IGC and TPP, which are partly informed by New Zealand's Treaty of Waitangi obligations, which are reflected in the Wai 262 Report. Overall, New Zealand seems to be dragging its feet on the matter. Arguably, because all roads seem to lead to the Treaty, New Zealand needs to first deal with the issues domestically – that is, address the Wai 262 Report – and go from there. Indeed, the Presiding Officer of the Wai 262 Claim and Report, Justice Joe Williams, specifically highlighted that New Zealand has the opportunity to be a 'first mover in international law reform' by addressing the Report and using this to lead the international debate, in his letter delivering the Report to the Crown (Wai 262, xxv–xxvi). What is more, the Government's international approach to TK, GRs and TCEs – namely, negotiating within the IGC with little consultation with or direct input from Māori – is arguably contrary to its Treaty obligations.

As stressed in the Wai 262 Report, though Article 1 of the Treaty granted the Crown the right to govern (*kāwanatanga*), which includes the right to represent New Zealand abroad and make foreign policy, Māori tribes were guaranteed Crown protection of their interests (Wai 262, 680). According to the Tribunal, the Crown and Māori are in a partnership (Wai 262, 14–19). The Report strongly stated that the Crown is consequently obliged to be in constant dialogue with Māori with respect to the negotiations of international instruments, because these affect Māori interests in trade and economic development, natural resources, the protection and transmission of Māori culture and TK, indigenous rights and environmental protection (Wai 262, 680–681). The level of engagement with Māori should be on a sliding scale, dependent on the importance of the instrument to Māori and the nature and extent of the likely impact on them. The Report states that 'Māori must have a say in identifying the interest and devising the protection' and, in some rare cases where Māori interests are so overwhelming and other interests narrow or limited, the Crown should delegate its negotiating and decision-making powers to Māori and 'place the Māori voice as the New Zealand voice in the international arena' (Wai 262, 681–682 and 685). There can be no doubt that the IGC negotiations are high on the sliding scale of importance to and impact on Māori, which at the least justifies a great deal of communication between the Crown and Māori, and co-decision-making and consensus on New Zealand's stance in the IGC.

Finally, one should not view the IGC as a means to circumvent the domestic debate. This is particularly because the instruments – being creatures of compromise – will not meet the interests of every indigenous or traditional community. Relatedly, the instruments contain broadly drafted minimum standards that could be implemented any number of ways. It is not even clear if the instruments will be binding. Even if they are, New Zealand might not sign them. While New Zealand supported the continuance of the normative mandate of the IGC in 2015 (Saez and New 2015), it has always expressed reservation about the need for binding instruments and concerns about loss of flexibility (e.g. WIPO IGC 2009, paras 47, 76 and 185). Moreover, even if they are binding and New Zealand does sign onto them, New Zealand arguably either already meets the encapsulated standards or would meet them if it were to implement the Wai 262 recommendations,[10] such that one cannot maintain that New Zealand must wait until the finalisation of the WIPO instruments before dealing with national law and policy.

Conclusion

New Zealand has partaken in the IGC since its inception, despite the Wai 262 Claim and resultant Report. This is in itself not problematic. As discussed in this chapter, New Zealand has much to gain from international cooperation and protection. However, although New Zealand has proven that it can be avant-garde at the national level, rather than using the international stage to emphasise the partnership that the Treaty created and being a world leader in the debate on TK, GRs and TCEs, the Government seems to be using its international negotiations to avoid dealing with domestic law and policy. Furthermore, it is arguably doing so in a manner inconsistent with the partnership that was meant to be created between Māori and the Crown by the Treaty. This is arguably disingenuous and futile, as the Wai 262 Report is not going away and New Zealand is missing the opportunity to act as an example to the world of how a post-colonial nation can move beyond talk of grievances towards multicultural unity.

Notes

1 Though only recommendations, governments have usually taken the reports seriously. However, the Crown has rejected a number of Waitangi Tribunal reports; see Charters 2006.

2 For the English and modern translation of the Māori text of the Treaty of Waitangi, see MCH (with explanatory footnotes by Professor Hugh Kawharu).

3 The history of the establishment of the Committee is described in IPONZ 2010, Chapter 16.2.

4 The TPP was finalised on 5 October 2015 and signed on 4 February 2016.

5 E.g. New Zealand–Hong Kong, China Closer Economic Partnership (CEP) (signed on 29 March 2010, entered into force 1 January 2011), Chapter11 Art. 8 and Chapter 19 Art. 3; New Zealand–Malaysia FTA (signed on 26 October 2009), Arts 11.4, 11.6 and 17.6; ASEAN, Australia–New Zealand FTA (signed on 27 February 2009, entered into force 1 January 2010), Chapter 13 Art. 8 and Chapter 15 Art. 5; New Zealand–China FTA (signed on 7 April 2008, entered into force 1 October 2008), Art. 165 and 205; New Zealand–Thailand CEP (signed on 19 April 2005, entered into force 1 July 2005), Arts 12.5(d) and 15.8.

6 The IGC has played an important role as a forum for indigenous peoples to air their concerns and to place their hope. This has not gone without controversy, with many indigenous peoples unhappy with only having observer status and states dominating the debates and drafting (Gordon 2014; Saez 2012), and the economic-centric and IP-thematic nature of the drafts (Morgan 2014, 321–323).

7 Marrakesh Agreement Establishing the World Trade Organization, Annex 1C, 1869 UNTS 299; 33 ILM 1197 (adopted on 15 April 1994, entered into force 1 January 1995). TRIPS contains minimum standards for IP.

8 1760 UNTS 79; 31 ILM 818 (opened for signature 5 June 1992, entered into force 29 December 1993). The CBD contains obligations for TK relevant for the conservation and sustainable use of biological diversity (Art. 8(j)).

9 The Nagoya Protocol contains stronger obligations for TK relating to GRs and GRs per se (Arts 5.2, 7 and 12).

10 This is of course extremely difficult to say with any certainty due to the constant changes made to the drafts, and all the alternatives and bracketed text.

Bibliography

Chapman, Kate and McCammon, Belinda. 2011, 4 July. 'Govt Take Time to Considering Wai 262 Report', *stuff.co.nz National*, available at www.stuff.co.nz/national/politics/5228332/Govt-to-take-time-considering-Wai-262-report (accessed on 10/1/2017).

Charters, Claire 2006. 'An Imbalance of Powers: Māori Land Claims and an Unchecked Parliament', *Cultural Survival: Bridging the Gap Between law and Reality*, 30(1), available at www.culturalsurvival.org/publications/cultural-survival-quarterly/imbalance-powers-maori-land-claims-and-unchecked (accessed on 10/1/2017).

Finlayson, Christopher (Attorney General). 2011, 2 July. 'Government Receives Waitangi Tribunal Report', *Press Release: Government*, available at www.beehive.govt.nz/release/government-receives-waitangi-tribunal-report (accessed on 10/1/2017).

Frankel, Susy. 2005. 'Third-Party Trade Marks as a Violation of Indigenous Cultural Property: A New Statutory Safeguard', *Journal of World Intellectual Property*, 8, 83–98.

Frankel, Susy. 2011. *Intellectual Property in New Zealand*, Wellington, New Zealand: LexisNexis New Zealand.

Frankel, Susy. 2012. 'A New Zealand Perspective on the Protection of Mātauranga Māori (Traditional Knowledge)', in Christoph B. Graber, Karolina Kuprecht and Jessica C. Lai (eds), *International Trade in Indigenous Cultural Heritage: Legal and Policy Issues*, Cheltenham, UK: Edward Elgar, 439–459.

Gordon, Veronica. 2014. 'Appropriation without Representation? The Limited Role of Indigenous Groups in WIPO's Intergovernmental Committee on Intellectual Property and Genetic Resources, Traditional Knowledge and Folklore', *Vanderbilt Journal of Entertainment and Technology Law*, 6, 629–667.

Gray, Earl. 2005. 'Māori Culture and Trade Mark Law in New Zealand', in Christopher Heath and Anselm Kamperman Sanders (eds), *New Frontiers of Intellectual Property. IP and Cultural Heritage – Geographical Indicators – Enforcement – Overprotection*, Oxford and Portland, Oregon: Hart Publishing, 71–96.

Heath, Christopher and Weidlich, Sabine. 2003. 'Intellectual Property: Suitable for Protecting Traditional Medicine?', Intellectual Property Quarterly, 1, 69–96.

ICTSD (International Centre for Trade and Sustainable Development). 2009, 17 June, 'Way Forward on IP Issues at the WTO Still Unclear', *Bridges*, 13(22), 3–5.

IPONZ (Intellectual Property Office of New Zealand). 2010, 26 January. Trade Marks Practice Guidelines 2010.

Lai, Jessica C. 2013. 'Bicultural Art: Offensive to the Māori or Acceptable Freedom of Expression? Wai 262, the CCPR and NZBORA', *Australian Journal of Human Rights*, 19(2), 47–92.

Lai, Jessica C. 2014. *Indigenous Cultural Heritage and Intellectual Property Rights: Learning from the New Zealand Experience?* Heidelberg: Springer.

MCH (Ministry of Cultural Heritage) (with explanatory footnotes by Professor Hugh Kawharu). 'Te Tiriti o Waitangi – The Treaty of Waitangi', available at www.tiritiowaitangi.govt.nz/treaty/translation.pdf (accessed on 19 April 2012).

MFAT (Ministry of Foreign Affairs and Trade). 2002, 17–19 September. 'New Zealand and the World Trade Organisation. Intellectual Property. The Protection of Traditional Knowledge. Intervention by New Zealand at the TRIPS Council'.

MFAT, 'Convention on Biological Diversity – Nagoya Protocol on Access to Genetic Resources and the Fair and Equitable Sharing of the Benefits Arising from their Utilization', available at www.treaties.mfat.govt.nz/search/details/p/41/1190 (accessed on 10/1/2017).

Morgan, Owen. 2014. 'Assessing the Work of the WIPO IGC', *European Intellectual Property Review*, 36(5), 319–326.

Nagoya Protocol on Access to Genetic Resources and the Fair and Equitable Sharing of Benefits Arising from their Utilization to the Convention on Biological Diversity (UN Doc. UNEP/CBD/COP/DEC/X/1) (adopted on 29 October 2010).

New Zealand. 2002. 'Specific Legislation for the Legal Protection of Traditional Cultural Expressions – Experiences and Perspectives of New Zealand', in WIPO IGC, Secretariat, 'Presentations on National and Regional Experiences with Specific Legislation for the

Legal Protection of Traditional Cultural Expressions (Expressions of Folklore)' (WIPO Doc. WIPO/GRTKF/IC/4/INF/2, 2002), Annex II.

New Zealand. 2007. 'Written Comments on the List of Issues for the Protection of Traditional Knowledge and Traditional Cultural Expressions/Expressions of Folklore' (WIPO/GRTKF/IC/10/7 Prov.), Annex I.

New Zealand. 2015. 'Opening Statement of New Zealand', Assemblies of the Member States of WIPO, 55th Series of Meetings.

Orange, Claudia. 1987. *The Treaty of Waitangi*, Wellington, New Zealand: Bridget Williams Books Limited.

Saez, Catherine. 2012, 22 February. 'Indigenous Peoples Walk Out Of WIPO Committee On Genetic Resources', *Intellectual Property Watch*, available at www.ip-watch.org/2012/02/22/indigenous-peoples-walk-out-of-wipo-committee-on-genetic-resources/ (accessed on 26 October 2015).

Saez, Catherine and New, William. 2015, 15 October. 'WIPO Assembly Extends Talks On Traditional Knowledge, Design; Sets Policy For New Offices', *Intellectual Property Watch*, available at http://www.ip-watch.org/2015/10/15/wipo-assembly-extends-talks-on-traditional-knowledge-design-sets-policy-for-new-offices/ (accessed on 5 February 2016).

Smith, Huhana. 2003. 'A Comment on *The Politics of Māori Image and Design*', *He Pukenga Kōrero*, 7 (1), 38–41.

Solomon, Maui. 2005a. 'Protecting Māori Heritage in New Zealand', in Barbara T. Hoffmann (ed.), *Art and Cultural Heritage: Law, Policy, and Practice*, New York: Cambridge University Press, 352–362.

Solomon, Maui. 2005b. 'Peer Review Report on WIPO Documents: The Protection of Traditional Cultural Expressions/Expression of Folklore: Revised Objectives and Principles' (WIPO/GRTKF/IC/8/4); and 'The Protection of Traditional Knowledge: Revised Objectives and Principles' (WIPO/GRTKF/IC/8/5).

Waitangi Tribunal. 'Treaty of Waitangi', available at www.waitangi-tribunal.govt.nz/treaty/ (accessed on 8 June 2012).

Wai 38, 1992. *Report of the Waitangi Tribunal on Te Roroa.*

Wai 262, 2011. *Ko Aotearoa Tēnei. Report of the Waitangi Tribunal on Claims Concerning New Zealand Law and Policy Concerning New Zealand Law and Policy Affecting Māori Culture and Identity.*

WIPO General Assembly. 2015. 'Matters Concerning the Intergovernmental Committee on Intellectual Property and Genetic Resources, Traditional Knowledge and Folklore (IGC): Proposal of the Holy See, Kenya, Mozambique, New Zealand, Norway and Switzerland to the WIPO General Assembly', 47th (22nd Ordinary) Session (WO/GA/47/18).

WIPO IGC. 2004. 'Report', 6th Session (WIPO/GRTKF/IC/6/14).

WIPO IGC. 2009. 'Report', 14th Session (WIPO/GRTKF/IC/14/12).

WIPO IGC. 2012. 'Report', 19th Session (WIPO/GRTKF/IC/19/12).

WIPO IGC. 2014a. 'The Protection of Traditional Cultural Expressions: Draft Articles', 28th Session (WIPO Doc. WIPO/GRTKF/IC/28/6).

WIPO IGC. 2014b. 'The Protection of Traditional Knowledge: Draft Articles', 28th Session (WIPO Doc. WIPO/GRTKF/IC/28/5).

WIPO IGC. 2014c. 'Consolidated Document Relating to Intellectual Property and Genetic Resources', 28th Session (WIPO Doc. WIPO/GRTKF/IC/28/4).

WIPO Program and Budget Committee. 'Participation of Representatives of Accredited Indigenous and Local Communities in the Work of the Intergovernmental Committee on

Intellectual Property and Genetic Resources, Traditional Knowledge and Folklore (IGC): Proposal for Subsidiary Funding from the Regular Budget of WIPO', 22nd Session (WO/PBC/22/24).

Zografos, Daphne. 2005. 'New Perspectives for the Protection of Traditional Cultural Expressions in New Zealand', *International Review of Intellectual Property and Competition Law*, 36(8), 928–952.

Zografos, Daphne. 2010. *Intellectual Property and Traditional Cultural Expressions*, Cheltenham, UK: Edward Elgar.

Part 4

The future of the IGC

19 Preferential trade agreements and the protection of genetic resources and associated traditional knowledge

Pedro Roffe

The World Intellectual Property Organization (WIPO)'s Intergovernmental Committee on Intellectual Property and Genetic Resources, Traditional Knowledge and Folklore (IGC) plays a role in the delicate search for new multilateral principles and rules on genetic resources and associated traditional knowledge (GRAATK). Since its establishment in 2000, however, important normative activity on intellectual property has taken place in bilateral and plurilateral negotiations. Countries advocating the need for multilateral rules on GRAATK at the IGC have also entered in what we could generally characterise as preferential trade agreements (PTAs)[1] incorporating dedicated chapters on intellectual property (IP) rights.

In contrast to the slow or sporadic progress made at the multilateral level on IP normative instruments, particularly since entry into force of the WTO's Agreement on Trade-Related Aspects of Intellectual Property Rights (TRIPS) in 1995, a large number of PTAs with IP provisions have been negotiated in the last two decades. PTAs are today probably the most active venue in the formation of new IP principles and norms and, in most cases, exceed what was agreed on protection and enforcement in the TRIPS Agreement.

Major trading partners, particularly the USA and the EU, are among those most frequently including IP chapters in trade negotiations. The European Free Trade Association (EFTA), Japan, China and Taiwan have also been dynamic players in including IP as an important component of their trade agreements. PTAs have dealt occasionally with GRAATK. In more recent agreements, particularly those where the EU is a party, PTAs display receptiveness to some of the concerns raised by developing countries multilaterally on the relationship principally between environmental agreements and the international IP system. It is difficult, however, to discern a common pattern in diversity of PTAs negotiated in recent decades (Vivas and Oliva, 2010).

This chapter looks at the approaches followed in recent PTAs in dealing with IGC associated issues and particularly with respect to the nexus between IP rights, genetic resources, and traditional knowledge. It will review trends and the approaches in selected PTAs on GRAATK issues. A major question is how relevant and how harmonious the treatment of these issues is compared to the aspirations raised by developing countries in dealing with GRAATK at the multilateral level.

The United States of America (USA)

The USA has systematically sought since the early 1970s to place the respect of intellectual property rights of its own citizens high on the international trade agenda. Contemporaneous to the TRIPS Agreement, the USA negotiated the North America Free Trade Agreement (NAFTA) with Canada and Mexico where the IP chapter makes the former's provisions sharper and more specific. Since NAFTA, the US has negotiated more than twenty PTAs (USTR, 2016) with a range of countries where the IP chapters are intended to reflect the level of protection found in US legislation. Another example in this vein is the recently concluded Trans-Pacific Partnership (TPP) agreement.

Genetic resources and traditional knowledge have not featured prominently in most of the trade agreements where the USA is a party. However, in discussions with Colombia, Panama, and particularly with Peru, the treatment of GRAATK became an important subject in negotiations of their respective PTAs. Peru has been a traditional *demandeur* on these issues at the multilateral front both in the WTO and the WIPO IGC. While genetic resources and associated traditional knowledge were not finally incorporated in the IP chapters concluded between the US and these three Latin American countries, these are the subject of specific annexes in the form of side letters that gave credit to the significance of these issues particularly for the *demandeur* countries.

The joint understanding with Colombia and Peru recognises 'the importance of traditional knowledge and biodiversity, as well as the potential contribution of traditional knowledge and biodiversity to cultural, economic, and social development'. It further recognises in this context – even if the USA is not party to the Convention on Biological Diversity (CBD) and its accompanying Nagoya Protocol – the need for: (i) obtaining informed consent from the appropriate authority prior to accessing genetic resources under the control of such authority; and (ii) the equitable sharing of the benefits arising from the use of GRAATK.

The joint understanding also highlights that promoting patent quality examination would ensure that the conditions of patentability are satisfied. At the same time it stresses that 'access to genetic resources or traditional knowledge, as well as the equitable sharing of benefits that may result from use of those resources or that knowledge, can be adequately addressed through contracts that reflect mutually agreed terms between users and providers' (Understandings Regarding Biodiversity and Traditional Knowledge, SICE, 2016).[2]

It is worthwhile noting the different approach followed in the case of the USA–Panama agreement in the form of a side letter attached to the agreement. The side letter – less explicit in terms of principles than in the case of Colombia and Peru – recognises the importance of traditional knowledge and folklore and to that end the parties 'seek to work together in consulting on issues and positions in the World Intellectual Property Organization Intergovernmental Committee on Intellectual Property and Genetic Resources, Traditional Knowledge, and Folklore in addressing matters related to traditional knowledge and folklore' (Letter on Traditional Knowledge, SICE, 2016).[3] It is also interesting to compare this approach with the PTA signed between Panama and Taiwan, discussed later.

The recently negotiated mega-regional TPP includes in its IP chapter provisions on traditional knowledge under the rubric of 'cooperation'. It is less aspiring than the joint understanding reached with Peru, which is also a Party to the TPP. It includes a general recognition of 'the relevance of intellectual property systems and traditional knowledge associated with genetic resources [to the Parties] when that traditional knowledge is related to those intellectual property systems' (USTR, 2016).

The IP chapter of the TPP needs to be seen also in the light of provisions in the same agreement on the environment (Chapter 20). The latter in dealing with trade and biodiversity recognises the importance of facilitating access to genetic resources within the respective national jurisdictions of the Parties, 'consistent with each Party's international obligations'. In this context the TPP acknowledges 'that some Parties require, through national measures, prior informed consent to access such genetic resources in accordance with national measures and, where such access is granted, the establishment of mutually agreed terms, including with respect to sharing of benefits from the use of such genetic resources, between users and providers' (Article 20.13.4, TPP, USTR, 2016). In other words, the TPP recognises the significance of facilitating access to genetic resources but it limits itself to reiterate that some – not all – members are bound by international obligations, of prior informed consent and mutually agreed terms and benefit sharing in accordance with national measures. This, however, is by itself an important acknowledgement for *demandeurs* in the TPP.

In more precise terms the TPP places emphasis on improving the quality of patent examination by taking into consideration prior art found in relevant publicly available documented information related to traditional knowledge associated with genetic resources. The TPP would also see parties endeavour to provide an opportunity for third parties to cite, in writing, the competent examining authority prior art disclosures that may have a bearing on patentability, including prior art disclosures related to traditional knowledge associated with genetic resources. Databases or digital libraries containing traditional knowledge associated with genetic resources should be used if applicable and cooperation undertaken in the training of patent examiners in the examination of patent applications related to traditional knowledge associated with genetic resources. (See Art. 18.16.3, TPP, USTR, 2016.)(See also Chapter 6 by Robinson and Chiarolla on databases, in this volume.)

Following the adoption of the TPP, Australia exchanged letters, with Malaysia and Peru underlining the recognition that access to traditional knowledge associated with genetic resources, as well as the equitable sharing of benefits that may result from the utilisation of that traditional knowledge, 'may be addressed through contracts that reflect mutually agreed terms between users and providers.' The exchange of letters made in February 2016 at the time of signing of the TPP in Auckland, New Zealand, constitutes a Memorandum of Understanding between these parties.[4] These two memorandums bear a resemblance to the joint understanding between Peru with the USA, as discussed earlier.

European Union

The European Union has been aligning its policies towards third countries with respect to trade agreements and IP. In earlier agreements negotiated around the same time as the TRIPS Agreement, the main objective was to demand from trading partners 'conformity with the highest international standards' of IP, and adherence to a number of international treaties administered by WIPO (European Union, 2016). In most recent years, following the new EU strategy towards third countries (Intellectual Property Strategy, 2016), a more aggressive agenda on IP enforcement has prevailed that in many respects mirrors the style of the US. In the case of GRAATK, however, a more sympathetic approach to the concerns of trading partners from developing countries may be surmised most notably through references to the disclosure requirement of the origin or source of genetic resources and associated traditional knowledge used in patent applications.

The CARIFORUM-EU free trade agreement was the first of a new generation of trade agreements negotiated by the EU with third countries that acknowledges the need to take into account trading partners' national approaches with respect to the defensive and positive protection of traditional knowledge. Signed in 2008, the CARIFORUM Agreement states that 'subject to their domestic legislation' signatory parties:

> respect, preserve and maintain knowledge, innovations and practices of indigenous and local communities embodying traditional lifestyles relevant for the conservation and sustainable use of biological diversity and promote their wider application with the involvement and approval of the holders of such knowledge, innovations and practices and encourage the equitable sharing of the benefits arising from the utilization of such knowledge, innovations and practices.
>
> (Article 150.1, CARIFORUM-EU, SICE, 2016)

In this context, CARIFORUM recognises the importance of adopting measures – subject to national legislation – to preserve such traditional knowledge, and that patent provisions in the agreement and the CBD shall be implemented in a mutually supportive way.

In a more categorical way, the CARIFORUM-EU Agreement acknowledges that parties may demand in patent applications using biological material 'as a necessary aspect of the invention' the disclosure or identification of the sources of the biological material used in the invention and described in the patent application.

With respect to the evolution of international norms in this area, the parties agree to work towards the 'development of internationally agreed sui generis models for the legal protection of traditional knowledge' and to regularly exchange views and cooperation in the framework of WIPO IGC. As noted above, a similar kind of cooperation is embedded in the PTA signed by the USA and Panama. However, one difference is the reference in CARIFORUM that comparable cooperation should also prevail in the case of the WTO 'on the issues related to the relationship between the TRIPS Agreement and the Convention on Biological Diversity, the protection of traditional knowledge and folklore' (CARIFORUM, Article 150.5, SICE 2016).

As in the case of the CARIFORUM-EU Agreement, subsequent PTAs negotiated with Peru, Colombia and more recently with Ecuador (European Union, 2016), the respective IP chapters provide for the link with the CBD on fair and equitable benefit sharing in relation to intellectual property rights arising from the use of genetic resources and associated traditional knowledge. The trade agreements also recognise, for the sake of transparency, the usefulness of the disclosure requirement of the origin or source of genetic resources and associated traditional knowledge used in patent applications (see Chapter 5 by Margo A. Bagley in this volume for more details on a disclosure of origin patent requirement).

More explicit than in the case of the CARIFORUM-EU Agreement, the latest PTAs with Colombia and Peru provide that with respect to the effects of non-disclosure this should be subject to the domestic law of the respective party (SICE, 2016).[5]

In 2012 the EU signed a free trade agreement with the Central American countries (SICE, 2016) that, contrary to the previous efforts, dealt with genetic resources and traditional knowledge in the introductory provisions of the IP chapter (nature and scope of obligations). A brief statement reiterates the recognition of states' sovereign rights over natural resources and access to their genetic resources in accordance with the CBD. To this end, nothing should prevent parties from 'adopting or maintaining measures to promote the conservation of biological diversity, the sustainable utilisation of its components and the fair and equitable participation in the benefits arising from the utilisation of genetic resources' (Article 229.4, EU-Central American countries, SICE, 2016).[6]

European Free Trade Association (EFTA)

EFTA has concluded twenty-five free trade agreements including a good number of developing countries (EFTA, 2016). While these agreements incorporate provisions dealing with IP, only in agreements negotiated after 2010 can one detect provisions dealing directly with GRAATK. In these latest PTAs, EFTA has in general followed the approach taken more recently by the EU, as discussed above. EFTA's PTA with Colombia, for example, uses a similar model found in the EU's PTA with the same country. The EFTA-Colombia agreement refers specifically to the legitimacy, according to national law, of the declaration of the origin or source of a genetic resource, to which the inventor or the patent applicant has had access. It emphasises also that, in accordance with national law, each Party has sovereign rights to provide for administrative, civil or criminal sanctions if the inventor or the patent applicant 'wilfully makes a wrongful or misleading declaration of the origin or source'.

Contrary to provisions in some US and EU PTAs, the EFTA agreements do not make specific reference to the ongoing IGC process. (See for details Box 19.1 on the EFTA-Colombia trade agreement, SICE 2016.)

Box 19.1: EFTA-Colombia

Article 6.5

Measures Related to biodiversity

1 The Parties reaffirm their sovereign rights over their natural resources and recognise their rights and obligations as established by the Convention on Biological Diversity with respect to access to genetic resources, and to the fair and equitable sharing of benefits arising out of the utilisation of these genetic resources.
2 The Parties recognise the importance and the value of their biological diversity and of the associated traditional knowledge, innovations and practices of indigenous and local communities. Each Party shall determine the access conditions to its genetic resources in accordance with the principles and provisions contained in applicable national and international law.
3 The Parties recognise past, present and future contributions of indigenous and local communities and their knowledge, innovations and practices to the conservation and sustainable use of biological and genetic resources and in general the contribution of the traditional knowledge of their indigenous and local communities to the culture and economic and social development of nations.
4 The Parties shall consider collaborating in cases regarding non-compliance with applicable legal provisions on access to genetic resources and traditional knowledge, innovations and practices.
5 According to their national law, the Parties shall require that patent applications contain a declaration of the origin or source of a genetic resource, to which the inventor or the patent applicant has had access. As far as provided for in their national legislation, the Parties will also require the fulfilment of prior informed consent (PIC) and they will apply the provisions set out in this Article to traditional knowledge as applicable.
6 The Parties, in accordance with their national laws, shall provide for administrative, civil or criminal sanctions if the inventor or the patent applicant wilfully make a wrongful or misleading declaration of the origin or source. The judge may order the publication of the ruling.
7 If the law of a Party so provides:

(a) access to genetic resources shall be subject to the prior informed consent of the Party that is the Party providing the genetic resources; and
(b) access to traditional knowledge of indigenous and local communities associated to these resources shall be subject to the approval and involvement of these communities.

8 Each Party shall take policy, legal and administrative measures, with the aim of facilitating the fulfillment of terms and conditions for access established by the Parties for such genetic resources.
9 The Parties shall take legislative, administrative or policy measures, as appropriate, with the aim of ensuring the fair and equitable sharing of the benefits arising from the use of genetic resources or associated traditional knowledge. Such sharing shall be based on mutually agreed terms.

EFTA's PTA with Peru includes identical principles to those found in its deal with Colombia, with the sole exception of an additional provision in the former affirming and recognising existing parties' rights and obligations under the International Treaty on Plant Genetic Resources for Food and Agriculture of the Food and Agriculture Organization.[7]

EFTA has also more recently (2013–2015) negotiated a PTA with Central American states Costa Rica, Guatemala and Panama. The agreement, following an earlier model, includes provisions on IP in Annex (XIX). Article 9 of the Annex, dealing with measures related to the protection of biodiversity and traditional knowledge, is not as detailed as in the case of the EFTA-Colombia and Peru. But, it reaffirms state sovereign rights over natural resources and recognises the rights and obligations under 'the Convention on Biological Diversity with respect to access to genetic resources and to the fair and equitable sharing of benefits arising out of the utilisation of these genetic resources' (EFTA, 2016).

The conditions of access and benefit-sharing are subject to the principles and legal framework of 'the relevant domestic legislation, international provisions and instruments' (Article 9.2, Annex XIX, EFTA-Central American States, SICE, 2016). The parties acknowledge the importance and value of knowledge, innovations and practices of indigenous and local communities, and their past, present and future contributions to the conservation and sustainable use of biological and genetic resources. The PTA is not, however, specific on measures that could be adopted to protect biodiversity but reiterates the commitment, in accordance with domestic law, 'to respect, preserve and maintain knowledge, innovations and practices of indigenous and local communities'.

There are no provisions dealing with GRAATK in other PTAs negotiated by EFTA in the past, such as in the case of Morocco (1999), Jordan (2002), Egypt (2007), and the Southern African Customs Union (SACU) States (2008) (EFTA, 2016).

Other major trading partners

In agreements negotiated by China, New Zealand and Taiwan, references are made to GRAATK, but in restrained terms, cautious not to encroach on ongoing international processes.

China has signed a number of PTAs that include provisions on IP. According to the China FTA Network, at the time of writing the country has negotiated thirteen FTAs and eight are under negotiation (Mofcom, 2016). As noted by Peter Yu, China PTAs 'are very different from EPAs or the FTAs that the European Union and the United States established. While [PTAs] do not seek to export Chinese laws to the signatory countries, China developed the agreements with specific objectives and its self-interests in mind' (Yu, 2011, 1027). Again, in the case of agreements signed with Latin American countries, reference is included to the treatment of GRAATK.

The agreement with Chile[8] includes a reference to GRAATK on IP-related cooperation matters. More explicit provisions address the subject matter in recent deals negotiated with Costa Rica and Peru. In both agreements the narrative is almost identical: reaffirmation of adherence to the CBD and the need to establish a supportive relationship with the TRIPS Agreement. The PTAs are not explicit on disclosure of origin or source, however, leaving the matter to future developments of domestic laws and the 'outcome of negotiations in multilateral fora'. (See, as an example, the 2010 agreement with Costa Rica, Box 19.2.)

New Zealand has concluded sixteen PTAs with members of the WTO (New Zealand, 2016). Reference to GRAATK is only found in one of these.

Box 19.2: China–Costa Rica

Article 111

Genetic resources, traditional knowledge and folklore

1 The Parties recognize the contribution made by genetic resources, traditional knowledge and folklore to scientific, cultural and economic development.
2 The Parties acknowledge and reaffirm the principles and provisions established in the Convention on Biological Diversity adopted on 5th June 1992 and encourage the effort to establish a mutually supportive relationship between the TRIPS Agreement and the Convention on Biological Diversity, regarding genetic resources and the protection of traditional knowledge and folklore.
3 Subject to each Party's international obligations and domestic laws, the Parties may adopt or maintain measures to promote the conservation of biological diversity, share equitably benefits arising from the use of traditional knowledge, innovations and practices relevant to the conservation of biological diversity and the sustainable use of its components in conformity with what is established in the Convention on Biological Diversity.

4 Subject to future developments of domestic laws and the outcome of negotiations in multilateral fora, the Parties agree to further discuss the disclosure of origin or source of genetic resources; and/or prior informed consent obligations in patent applications; and the grant of a patent for an invention that involves or relies on genetic resources, when such resources were acquired or exploited without complying with the relevant domestic laws or regulations.

In the New Zealand–Thailand PTA, parties agree to cooperate to enhance understanding in new areas of intellectual property, with traditional knowledge, genetic resources, and folklore cited as an example. It is further recognised that, consistent with obligations under the TRIPS Agreement, appropriate measures could be established to protect GRAATK (New Zealand, 2016).[9]

Taiwan provides a particularly interesting case in that five PTAs negotiated with Central American Countries (El Salvador, Honduras, Guatemala, Nicaragua and Panama) (SICE, 2016) have included detailed provisions on GRAATK. The provisions are similar across the board. They are essentially of an aspirational nature and refer to each country's commitment to protect the conservation and sustainable use of biological diversity and access to its GRAATK developed by indigenous peoples. The PTAs with Taiwan further recognise, without referring to the CBD explicitly, the fair and equitable participation in the benefits derived from access to the GRAATK associated with these resources by the indigenous peoples and local and ethnic communities. The agreements do not refer specifically to the issue of disclosure of source or origin but validate the fact that the licensing of patents developed from material obtained from biological and genetic heritage or traditional knowledge are subject to the condition that this material was acquired according to relevant national and international laws and regulations. See Box 19.3 reproducing relevant sections of the Taiwan–Panama agreement (SICE, 2016).

Box 19.3: Taiwan–Panama

Article 16.05

Protection of Traditional Knowledge

1 Each Party shall protect the collective intellectual property rights and the traditional knowledge of indigenous people on their creations, subject to commercial use, through a special system of registration, promotion and marketing of their rights, aiming at emphasizing the indigenous sociological and cultural values of the indigenous people and the local communities and bring to them social justice.

2 Each Party shall recognize that the customs, traditions, beliefs, spirituality, religiosity, cosmos vision, folklore expressions, artistic manifestations, traditional skills and any other form of traditional expression of the indigenous people and local communities are a part of their cultural heritage.
3 The cultural heritage shall not be subject to any form of exclusivity by unauthorized third parties applying the intellectual property system, unless the request is done by the indigenous people and local communities or by third parties with their authorization.

Article 16.07

Relation between Access to Genetic Resources and Intellectual Property

1 Each Party shall protect the access to its genetic resources and the traditional knowledge developed by indigenous people and local communities on the uses of the biological resources containing these genetic resources, against the indiscriminate use of biological diversity, as well as ensuring that the Party will participate in benefits derived from the use of its genetic resources.
2 Each Party shall accord a fair and equitable participation in the benefits derived from the access to its genetic resources and from the uses of its traditional knowledge and folklore expressions.
3 Each Party shall ensure that the protection accorded to the industrial property shall safeguard its biological and genetic heritage. Consequently, the licensing of patents on inventions developed from material obtained from such heritage or traditional knowledge shall be subject to the condition that this material was acquired according to relevant national and international laws and regulations make use of such heritage, in the understanding that this shall not be an exclusive right.

Discernible trends and conclusions

The chapter has selectively examined how PTAs by major trading nations have dealt with GRAATK and their relationship with multilateral normative processes. It appears that most existing PTAs have not included major commitments to the protection of GRAATK even in cases where biodiverse developing countries, major *demandeurs* in the context of the IGC, are parties to these PTAs. This analysis nevertheless shows that recent agreements where Latin American countries are parties do incorporate some provisions on the subject. The role played by Peru in PTAs signed with major trading partners is noticeable in this respect.[10] In other agreements negotiated with major trading partners the treatment of GRAATK is rather trivial.

The EU and EFTA seem to be more responsive to concerns voiced by developing countries in the framework of multilateral institutions, particularly in the case of the IGC, including around disclosure of origin or source that continues to be controversial and elusive in multilateral negotiations. It should be noted that the position of these trading partners in the case of disclosure and consequences is to leave the adoption of specific measures to the sovereign right of each country. In the case of EFTA for example, this is consistent with the position of member countries such as Switzerland (see Chapter 14 in this volume), which provides in its domestic law for disclosure of origin or source but with no consequences of revocation of the patent as has been advocated by some developing countries in the context of WTO TRIPS submissions (Draft Decision)[11] and WIPO IGC negotiations.

Could the EU position, with respect to disclosure of origin or source, be considered a new trend and/or consistent with European law on this matter? Reference is often made to the 1998 European Parliament Directive on biotechnological inventions and its preamble recognising that 'if an invention is based on biological material of plant or animal origin or if it uses such material, the patent application should, where appropriate, include information on the geographical origin of such material, if known; whereas this is without prejudice to the processing of patent applications or the validity of rights arising from granted patents' (27, Preamble, Directive 98/44/EC, 2016). The directive is not mandatory on the disclosure requirement, however, instead making it optional ('where appropriate') and without prejudice to the consequences of non-disclosure on the validity of the patent. As noted above, the PTAs correspondingly leave these matters to parties' national laws.

Furthermore, the position on GRAATK of the EU as well as of most countries of EFTA in recent PTAs, appear fully consistent with the position adopted in WTO in the joint submission made in 2008 together with IGC *demandeurs* (e.g., Brazil, Colombia, Ecuador, India, Indonesia, Pakistan, Peru, Sri Lanka and Thailand) on the discussion of TRIPS of three related outstanding issues under the Doha Work Programme: the geographical indications register, TRIPS disclosure requirement and geographical indications extension (Draft Modalities for TRIPS Related Issues, 2016).

On the relationship between IP and genetic resources and related traditional knowledge, the USA's PTAs consistently identify this as a matter related to patent quality examination, as well as to the use of databases or digital libraries containing traditional knowledge associated with genetic resources (see also Chapter 16 in this volume). The latest expression of this position, endorsed by eleven other countries, is the recently negotiated TPP agreement. Ideally for the USA the best way to deal with access to genetic resources and equitable sharing of benefits is through contracts that 'reflect mutually agreed terms between users and providers', as found in side letters to trade agreements, signed with Colombia and Peru and, more recently, by Australia with Malaysia and Peru (see Chapter 6 by Robinson and Chiarolla on contracts in this volume).

Most of the PTAs examined in this chapter reiterate the importance of traditional knowledge and biodiversity, as well as the potential contribution of traditional knowledge and biodiversity to cultural, economic and social development.

However, they do not agree on the modalities for its protection, or on the measures that might be adopted to prevent misuse or misappropriation. In a number of cases an allusion is made to the ongoing deliberations at the IGC. This could be perceived as recognition of the legitimacy of the ongoing WIPO process as well as understanding that the multilateral body is the sole competent institution to continue dealing with the subject in normative or simply deliberative terms. PTA provisions examined do not commit parties to the nature of the outcome of these deliberations.

In the realm of bilateral and regional trade agreements, *demandeurs* in WTO and WIPO might consider pursuing a parallel positive agenda on GRAATK, particularly when dealing with other developing *demandeur* countries; for example by including mutual recognition of traditional knowledge and the legitimacy of defensive measures such as disclosure. Developments of this type could set important precedents and thus establish state practices that one day could find a place in international law. An interesting precedent is the Swakopmund Protocol on the Protection of Traditional Knowledge and Expressions of Folklore concluded in 2010 in the context of the African Regional Intellectual Property Organization (ARIPO, 2016) according to which foreign holders of traditional knowledge enjoy benefits of protection under a national treatment principle.[12] (See also Chapter 21 by Frankel, in this volume.)

This chapter has not attempted to assess the actual impact of provisions in PTAs on GRAATK but tried to understand what could be the possible synergies on related questions with multilateral processes such as the IGC. One conclusion from this analysis is that PTAs, as one of the most significant sources of today's new emerging IP normative trends, do pay attention to GRAATK as a matter that cannot be eluded but do not provide clear orientation as to what agreed multilateral standards might emerge in this area. A related observation is that *demandeurs* in WIPO and WTO for defensive as well as positive measures for the protection of GRAATK have not been able to make their positions fully acceptable by their trading partners in PTAs. The proper interaction, thus, between multilateral normative processes and PTAs appears still to be work in progress.

Notes

1 Preferential trade agreements adopt different denominations (e.g., free trade agreements, trade promotion agreements, economic partnership agreements). We refer, hereinafter, to PTA to include also these denominations.

2 The Understanding adds that 'Each Party shall endeavour to seek ways to share information that may have a bearing on the patentability of inventions based on traditional knowledge or genetic resources by providing:
   (a) publicly accessible databases that contain relevant information; and
   (b) an opportunity to cite, in writing, to the appropriate examining authority prior art that may have a bearing on patentability.'

3 The Letter concludes by stating that 'If the United States and another government sign a free trade agreement that contains provisions addressing traditional knowledge or folklore, the United States and Panama shall promptly consult after that agreement enters into force on whether to apply similar provisions, as appropriate, between

the United States and Panama.' It is also interesting to note the sole reference to consultations in the case of the WIPO IGC and not to the WTO process.

4 Memorandum of Understanding with Peru available at http://dfat.gov.au/trade/agreements/tpp/official-documents/Documents/australia-peru-traditional-knowledge.PDF (2016) (accessed 10/01/2017).

5 'Parties will provide, in accordance with their domestic law, for applicable effects of any such requirement so as to support compliance with the provisions regulating access to genetic resources and associated TK.' EU-Peru-Colombia-Ecuador Agreements.

6 It further provides that 'The Parties recognise the importance of respecting, preserving and maintaining the indigenous and local communities' knowledge, innovations and practices that involve traditional practices related to the preservation and the sustainable use of biological diversity.' Art. 229, 4-5.

7 Art. 6.5 of FTA EFTA-Peru, SICE.

8 http://fta.mofcom.gov.cn/chile/xieyi/freetradexieding2.pdf (accessed 10/01/2017).

9 See Article 12.5, FTA NZ-Thailand at www.mfat.govt.nz/assets/_securedfiles/FTAs-agreements-in-force/Thailand-FTA/thainzcep-agreement.pdf (accessed 10/01/2017).

10 Interesting to note in the case of Peru that in a free trade agreement signed with Canada (2008) where there is no IP specific chapter, the environmental provisions do recognise 'that each Party has sovereign rights and responsibilities to conserve and protect its environment, and affirm their environmental obligations under their domestic law, as well as their international obligations under multilateral environmental agreements (Article 1701). In this context, 'the Parties have set out their mutual obligations in an Agreement on the Environment that addresses inter alia: conservation and sustainable use of biological diversity and protection and preservation of traditional knowledge' (Article 1702 c) (SICE, 2016).

11 'If it is discovered after the grant of a patent that the applicant failed to disclose the information set out in paragraph 2 of this Article, or submitted false and fraudulent information, or it is demonstrated by the evidence that the access and utilization of genetic resources and/or associated traditional knowledge violated the relevant national legislation of the country providing genetic resources and/or associated traditional knowledge, that is, the country of origin of such resources or a country that has acquired the genetic resources and/or associated traditional knowledge in accordance with the CBD, Members shall impose sanctions, which may include administrative sanctions, criminal sanctions, fines and adequate compensation for damages. Members may take other measures and sanctions, including revocation, against the violation of the obligations set out in paragraph 2.' See WTO DRAFT DECISION, 2011.

12 Regional Protection

24.1. Eligible foreign holders of traditional knowledge and expressions of folklore shall enjoy benefits of protection to the same level as holders of traditional knowledge and expressions of folklore who are nationals of the country of protection, taking into account as far as possible the customary laws and protocols applicable to the traditional knowledge or expressions of folklore concerned.

24.2. Measures should be established by the national competent authority and ARIPO Office to facilitate as far as possible the acquisition, management and enforcement of such protection for the benefit of the holders of traditional knowledge and expressions of folklore from foreign countries.

24.3. ARIPO may be entrusted with the task of settling cases of concurrent claims from communities of different countries with regard to traditional knowledge or expressions of folklore; to this end, ARIPO shall make use of customary law, local information sources, alternative dispute resolution mechanisms, and any other practical mechanism of this kind, which might prove necessary.

References

ARIPO, Laws & Protocols, www.aripo.org/resources/laws-protocols (last accessed 25/04/2016)

Directive 98/44/EC of the European Parliament and of the Council of 6 July 1998 on the legal protection of biotechnological inventions, http://eur-lex.europa.eu/legal-content/EN/TXT/HTML/?uri=CELEX:31998L0044&from=EN (last accessed 25/04/2016)

Draft Modalities for TRIPS Related Issues, *Communication from Albania, Brazil, China, Colombia, Ecuador, the European Communities, Iceland, India, Indonesia, the Kyrgyz Republic, Liechtenstein, the Former Yugoslav Republic of Macedonia, Pakistan, Peru, Sri Lanka, Switzerland, Thailand, Turkey, the ACP Group and the African Group*, TN/CW/52, 19 July 2008, https://docs.wto.org/dol2fe/Pages/FE_Search/FE_S_S009-DP.aspx?language=E&CatalogueIdList=80404,69295,91464,92094&CurrentCatalogueIdIndex=3&FullTextHash= (last accessed 25/04/2016)

EFTA, Free Trade Agreements, www.efta.int/free-trade/free-trade-agreements (last accessed 25/04/2016)

European Union, Trade Agreements, http://ec.europa.eu/trade/policy/countries-and-regions/agreements/ (last accessed 25/04/2016)

Intellectual property – Strategy for the protection and enforcement of IP rights in third countries, European Union, 2014, http://trade.ec.europa.eu/civilsoc/meetdetails.cfm?meet=11435 (last accessed 25/04/2016)

Mofcom, China FTA Network, http://fta.mofcom.gov.cn/topic/enchile.shtml (last accessed 25/04/2016)

New Zealand Foreign Affairs and Trade, Free Trade Agreements in force, www.mfat.govt.nz/en/trade/free-trade-agreements/free-trade-agreements-in-force/ (last accessed 25/04/2016)

SICE, Foreign Trade Information System, OAS, Trade Agreements, www.sice.oas.org/agreements_e.asp (last accessed 25/04/2016)

USTR, Office of the United States Trade Representatives, Free trade agreements, https://ustr.gov/trade-agreements/free-trade-agreements (last accessed 25/04/2016)

Vivas, David and Julia Oliva, (2010) 'Biodiversity Related Intellectual Property Provisions in Free Trade Agreements', *ICTSD*, Issue Paper No. 4, September 2010, available at www.google.ch/?gws_rd=ssl#q=Vivas%2C+David+and+Julia+Oliva%2C+%282010%29+Biodiversity+Related+Intellectual+Property+Provisions+in+Free+Trade+Agreements%2C+ICTSD%2C+Issue+Paper+No.+4%2C+September+2010 (last accessed 25/04/2016)

Yu, Peter, (2011) 'SINIC Trade Agreements', *UC Davis Law Review*, Vol 44, No. 3, pp. 953–1028.

WTO draft decision to enhance mutual supportiveness between the TRIPS agreement and the Convention on Biological Diversity, Communication from Brazil, China, Colombia, Ecuador, India, Indonesia, Peru, Thailand, the ACP Group, and the African Group, TN/C/W/59, 19 April 2011, https://webcache.googleusercontent.com/search?q=cache:3f0B2OaU04AJ:https://docsonline.wto.org/dol2fe/Pages/SS/DirectDoc.aspx%3Ffilename%3Dt%253A%252Ftn%252Fc%252Fw59.doc%26+&cd=2&hl=en&ct=clnk&gl=ch (last accessed 25/04/2016)

20 WIPO and the traditional knowledge conundrum

Ahmed Abdel-Latif[1]

Most of the chapters of this book have focused on the substantive issues the IGC has been grappling with in the areas of genetic resources, traditional knowledge and traditional cultural expressions and the perspectives of different countries and stakeholders regarding them. This chapter seeks to reflect on the IGC's work and its future from the perspective of its significance and implications for WIPO and more broadly for the global intellectual property system. It argues that the traditional knowledge 'conundrum' is the biggest policy and normative challenge that WIPO has ever faced and that the future of the global intellectual property system hinges upon its satisfactory resolution.

The protection of TK: a test for the universality and legitimacy of the global IP regime

The IGC's discussions, which have lasted now nearly seventeen years, may well be the longest ongoing intergovernmental negotiations WIPO has ever witnessed which are not taking place under the aegis of any of the established treaties it administers. They present WIPO with the biggest policy and normative challenge it has ever faced. At stake is a potentially unprecedented expansion in the realm of intellectual property protection and the emergence of an entirely new category of beneficiaries. Most of the previous norm setting under WIPO had been confined to adjusting the levels of protection enshrined in the two existing international treaties: *the Paris Convention for the Protection of Industrial Property* and *the Berne Convention for the Protection of Literary and Artistic Works*.

To understand the magnitude of this challenge, it is important to situate the IGC process in the context of WIPO's history and longer term institutional development. For several reasons, *universality* and *legitimacy* are two fundamental objectives that WIPO has actively pursued since its formal establishment as a modern international organisation in 1967. First, WIPO's key treaties – the Paris and Berne Conventions – were concluded in the nineteenth century between a small group of

1 The chapter is written by the author in his personal capacity and does not reflect the views of any institution he might be affiliated with.

developed countries – mostly European – at a time when much of the developing world was under colonial domination. Second, the technological firms and cultural industries of industrialised countries have been the greatest users and beneficiaries of the system designed by these countries.

Therefore, in 1967, WIPO faced defiance among developing countries towards an intellectual property system perceived as imbalanced – particularly in the area of patents – and which they were seeking to reform in the context of their efforts to establish a new international economic order. WIPO had few developing country members at that time. It also faced potential rivals in the United Nations system such as the United Nations Conference on Trade and Development (UNCTAD), established in 1964 and which had the favours of developing countries, and the United Nations Education, Science and Culture Organization (UNESCO) which was preferred by other countries, including the United States, which was not a party to the Berne Convention at the time.

In face of this, WIPO's Secretariat pursued membership of the UN system which it secured in 1974. The Agreement between the UN and WIPO highlighted a number of priority issues for developing countries, such as technology transfer and technical assistance. With UN status, the number of developing countries members of WIPO steadily increased, though many of them did not join either the Paris or Berne Conventions. The 1970s witnessed an effort by some key developing countries to reform these conventions and relax their rules so as to take into account differences in levels of developments and the socio-economic circumstances prevailing in the developing world. However, these efforts were not successful and accelerated the forum shift by industrialised countries to the GATT resulting in the conclusion of the TRIPS Agreement.

While the adoption of the TRIPS Agreement brought competition to WIPO in terms of global norm setting, it also brought along new opportunities for its engagement with developing countries. Given that the TRIPS Agreement incorporated key provisions of the Paris and Berne Conventions, many developing countries which were WTO members also became parties to these two conventions. More importantly, the WTO and WIPO concluded a cooperation agreement in 1996 which made WIPO the main multilateral technical assistance provider for the implementation of the TRIPS Agreement. This gave further impetus to developing countries in need of such assistance to become members of WIPO. By the end of the 1990s/early 2000s, WIPO had achieved its longstanding objective of near universal membership, encompassing most of the developing world.

In support of its efforts to attract a greater number of developing countries, the WIPO Secretariat worked relentlessly for decades to correct the perception that the IP system was imbalanced, and favoured industries of developed countries, and to anchor the 'legitimacy' of the IP system in the developing world. Through its technical assistance and advocacy, it showcased how the IP system could also be beneficial for developing countries as a 'power tool' for growth and socio-economic development (Idris, 2003). The establishment of the IGC and the prospects of elaborating new international rules for the protection of TK, GRs and TCEs, which are mainly found in developing countries, should be seen as a continuation of WIPO's long-term efforts to 'persuade' developing countries that they too can have an important stake in the system. Thus a failure by the IGC to reach tangible binding outcomes would constitute an important setback for WIPO's long-term efforts particularly as developing countries also became more vocal about the need for the organisation to be more responsive to their demands through initiatives such as the WIPO Development Agenda launched in 2004 (see also Chapter 23 by Oguamanam).

Such a setback would come at a particularly sensitive time when the IP system has come under criticism, during the past two decades, by public interest-oriented NGOs for not being sufficiently supportive of a range of public policy objectives. These include public health and access to medicines, access to knowledge and food security. An inability of the IP system to deliver new international measures and rules to combat misappropriation of GRs, TK and TCEs could further strengthen such criticisms.

Beyond the relationship with developing countries, the protection of GRs, TK and TCEs touches upon the fundamental tenets of the IP system: a system which projects itself as a set of universal principles and rules that protects and rewards human creativity in any time and place. Its reach is thus virtually unlimited. This assumption has been often invoked by developing countries and the WIPO Secretariat to justify the rationale for TK and TCEs protection in particular. Reference is made in this context to Article 2 of the WIPO (1967) convention which, when defining intellectual property, states that it encompasses 'all other rights resulting from intellectual activity in the industrial, scientific, literary or artistic fields'.

In this regard, the IP system has shown, in recent years, its ability to extend to new subject matter in fields such as biotechnology and computer software. It has also been capable of devising new rules, and tinkering with existing ones when needed, as exemplified by the 'invention' of protection systems for geographical indications and plant varieties. Adaptability and resilience are key features of the IP system which has enabled it to successfully overcome many challenges it has faced. In addition, the system is premised on the belief that the solutions to these challenges, whether technical or normative, can come from within. The way the debate in the IGC has evolved is emblematic of this. It progressed from being about how measures could be taken in the patent system to prevent it from misappropriating GRs and associated TK to being about how TK can be protected through an IP-like system and become the subject of exclusive rights. In any case, the inability of the IGC to deliver new rules to protect GRs, TK and TCEs would be a severe blow to the universalist aspirations of the IP system and to its self-perceived capacity to adapt and accommodate the new demands it is confronted with.

The key role of the WIPO Secretariat: opportunities and limitations

The WIPO Secretariat is fully cognisant that a successful conclusion of the IGC's work and the adoption of new international instrument(s) for the protection of

GRs, TK and TCEs would strengthen the universality and legitimacy of WIPO and of the global IP regime, in particular vis-à-vis developing countries. It would also enhance WIPO's pre-eminence in global norm setting in relation to its competitors, such as the WTO. It could also open new opportunities for the Secretariat to play a strategic role in the administration and management of the new global architecture for the protection of GRs, TK and TCEs, for instance, in areas such as TK databases. The protection of GRs, TK and TCEs could thus potentially herald a massive expansion of WIPO's role in global IP and knowledge governance. The WIPO Secretariat is also hopeful that a successful outcome at the IGC can help move the rest of WIPO's norm-setting agenda, particularly in the area of patents where it has been at a standstill for many years but also in the area of industrial designs where the African Group has put forward a proposal to include a disclosure requirement for GRs and TK in discussions on a procedural treaty on industrial designs which have remained inconclusive.

For these reasons, the WIPO Secretariat has been supportive of the IGC process in general and has played a key role in steering it forward. As we saw in Chapter 2, WIPO's Director General was instrumental in the creation of the IGC. The analytical work undertaken by the Secretariat, as reflected in the FFMs, provided the intellectual foundation of much of the IGC's work, particularly in its early years. In 2004, the Secretariat took it upon itself to table the first documents ever in the IGC, which contained draft objectives and principles on GR, TK and TCEs, attracting criticisms from a number of industrialised countries. The Secretariat contributed to keeping the IGC process alive even in periods in which it lacked momentum (see also chapter 3 by Wendland).

Although it is in favour of a successful outcome for the IGCs deliberations, at the same time, the WIPO Secretariat faces important limitations in playing a more active role towards achieving this. First, it is supposed to be neutral in ongoing negotiations between its member states. Second, and more importantly, the adoption of binding international rules to protect GRs, TK and TCEs with exclusive rights attached to them, faces staunch opposition from a number of key industrialised countries and industry associations representing patent holders. This puts the Secretariat in a difficult position in view of its traditional closeness with industry and, more importantly, in view of its dependence for most of its income on fees paid by patent holders to registration systems it administers, mainly the Patent Cooperation Treaty (PCT). In reality, the funding of WIPO by patent holders through these registration systems creates a potential conflict of interest with the organisation's general orientations and norm-setting functions. This has been underlined in recent years in the context of efforts and proposals to reform WIPO's governance and suggestions have been made to separate these functions.

At the same time, in order for the IGC to reach a positive conclusion, the WIPO Secretariat needs to exert its maximum efforts to persuade these stakeholders that such an outcome strengthens the credibility, legitimacy and universality of the IP system in a way that supersedes any losses they might incur. Indeed, even if international instruments for the protection of GRs, TKs and TCEs are adopted at

WIPO, the technological and cultural industries of developed countries will remain for many years the main beneficiaries of the IP system. The reason is that this system is not only about granting rights but also about the capacity to enforce those rights, and these industries have huge means at their disposal to do so, which is not the case of indigenous peoples or local communities.

Another important consideration for the WIPO Secretariat to take into account is that industrialised countries do not represent the monolithic bloc they once did when it comes to these issues. Two groups have clearly emerged. On the one hand, Switzerland and European countries have accepted the disclosure requirement in principle – though without strong sanctions for non-compliance – and have shown signs that they could live with some sort of TK protection regime. On the other hand, only a few hardliner industrialised countries continue to vehemently oppose both.

The IGC negotiations as they advance place the WIPO Secretariat under increased tension as it is caught between, on one side, the universalist aspirations of the IP regime, of which it is a key custodian, and on the other side the interests of the stakeholders which have been the traditional beneficiaries of this regime and the funders of the organisation and which, for the first time, could stand to lose from the expansion of IP protection. The Secretariat's ability to reconcile these two dynamics will be key in determining the future of the IGC negotiations and the nature of the outcome it could reach.

The IGC in the context of the challenges facing IP treaty making at WIPO and beyond

Efforts to reach new international rules to protect GRs, TK and TCEs face an exceptional number of challenges from the perspective of norm setting.

First, these efforts are taking place against the backdrop of a severe crisis in multilateral treaty making. With an ever increasing number of countries and a growing diversity of stakeholders and interests at play, it has become challenging for the international community to agree on global multilateral norms to address issues which are becoming more complex and that straddle different areas of international regulation. This is even more true for rule making on GRs, TK and TCEs, which also have important implications for domestic legal orders in many countries and that touch upon the very cultural fabric of societies. However, it is important to note that the crisis in IP rule making is not limited to multilateralism. The Anti-Counterfeiting Trade Agreement (ACTA) is an example of a plurilateral IP agreement that was negotiated and concluded by a group of mostly like-minded industrialised countries but which ultimately did not come into force as a result of a number of factors, including civil society mobilisation in these countries and lack of parliamentary consent.

Second, there has been some scepticism all along the IGC process about whether states negotiating in an intergovernmental setting can devise rules that cater and sufficiently take into account the specificities of TK and TCEs held by indigenous peoples and local communities. This view has been echoed by some representatives of indigenous peoples and also by some observers (see Chapter 22 by Dutfield in this volume). At the same time, it is important to note that some of the recurring questions that the IGC has grappled with unsuccessfully, such as the definition of TK, have been tackled with success in other forums. For instance, Article 2 of the 2003 UNESCO Convention on Intangible Cultural Heritage provides a definition of intangible cultural heritage. While the UNESCO convention requires state parties to take the necessary measures to safeguard intangible cultural heritage (Article 11), it leaves them with a certain latitude about how to do that. Although the UNESCO Convention is more about preservation than protection, and leaves out the intellectual property dimension, it shows that multilateral treaty making in relation to TK and TCEs is possible.

Third, in the absence of a 'single undertaking' principle like the one in the WTO, states engaged in WIPO norm setting don't have any real incentives to agree to new international norms which do not advance their interests. This issue has been highlighted by a number of scholars (Okediji, 2008) who have argued that, for this reason, the WTO might be a more attractive forum for global IP norm setting.

In addition to the absence of a 'grand bargain approach', there is another more recent feature of WIPO's norm-setting activities which is important to underline. For many years, WIPO Member States were willing to accept the conclusion of international IP treaties under WIPO even though they would subsequently not necessarily become party to them, as adherence to WIPO treaties is voluntary. WIPO's norm-setting history is replete with treaties which have been negotiated with the participation of all WIPO Member States but have no more than a handful of states that are party to them. However, this has changed after the adoption of the TRIPS agreement. The stakes around IP rules have become so high and the polarisation so intense that countries are now even reluctant to agree to the adoption of multilateral instruments even if they will not adhere to them. This could result from an apprehension that these instruments might set a precedent which could be used or imported to other settings, in particular bilateral and regional free trade agreements or even domestic legislation.

Thus, the IGC deliberations take place against the backdrop of already grim prospects facing IP treaty making in general at WIPO and elsewhere. As a result, few treaties have been adopted at WIPO in the past two decades, the most notable one being the 2013 *Marrakesh Treaty to Facilitate Access to Published Works for Persons Who Are Blind, Visually Impaired or Otherwise Print Disabled*. However, the Marrakesh Treaty has a rather narrow scope and responds to a clearly humanitarian imperative. In addition, there was a clear cohesive group of countries championing it, mainly Latin-American countries at the initial stages and then a broader group of countries, in tandem with a number of key NGOs such as the World Blind Union (WBU).

Such a dynamic seems to be lacking in the IGC deliberations. Latin American countries do not exert the prominent role they initially played. African countries appear as the most active players with the LMG group of developing countries. Indigenous peoples groups have contrasting views on what should be the outcome of the Committee's deliberations. The process clearly lacks the political momentum and coalitions which could bring it to a fruitful completion. In other fora, the political push to compromise and reach agreements often comes from ministers who are involved in the negotiations. This is seldom the case at WIPO. One possibility would be to consider convening a ministerial level meeting on the IGC to help advance the process in a decisive manner.

The IGC: towards what outcome?

Though the IGC's deliberations have lasted sixteen years, this is not so exceptional (taking into account that the first norm-setting mandate of the IGC was adopted in 2009). Just as a measure for comparison, discussions about a treaty on broadcasting organisations has been on the agenda of the Standing Committee on Copyright and Related Rights (SCCR) at WIPO since its first session in 1998 and up until today. It took twelve years to reach an agreement on one single provision in the treaty on the protection of audiovisual performances after the first diplomatic conference convened to adopt the treaty in 2000 failed to do so.

Despite the long time period elapsed since it began, it seems difficult to envisage the IGC not reaching an outcome. Too much intellectual and political capital has been invested by countries in these negotiations. Failure would deal a severe blow to the universality and legitimacy of the global IP regime which WIPO has been building up for several decades. It would show that the IP system is not able to address the grievances and demands of vulnerable groups such as indigenous peoples. At the same time, the IGC deliberations could last many more years in view of the wide gap in positions remaining between countries on the different provisions in the current texts and despite some progress with regard, for instance, to a 'tiered' approach to TK protection.

One possible way to disentangle this situation would be to have the draft instruments limit themselves to defining important principles and general obligations that countries would have to pursue to protect GRs, TK and TCEs, while leaving a broad latitude to national implementation schemes with mutual recognition between. This is the model that has proven successful in some multilateral negotiations recently, such as in the context of the United Nations Framework Convention on Climate Change (UNFCCC) leading to the adoption of the Paris Agreement. After years of pursuing a top-down approach, which would have countries take up extensive detailed obligations and targets, climate negotiations took a bottom-up approach in which countries agree to general principles and targets and undertake national actions to meet them with mechanisms in place for raising the level of ambition of these actions over time. A transparency system is also established to monitor these actions akin to the WTO's Trade Policy Review (TPR). Elements of such an approach for TK protection, most notably the transparency regime, have been proposed by some, most notably Drahos (2004) in the context of reflections on 'nodal governance'. This path would seem a viable option in order to avoid potentially more years of wrangling over the scope of TK protection and the rights attached to it while, at the same time, concluding a legally binding instrument.

Conclusion

The question is thus not whether the IGC will reach an outcome but rather what kind of outcome will it be: a diluted one which leaves the patent system essentially untouched and provides weak protection for GRs, TK and TCES, or a more ambitious one which entails measures to ensure that the patent system is supportive of biodiversity and provides for a strong protection of GRs TK, and TCEs? In any case, the window of opportunity for adopting soft norms might have passed, superseded by the length of the negotiations and the hardening of country positions.

Developing countries and indigenous peoples on their part need to weigh carefully the costs and benefits of any outcome at the IGC, taking into account a variety of factors such as: outcomes reached at WIPO are not backed by dispute settlement, adherence to new WIPO treaties is voluntary and the funding of WIPO by patent holders makes it difficult to reach outcomes which are detrimental in any significant manner to the interests of patent holders. In any case, developing countries should also continue to pursue their proposals at the WTO and other WIPO bodies. Ultimately, only time will tell if WIPO is capable of surmounting the TK 'conundrum' and how this will impact the universality and legitimacy of the global IP regime.

References

Drahos, P. (2004). Intellectual property and pharmaceutical markets: A nodal governance approach, *Temple Law Review*, 77: 401–424.

Idris, K. (2003). *Intellectual Property: A Power Tool for Economic Growth*, World Intellectual Property Organisation, Geneva.

Okediji, R. L. (2008). WIPO-WTO relations and the future of global intellectual property norms, *Netherlands Yearbook of International Law*, 39(1): 69–125.

21 The challenge of cross-border protection of traditional knowledge

Susy Frankel

In 2015, at a WIPO seminar I spoke about cross-border mechanisms.[1] That seminar took place while the mandate of the WIPO-IGC to negotiate an international legal instrument for the protection of traditional knowledge (and related traditional cultural expressions and genetic resources)[2] had expired and the new mandate had not yet been agreed. A key theme from that presentation was that without national laws protecting traditional knowledge not only is international protection hard to develop,[3] but also the protection of traditional knowledge in cross-border disputes is virtually impossible.

Much traditional knowledge is place-based, both as a matter of origin and the manner in which the knowledge is developed and used. As many indigenous peoples and local communities have explained, their traditional knowledge is often part of a connection with the land and the environment in which they live.[4] The challenge addressed here is the legal framework for achieving an appropriate sort of protection in places other than where the traditional knowledge originates. Details about possible modes of protection are discussed in other chapters in this volume (see Chapter 5 by Bagley, Chapter 6 by Robinson and Chiarolla, Chapter 7 by Ruiz Muller and Chapter 9 by Perlman). Existing intellectual property (IP) and trade mechanisms that enable cross-border protection provide some guidance, even though it is well-recognised that IP law does not provide the right mechanisms or framework for the protection of most traditional knowledge.[5] The experience of trans-border IP protection is useful because much of IP law is territorial and international agreements have created ways to address some cross-border issues.

National treatment and minimum standards of protection

A cornerstone of international IP agreements, including the Berne Convention, the Paris Convention and the TRIPS Agreement is the non-discrimination principle of national treatment.[6] National treatment broadly means that foreign creators and owners of IP are treated as well as national creators and owners in domestic legal systems. In IP law the point of attachment of national treatment is the creator, author or rights owner. In goods (tangible properly) the point of attachment is the goods.

An example of a national treatment clause, found in a regional instrument for protection of traditional knowledge, is the Swakopmund Protocol,[7] which provides that:

> Eligible foreign holders of traditional knowledge and expressions of folklore shall enjoy benefits of protection to the same level as holders of traditional knowledge and expressions of folklore who are nationals of the country of protection, taking into account as far as possible the customary laws and protocols applicable to the traditional knowledge or expressions of folklore concerned.

The rationales and goals of national treatment include economic and cultural justifications. The aim of national treatment is non-discrimination and it is, therefore, a mechanism for the assimilation of foreign works and owners into national systems. This is aimed at creating a level playing field. Importantly, however, national treatment cannot work without local laws. In international instruments, national treatment works together with minimum standards to form the structure of international norms. These two structural devices embed international norms in domestic legal systems.

The nature of minimum standards at the international level means that the laws can look different in domestic legal systems. There will often be more than one way to meet a minimum standard. Thus, there is no complete harmonisation in IP systems. Some IP norms are more similar among jurisdictions than others, as the international agreements provide more details of some standards.[8] This structure of national treatment and minimum standards explicitly recognises difference and allows it within a framework. Put differently, the structure can allow greater co-operation but provides the appropriate domestic policy space for difference. In the context of traditional knowledge, differences in protection are very important because the rational basis for its protection and the traditions, at the heart of what needs protection are inherently local; as noted above, they are place-based. Thus, a common framework through agreed minimum standards in areas of exclusive rights, such as protection from misappropriation, is desirable, but the parameters of rights may differ when enacted in domestic law according to the local policy. In some jurisdictions the scope of protection of traditional knowledge may even be determined by customary law, where a pluralistic legal system exists or is part of the culture. In other places the parameters may be determined by a governing body that is external to the traditional knowledge holders.

How national treatment applies to rights and exceptions

Importantly, national treatment provisions in existing international IP agreements refer to rights; in the Berne Convention the term ‘rights’ is used and in the Paris Convention, Rome Convention and TRIPS Agreement the term is ‘protection’. National treatment functions when there are rights for protection at national law. One country must treat foreign nationals the same as domestic nationals. Much IP functions with domestic law exceptions to rights. For the protection of traditional knowledge there need to be rights, exceptions may be relevant to shaping those rights, but as exceptions are designed according to local needs the role of national treatment is different. 'Authors' and 'owners' are usually not claiming exceptions. Exceptions are also relevant to traditional knowledge interfacing with existing IP regimes.

Consider an example of national legislation, which affords some protection for Māori culture via an objection system, to see how national treatment works. Anyone, from any Paris Convention or TRIPS Agreement member jurisdiction, can apply for a trade mark in New Zealand on the basis that the trade mark is used or proposed to be used. New Zealand's trade mark law also provides:[9]

> The Commissioner must not register as a trade mark or part of a trade mark any matter the use or registration of which would, in the opinion of the Commissioner, be likely to offend a significant section of the community, including Māori.

This is a provision for trade mark opposition.[10] Applying national treatment to this sort of opposition provision is different from applying it to rights granted to trade mark owners or the opportunity for anyone to apply for a trade mark. Foreign trade mark applicants can also expect their application to be opposed if it 'would be likely to offend a significant section of the community'. Note that the objection can be that the trade mark offends any significant section of the community. Such an objection process could be extended to other indigenous or local communities in the event of an international agreement to do so (that is a different process from national treatment applicable to rights). An international agreement on protection would ideally require other sorts of provisions, but it is useful to consider how the above provision could be amended. 'Significant section of the community' is not defined. It could be redefined to include communities from outside New Zealand.[11]

The strength of national treatment compared to other cross-border mechanisms

National treatment is a strong mechanism because of its anti-discrimination goals and characteristics. It also has potential for more robust outcomes than other cross-border frameworks. An alternative to national treatment is material reciprocity (foreigners get the same protection as they would receive in their country of origin). That is sometimes used in IP, and can be a starting point, but it is not the preferred approach of international IP because it entrenches differences.[12]

Another trans-border mechanism is that of mutual recognition, which is often used for goods (primarily for enabling market access). It provides a mode of recognition of others' standards. Possible venues for such recognition are registration processes, border controls, tribunals and courts.

In trade mark law the *telle quelle* (as is) rule, of the Paris Convention, has elements of (but not complete) mutual recognition. It provides:[13] 'every trademark duly registered in the country of origin shall be accepted for filing and protected as is in the other countries of the Union'.

When used in connection with IP, mutual recognition is most obviously applicable to registration procedures, but it has not been a preferred method. Patent offices, for example, prefer to do their own examination reflecting the territorial nature of patents and concerns about sovereignty.[14] Training and technocratic trust between IP offices can lead to some regulatory co-operation, as is found in the EU and between Australia and New Zealand. However, even those processes between countries with close relationships fall short of full mutual recognition.

Where registration systems are used for traditional knowledge (although I do not suggest they are a solution for protection) then mutual recognition of such registrations is a possible way to internationalise such information in order that it is available to examiners and applicants around the world.[15] The method of such uses would need to be appropriately and carefully designed.

Mutual recognition allows acceptance of others' standards, but where those standards are diverse mutual recognition is not an effective option because it requires a common procedure and cultural approach. It is, therefore, a flawed mechanism for achieving effective cross-border protection. National treatment has a better framework to achieve the goals of cross-border protection combined with appropriate national autonomy to create policies that support cultural diversity.

National treatment, private international law and minimum standards

It is important to appreciate, however, what national treatment does *not* do. It is a public international law rule that functions in state-to-state agreements. National treatment does not provide private international law rules, which are applicable to disputes between private parties, relating to jurisdiction or choice of law rules (applicable law in a dispute). These private international law rules will be important in cross-border traditional knowledge disputes. That is made more complex because there is no international agreement on these procedural rules applicable to IP broadly. However, the operation of such rules in national legal systems may very well provide greater coordination around resolving disputes where substantive law cannot do so because of territorial differences.[16] If, for example, country A provides protection for traditional knowledge against some sorts of misappropriation (and let's assume that it does so on the basis of an international agreement) then the courts of country A should have the jurisdiction to hear a dispute that the traditional knowledge, of a foreign local community, has been misappropriated in country A. However, it may not be practical for the local community to bring that dispute in a foreign country and so it might be important that there is jurisdiction to bring it in its own country, even though the infringement occurred overseas. Courts often cannot hear cases against foreigners who have no connection with the court's jurisdiction, but if the act of infringement has a subject matter connection to the place where the court sits some courts will take jurisdiction.[17] Broadly, such matters are some of the difficulties of international litigation which are not resolved in IP law either (although there are many detailed studies proposing solutions).[18]

If a foreign infringement claim is heard in a local court or a foreign court, in either case the court must decide what law it applies to the dispute. Courts do apply foreign law, although they can be reluctant to do so in some circumstances. In the situation of the protection of traditional knowledge it makes most sense to apply the law of the local community seeking protection as, after all, that is the rational basis for protection. However, as that concept may be alien to a foreign court it may be reluctant to apply a foreign law (it may not entertain such pluralistic notions). Rather, such a court might prefer to apply its own standard of misappropriation. That in turn shows how it will be very important not only to have national norms about what amounts to an actionable misappropriation, but also why there must be some level of detail in the international norm.

Detailing what amounts to misappropriation (that is minimum standards) is undeniably both complex and very important. The concept of misappropriation exists in many areas of IP and there are varying degrees of precision around what is legitimate use rather than illegitimate misappropriation. In the IGC process, defining the contours of misappropriation should not require the same level of detail as the national statutes should have, but what instances of misappropriation matter will require international norms. Unless there are details of some minimum standards at the international level, some countries will not develop national norms, just as they do not exist now. The point has been shown in existing IP law. In copyright, for example, without agreed international norms through the Berne Convention there would be no national protection in some places. In the IGC process, an outcome that does not require minimum standards of protection would likely mean that the international process is really only partially international and some economies – probably those most likely to misappropriate – are out of the reach of any international obligations. This would be the reverse result from the purpose of the IGC negotiations.[19] It also would make rights optional, like exceptions, which is not only counterintuitive, but also inconsistent with the general framework of international IP law, particularly the combination of national treatment and minimum standards.

Conclusion

Even though local difference is to be valued, international minimum standards for the protection of traditional knowledge must reach a level so that national laws, which are compliant with the international standards, are capable of dealing with trans-border disputes.

National treatment as a trans-border mechanism has the greatest potential to ensure that some minimum standards of protection are global and non-discriminatory. An effective pluralism for the international protection of traditional knowledge can emerge through the combination of minimum standards with national treatment. It is important that the IGC process ensures that there are some minimum standards that would enable not only domestic protection, but also cross-border protection as a result of national treatment. In addition to national treatment, national rules about private international law will necessarily emerge once international norms in the form of minimum standards, are established. The minimum standards do not need to be too detailed, but if they are empty or too thin then cross-border protection, by any mechanism, is not possible. If the IGC process is to succeed then rights should not be wholly left to national autonomy, but must have at least basic norms which can be fleshed out with appropriate detail in national systems, according to national policy drivers and goals.

Notes

1 Susy Frankel, 'The Cross-Border Protection of Intellectual Property and its Relevance for the Protection of Traditional Knowledge and Traditional Cultural Expressions', WIPO Seminar, WIPO/IPTK/GE/2/15/PRESENTATION/SUSY FRANKEL, Geneva, 23 June 2015, available at, www.wipo.int/tk/en/news/tk/2015/news_0005.html (accessed 10/01/2017).
2 In this chapter I use traditional knowledge in the broad sense to encompass traditional knowledge and where that knowledge is associated with cultural expressions and biological resources.
3 Many international agreements are drafted using national norms and experience. This approach is not as easy for the protection of traditional knowledge because of the paucity of national norms, but not impossible because, after all, it is the cross-border use of traditional knowledge that in large part has given rise to the calls for its international protection.
4 See, for example, the discussion of *mātauranga Māori* (Māori knowledge) in, *Report of the Waitangi Tribunal, Ko Aotearoa Tēnei: A Report into Claims Concerning New Zealand Law and Policy Affecting Māori Culture and Identity* (2011), ('Ko Aotearoa Tēnei').
5 For a discussion of the importance of the interface between *sui generis* traditional knowledge protection and intellectual property see 'A New Zealand Perspective on the Protection Traditional Knowledge', in Christoph Graber, Karolina Kuprecht and Jessica Lai (eds., 2012) *International Trade in Indigenous Cultural Heritage: Legal and Policy Issues* (Edward Elgar), 439–459.
6 Berne Convention, art. 5(1), Paris Convention art. 2(1), Rome Convention, art. 2(1) and TRIPS Agreement, art. 3.
7 Swakopmund Protocol, African Regional Intellectual Property Organisation, 2010, art. 24.
8 The subject matter of trade marks is, for example, more prescribed in article 15 of TRIPS than the subject matter of patents is in article 27 of TRIPS.
9 Trade Marks Act 2002 (NZ), s17(1)(c).
10 When foreigners apply for trade marks, they are treated the same as domestic applicants.
11 In this chapter, questions of the standing of foreigners is noted as may be requiring further consideration.
12 Exceptions to national treatment, where material reciprocity can apply, are found in the Berne Convention, art. 5; Paris Convention, art. 3; Rome Conventions, art. 2 and the TRIPS Agreement, art. 3.
13 Paris Convention, article 6 *quinquies* (A)(1).
14 The United Kingdom has considered using mutual recognition to overcome patent backlog, see UKIPO, 'Patent Backlogs and Mutual Recognition', at www.gov.uk/government/publications/patent-backlogs-and-mutual-recognition (accessed 10/01/2017).
15 That may, however, raise its own problems about confidentiality and systems integrity.
16 For a discussion of the harmonising effects of these rules in relation to IP, see Marketa Trimble, 'Advancing National Intellectual Property Policies in a Transnational Context', (2014) 74 *Maryland Law Review* 203.

17 This is a complex matter that requires discussion beyond the scope of this chapter.

18 See François Dessemontet, Rochelle Dreyfuss and Jane Ginsburg (2008), *Intellectual Property Principles Governing Jurisdiction, Choice of Law, and Judgments in Transnational Disputes* (American Law Institute); European Max Planck Group on Conflict of Laws in Intellectual Property (2013), *Conflict of Laws in Intellectual Property: The Clip Principles and Commentary* (Oxford University Press).

19 It also arguably would not give effect to article 31 of the United Nations Declaration on the Rights of Indigenous Peoples, see James Anaya, Technical Review of Key Intellectual Property-Related Issues of the WIPO Draft Instruments, WIPO/GRTKF/IC/29/INF/10, January 11, 2016, Intergovernmental Committee on Intellectual Property and Genetic Resources, Traditional Knowledge and Folklore, Twenty-Ninth Session, Geneva, February 15 to 19, 2016, para. 25.

22 Should we protect turmeric lattes?

Graham Dutfield

Introduction

On the day of writing this chapter, a Google search of the words 'turmeric and latte and biopiracy' yielded 182 results. Evidently there are people, at least a few, who seriously believe this type of product is worthy of protection from biopiracy (see Box 22.1). This chapter argues contrariwise that a legal regime that pulls in *all* of what people might call 'traditional knowledge' is unlikely to ever be agreed and would not deserve to be if it were. Probably you can get away without a clear definition of TK for a defensive protection regime that seeks merely to prevent excessively broad patents – that is assuming such a regime is ever agreed, given the opposition to disclosure of origin from some powerful countries. For most developing countries disclosure of origin is an indispensable provision (see Chapter 5 by Bagley in this volume). But for positive protection, traditional knowledge is not a sufficiently stable and workable concept for us to construe its application with so much breadth and expect everything to work out fine. One thing that is certainly clear from the anthropological literature is that 'traditional' knowledge is not static so that its origin can more easily be pinpointed to a space and time. It evolves and hybridises and it moves around. It is invisible and weighs nothing. Plants are neither but pollen blows in the wind and migrating people have taken their seeds with them all around the world mostly for perfectly innocent reasons. So any effort to 'fix' TK in time or space and claim it as one and only one group's or country's property can create its own injustices (see Chapter 7 by Ruiz Muller on TK in this volume). This is not to say one should not do so, but that it can definitely lead to conflicts if not done properly. Intellectual property law does just that in its own fashion. But the fact that advocates of TK protection tend to criticise IP law for its alleged failure to do this correctly or fairly should force us to be sceptical about property-based solutions especially – but not only – when governments put themselves forward as the ones to claim the exclusive rights. We have to be very careful before we go ahead and accept a property regime that might legitimate the vesting of rights on the basis of false or incomplete attributions of knowledge to specific countries or ethnic groups.

Box 22.1: What is biopiracy?

Biopiracy has received sufficient acceptance to be an entry in the *Oxford English Dictionary*. Its online version defines biopiracy as: '[t]he practice of obtaining commercially valuable genetic material, biochemicals, etc., from plant or animal species without fairly rewarding the country or community of origin, esp. by obtaining patents that restrict future use'. This is in fact inaccurate. My own definition, which in my view better reflects the way the word has been used from the start, is: '(i) the theft, misappropriation of, or unfair free-riding on, genetic resources and/or traditional knowledge through the patent system; and (ii) the unauthorized and uncompensated collection for commercial ends of genetic resources and/or traditional knowledge'. Arguably, this definition might be extended to embrace unfair appropriation through copyright law.

Conceptual and practical concerns surrounding traditional knowledge

Since the 1990s, the biopiracy discourse has tended to treat traditional and modern (or scientific) as binary opposites. Not explicitly perhaps, because 'traditional knowledge' is rarely given a definition of any great clarity, but certainly implicitly. This line of reasoning that assumes tradition and modernity are opposite, pure and unadulterated by each other is persistent. Proponents of this view, which remains the conventional wisdom, may not all be aware that such an opposition between tradition and modernity is a fundamental assumption of post-Second World War modernisation theory which has largely been discredited, at least in the academy, on a number of grounds including its determinism and its failure to accommodate the idea of mixing and hybridisation and their potentially beneficial aspects also for the traditional sector.[1] At its crudest, modernisation theory saw social progress and economic development as the necessary transformation of traditional societies into modern ones. It saw 'tradition' as holding societies back. The more tradition you have, the more modernisation you need; the less tradition the better. Given especially that some of the basic assumptions of modernisation theory continue to shape many people's assumptions, this should immediately be cause for reflection: that the term has tended to be used negatively as something outmoded and in some quarters continues to be. I know we can do our best to counter the negative connotations but the contraction of 'knowledge, innovations and practices' in the Convention on Biological Diversity to just plain 'knowledge' hardly helps. Companies pick up on this when their spokespeople say that TK is old and is therefore in the public domain. The contraction also downplays the pragmatism and adaptiveness of indigenous groups of each generation, as well as other societies with tradition-based bodies of knowledge that they wish to protect but that have changed culturally and in other ways from the groups and societies they were in the past. In fact, systems of knowledge tend to be hybrids because they are generally open, and they tend to have two-way 'valves': knowledge and materials go both out and in. Chinese medicine, for example, was not 'traditional' until it was named as such a few decades ago largely for political reasons. Traditional Chinese medicine co-evolved with western scientific medicine and has accommodated elements of modern science, for example the germ theory of disease. As for experiment and trying things out in a systematic way, it is not just white-coated laboratory scientists who do this; many traditional healers and farmers, who often breed modern varieties with their own, do as well. Chinese 'traditional' medicine remains highly popular as do the classical traditional Indian systems such as Ayurveda, Siddha and Unani Tibb, not all of which originate entirely in the subcontinent anyway. They may have very deep historical roots, but they are hardly devoid of novelty or innovation. Isolated indigenous peoples in places like the Amazon possess localised knowledge of flora and ecosystems, enabling them to meet many of their subsistence needs. But again it is unlikely that all of the biota they exploit or the knowledge they apply are entirely local or have ever been. Although uncontacted groups still exist in the Amazon, most human societies do not stay rooted to one spot over centuries *and* over a substantial period turn their backs on the world outside their own little part of it.

Despite all this, the use, borrowing, appropriation, misappropriation, or whatever name one chooses to call the inclusion of information, knowledge, methods and materials from one system of health or agriculture into another different and more dominant system, tends to be seen as being unidirectional. As tradition wanes and modern waxes, people assume the latter takes bits of tradition, used in the form of informational leads or raw materials, and gives nothing back in return. Modern *appears* no less modern for doing this because there is a translation and repackaging which generally strips tradition of its origins and cultural and spiritual entanglements or else denies it entirely. Accordingly, modernity is parasitic on tradition. An understanding of global power disparities reinforces such a perspective, ahistorical as it largely is. I am not saying that this is entirely wrong, but that the reality is more nuanced and that the nuances matter and need to be taken into account when making laws and devising policies. For those who take what I have just explained as given, it follows that the way to respond is to counter the otherwise inevitable TK erosion, either by eliminating contact with knowledge-seeking scientists and others, or else by creating a market for knowledge transactions so that access is exchanged for benefits. The former is impossible and ill-advised anyway. Isolationist policies ultimately fail. Knowledge is inherently 'leaky', plants get around pretty easily, and duplicates of both are likely to exist elsewhere. The latter seems perfectly reasonable and sensible. But again, there are pitfalls as we will see.

The most fundamental pitfall is this: treating traditional knowledge as a 'raw' input for life science corporations is to devalue it, essentially to reduce it to a random compilation of leads, hints, hopes, errors, deceptions and *cul de sacs* from which the useful needs to be separated from the supposedly useless. We end up considering only the value of TK to others, and not to the holders themselves within their own communities and among others with which they socialise and otherwise interact. This has negative practical implications, and many of the approaches being considered will not help. Contracts and licences might work in specific cases but are not the ideal solution. Corporations engage in knowledge transactions because intellectual property rights make doing this possible, and such exchanges work for them because intellectual property rights are the common currency that they all use and understand and which fit into their business models and the governing commercial law legal frameworks. Generally speaking, indigenous peoples inhabiting their own territorial spaces and elsewhere and rural communities are not companies having suitable governance structures and business nous sufficient to negotiate business deals.

There are many who accept this point but their responses are themselves problematic. There are two main approaches they tend to proffer, which in actual practice are not very different from each other. The first is to impose regulatory regimes. Experience strongly suggests that after twenty years of trying, these have done far less to encourage fair exchange – or any exchange for that matter – than we expected.[2] The second is for governments to step in to correct the power disparities. But to do this properly governments have to know about the people whose rights they are supposed to be promoting, including the local customary norms that operate. Lawmakers and central government representatives may not necessarily know a great deal about indigenous groups in distant parts of the country. Given the North-South slant on the TK debate, which can have the effect of downplaying the claims of specific national subpopulations, some governments seem inclined to treat traditional knowledge to be as much national property as that of minority and indigenous groups. Whether biopiracy is by corporation or the state it is still biopiracy. This is a somewhat controversial position and one must admit that it is likely to be uncommon for national ownership of TK to be *explicitly* provided in a country's statutory law. However, implicit in much of the rhetoric and even some of the practices of national governments is the assumption that traditional knowledge existing in a country belongs to that country and that the government on that basis has the moral authority to pursue claims as if it had a property interest. When a government agency initiates a legal challenge against a wrongly granted patent claiming common knowledge of that country, as the Indian government did when it opposed a patent on the use of turmeric in wound healing, what lies behind an apparently straightforward legal request that the patent be revoked for lack of novelty is effectively an effort to enforce a right over the 'biopirated' knowledge. If this were not so, why would it take the trouble of challenging the patent? Another example: if a government sets up a competent authority to collect benefits from use of TK that cannot be traced to a specific indigenous group *and* which is common knowledge, is this not tantamount to *de facto* nationalisation of common knowledge whose distribution may not even be confined to that country anyway?[3]

TK broad or TK narrow?

Despite my concerns as detailed above, many delegates of the WIPO IGC seem very keen to go ahead and draw up a legal instrument. Traditional knowledge continues to be the operative term and that is the way it is. The question arises of how broad or narrow should the regime define TK so as to protect it effectively? Clearly breadth can be excessive. Where does traditional knowledge end if anything done to or with turmeric by non-Indians is deemed to be misappropriation? An excessively broad meaning will unreasonably lock up vast amounts of publicly available knowledge which no identifiable group of people or nation could make any credible claim to, and whose circulation can no longer realistically be controlled anyway. The way that farmers in South Asia use neem tree seeds to protect their crops, to give one example, really is public domain information and compensation is due to nobody. I have heard no compelling argument for saying it belongs to the farmers of this generation or to the government of India, none of whom actually came up with the idea of using neem this way. The custodianship argument cannot take us very far in this particular context.

Accordingly, constructing a moral case for compensation from others' commercial use is a challenge. Neither James Watson nor the families of Francis Crick or Rosalind Franklin or the UK government has any right to claim benefits from those depicting DNA as a double helix or taking advantage of this discovered fact of nature to make money. They did not have such an entitlement at the time of the discovery and do not two generations later. It is hard to find a moral case for the government of India or of any other country to claim that any knowledge that ever came from their country that people and businesses elsewhere found commercially useful should be compensated for even when it gets hybridised, altered or otherwise transformed.

A narrower meaning, on the other hand, might exclude much of what many countries would like to have protected. Even so, this would get us far closer to a workable approach. What if one confined the legal regime to the knowledge of indigenous peoples as defined internationally under the International Labour Organization Convention 169 Concerning Indigenous and Tribal Peoples in Independent Countries? This approach could be justified in at least two ways. First, these are often culturally quite distinct groups of people. Consequently for such people an item of traditional knowledge may be more attributable unequivocally to such people. In some parts of the world, their knowledge and 'traditionality' are relatively unadulterated by mainstream knowledge systems and technologies. Second, they may have functioning customary norms governing access and use of certain knowledge and resources they possess (see Chapter 10 by Tobin on customary laws in this volume).[4] This is very important: what is deemed to be public domain in the intellectual property sense should not automatically be considered to be freely open for others to appropriate because right and duties over knowledge even after its circulation may be a matter for customary law (see Chapter 8 by Okediji on public domain in this volume). Why should we not take into account their own laws?

There is a real dilemma here, though. By adopting this approach, which I am advocating, much of what some countries regard as being traditional knowledge would be excluded from protection, perhaps unfairly. Just because they may not be the originators of some valuable knowledge it does not necessarily follow that communities have no rights over it or deserve no compensation. We would still need to discuss this, but I do not think that an *international* instrument should have to deal with this. Also, we would still need to have a conversation about the innovations of this generation including those of individuals in communities that, in the words of the Convention on Biological Diversity, embody traditional lifestyles broadly construed. As for indigenous peoples, their levels of acculturation vary widely. Should we ignore this or does it raise difficulties we would have to face up to? What about those who no longer live in such communities, and not necessarily by choice? Should their knowledge, innovations and practices be protected? After all these years important questions remain.

Another reasonable and very basic concern that the approach I suggest cannot fully alleviate is that even with ‘indigenous peoples’, as so defined, attributing knowledge to one group and one group alone is potentially erroneous. To name one example, the use and knowledge of hoodia as a thirst and appetite suppressant is almost universally attributed to the San people of Southern Africa. However, recent research suggests the situation is not entirely clear. While the San may well be the original discoverers, many of them did not consume it, while various non-San and mixed populations have used it in recent centuries, and some of them cultivated it too.[5]

Yet another dilemma arises, which I do not yet know how best to resolve. On the one hand, as I have tried to explain, a legal regime for TK that focuses on culturally distinct indigenous communities appears to be the most realistic approach. However, for such people biopiracy is simply not the biggest problem that they face. Land rights and other economic, social and cultural rights may be far more important. Is there a point to seeking to protect their knowledge when their absence of legal title may be causing them much more harm? There *is* a point to parallel campaigns to promote both. However, progress on land rights is probably an essential condition for a knowledge protection regime to work. Politically, confining the application of the regime in this way is a lot less interesting for governments who may lose interest in negotiating a legal instrument because other than a few indigenous groups the national economy does not stand to benefit in any substantial sense.

Conclusion

We need to open the traditional knowledge black box and make some tough decisions on what a legal regime should and should not cover. No decision on the scope of the regime will please everybody. My hunch is that a narrow definition is the only one that could lead to a workable regime.

As for turmeric latte, my message to the IGC is: Let it go.

Notes

1 A number of mostly US-based social scientists in the post-war era identified fundamental social and cultural differences between traditional and modern societies and assigned to each a set of descriptive terms that were in opposition to each other. Accordingly, as

they saw it, social and cultural evolution could be best understood in terms of progress that would entail the replacement of terms applicable to traditional societies such as 'community', 'patron-client relationship', 'routine', and 'solidarity', with their modern polar opposite counterparts: 'individual', 'bureaucratic relationships', 'innovation' and 'competition', respectively. Since evidence of progress essentially entailed the latter terms applying rather than the former ones, there was little accommodation for hybridity including its positive aspects for both societies. Ironically, like the Marxism that modernisation theory sought to counter, both approaches were susceptible to somewhat deterministic evolutionary stages notions. See J.-P. Olivier de Sardan (2005) *Anthropology of Development: Understanding Contemporary Social Change.* London and New York: Zed Books, pp. 46–48.

2 Two early examples are (1) Philippines Executive Order 247 'Prescribing Guidelines and Establishing a Regulatory Framework for the Prospecting of Biological and Genetic Resources, Their By-Products and Derivatives, for Scientific and Commercial Purposes; and for Other Purposes'; (2) the Andean Community's Decision 391 'Common System on Access to Genetic Resources'.

3 Iraq's Draft Law on the Protection and Exchange of Genetic Resources for Food and Agriculture states that information related to plant genetic resources is owned by the state, and the clear implication is that this includes TK. N.M. Khalaf (2016) *Green Economy in the Arab Region: Regulating Agricultural Biodiversity and Plant Related Innovation in Egypt and Iraq.* PhD thesis, Bangor University.

4 B. Tobin (2014) *Indigenous Peoples, Customary Law and Human Rights: Why Living Law Matters*. London: Routledge.

5 A.D. Osseo-Asare (2014) *Bitter Roots: The Search for Healing Plants in Africa.* Chicago and London: University of Chicago Press.

23 Ramifications of the WIPO IGC for IP and development

Chidi Oguamanam

Introduction

The year 2016 marked over a decade and a half of existence for the World Intellectual Property Organization (WIPO) Intergovernmental Committee on Intellectual Property and Genetic Resources, Traditional Knowledge and Folklore (IGC). The IGC is a child of circumstance and opportunity.[1] Detailed historical accounts of its evolution link it with earlier initiatives, including WIPO's Model Provisions on National Laws on Folklore and the Draft Treaty for the Protection of Expressions of Folklore Against Illicit Exploitation and other Prejudicial Actions.[2] Also, the IGC was in part a response to the failure of the World Trade Organization (WTO) Agreement on Trade-Related Aspects of Intellectual Property Rights (TRIPS) to accommodate traditional knowledge (TK) and traditional cultural expressions (TCEs) as sources of intellectual property rights and knowledge production.

Since the Convention on Biological Diversity (CBD) came into effect in 1993, legal and policy actions for the protection of traditional knowledge across multiple and overlapping fora and regimes have intensified.[3] Despite strides in those fora and regimes separate from the IGC, and weakened expectations in respect of the IGC's mandate to produce a text-based instrument(s) to ensure the effective protection of genetic resources (GR), TK and TCEs, there is still acknowledgement, however grudging, that intellectual property (as a concept, jurisprudence and policy instrument) is, in at-times-ironic fashion, at the very core of both the exploitation and the protection of traditional knowledge. And there is also an anchoring perception that WIPO is the most apposite forum for shaping international intellectual property policy.[4]

Over the years, the idea of traditional knowledge has progressively unravelled as a traction point for complex issue linkages between intellectual property and, for example, genetic resources, biodiversity conservation, the rights of indigenous peoples and local communities (ILCs), food, nutritional and environmental security. Because traditional knowledge cuts across regimes and is crucial to the lived realities of the world's impoverished regions and peoples, it has become a site for engaging, testing, and even challenging the development imperative in intellectual property. In this short chapter I reflect on the ramifications of the WIPO

IGC process for intellectual property and development, and consider the implications of the wobbly nature of the seemingly interminable IGC process for developing countries.

Regime complexity

The IGC symbolizes an intensive policy space for continued exploration of the historical tension between conventional intellectual property and traditional knowledge systems. However, as indicated above, the IGC is but one of the variegated sites, established since the late twentieth century, in which the issues relating to traditional knowledge and its multiple interfaces have been, and are continuing to be, explored. The WIPO General Assembly's understanding of the IGC mandate is clear: that deliberations are 'without prejudice to the work pursued in other fora'.[5]

But there is a staggering litany of fora in which the issue of protection of TK is writ large, as well as its associations with GR and the protection of various aspects of TCEs (see Chapter 4 by Lawson, this volume). Discerning analysts have expressed concerns over how this regime complexity could potentially undermine efforts aimed at the protection of TK.[6] These analysts point out that where multiple regimes engage a specific subject matter, the more powerful actors are in a position to leverage such regimes in a selective manner to advance their interests, e.g. through deliberate patronage of preferred regimes and fora, and neglect of less-favourable ones. One gets the impression that the fifteen years without agreed treaty texts in terms of outcomes at the IGC reflects the dynamic of this 'regime game' at its worst. A few powerful states and, to some extent, the global North as a geopolitical bloc, seem bent on ensuring that no outcome at the IGC will be a good outcome.

The development imperative

The IGC is a crucial aspect of sustained historic pressure on WIPO and, certainly, the international economic order, to embrace the development imperative. In the intellectual property or broader knowledge governance arena, Peter Yu describes the development agenda specifically as 'an ongoing struggle by less developed countries to develop an innovation system that responds to their needs, concerns and local conditions'.[7] In 2007, WIPO formally outlined the current operative phase of its Development Agenda.[8] Essentially, the agenda aims at mainstreaming, in practical ways, the development imperative in national and international intellectual property law and policy. Specifically, Development Agenda Recommendation 18 urges 'the IGC to accelerate the process on the protection of genetic resources, traditional knowledge and folklore, without prejudice to any outcome, including the possible development of international instrument or instruments'.[9]

As a crucial aspect of the WIPO Development Agenda, the IGC's mandate makes a link across several practical-conceptual elements: intellectual property, genetic resources, traditional knowledge and traditional cultural expressions (folklore). The innovation landscape of the late twentieth century[10] demonstrates the concrete interactions of the above-enumerated elements. The shared perception among many in developing countries is that these interactions result in an outcome that consistently undermines the interests of developing countries and indigenous peoples and local communities across the world, especially through the phenomenon of biopiracy.[11] The IGC represents a crucial forum to negotiate how knowledge production at these complex intersections can advance the needs, concerns and local conditions of stakeholders, especially developing countries, indigenous peoples and amorphous local communities.

Hardening of positions

The fifteen years of IGC negotiations have reinforced and hardened positions among proponents and opponents of TK, along opportunistic, ideological and political lines – all of them bearing the hallmarks of North-South geopolitical power relations as an obvious and enduring undercurrent. As an outcome, these trends impugn expectations and chill hope by developing countries over the sincerity of their developed-world counterparts in nurturing the development agenda within and outside the WIPO committee processes.

A selective sample of lingering issues that have held up negotiations for too long reflects a mixed bag of the finicky, the semantic and the substantive. The policy objectives to be served by the anticipated IGC instrument or instruments, for example, range between, on the one hand, preventing 'misappropriation' of genetic resources and associated traditional knowledge, or misappropriation of traditional knowledge and associated genetic resources, and, on the other, preventing 'erroneous grant of patents', 'invalid patents' or 'biopiracy'. Delegates still bicker over what constitutes misappropriation and how it is to be defined. In addition, there is palpable reluctance to legitimize the concept of 'biopiracy' (perceived as a too-politically-loaded term) into legal text and jurisprudence. Consensus also remains elusive on whether the IGC text(s) should focus on patent qua patents as opposed to intellectual property in general.[12] And debate continues on whether the text would specifically provide for the protection of derivatives of TK and of GR, derivatives of TK associated with GR, or derivatives of GR associated with TK. In addition, experts have yet to agree on what constitutes the trigger (i.e. what form/quantity/quality of uses or association of GR, and/or TK, in an invention) for an obligation to arise under the proposed instrument(s). While some stakeholders insist that invention must be 'directly derived' from GR and associated TK, others insist that in the era of biotechnology, bioinformatics and synthetic biology, the use and appropriation of knowledge related to genetic information can happen without physical association with a specific genetic resource, rendering moot the concept of direct contact or derivation.

Among the more substantive sources of disquiet is the matter of whether there has to be disclosure (see Chapter 5 by Bagley, this volume), or notification of information on, GRs or TK utilized for intellectual-property-protected innovation (i.e. in a patent application), and if so, what the scope of the disclosure should be (i.e. should it be of the source, or the origin?). Likewise, the nature of sanctions that would result from failure to disclose remains a matter of testy debate. With the disclosure debate, the element has focused on source and origin, which is particularly complicated, with 'non-demandeurs' (who do not favour disclosure of source/origin)[13] generating political capital from their resistance. The complex, natural and anthropomorphically driven migrations of GRs, and the equally complex and varied cultural trajectories of their applications and appropriations across ecological and jurisdictional lines – not to mention their applications across fluid epistemic contours – show that there can be no simplistic demarcation between source and origin. Also contentious is the role of national patent or intellectual property offices in any potential bureaucracy that would be responsible for implementation of treaty text(s) emanating from the IGC negotiations. Perhaps the most significant logjam is the lack of consensus on the nature or uses of genetic resources that should be eligible for exceptions and limitations, or waiver from a disclosure requirement.

Lastly, there is no consensus on the role or modality for creating databases of TK (see Chapter 6 by Robinson and Chiarolla, this volume). Would such databases be complementary, defensive measures against the abuse of TK through the patent system (i.e. against biopiracy), or, rather, part of a stand-alone alternative scheme to the instrument(s) expected from the IGC?[14]

The TK construct

As in similar fora where traditional knowledge comes under critical scrutiny, there is ample resistance, ranging from the conceptual, and the definitional to the philosophical, against TK as a construct. Definitions of what constitutes TK tend to be open-ended, defying legal certainty (see Chapter 7 by Ruiz Muller, this volume). The very idea of demarcation of knowledge systems across traditional and other juxtaposing models tends to re-open the old debate – in sociology-of-science or sociology-of-knowledge circles, where the ideas of knowledge hybridization, knowledge fluidity and knowledge creolization are resurgent – around the suspect relationship between modernism and tradition.[15] Yet few antagonists of TK would deny that some associated uses and practices, in their interface with genetic resources, have deeper or earlier historical roots than others in a cultural, ecological region, or in a local community, and that such an acknowledgement can be made without prejudice to the migratory and co-dependent nature of knowledge evolution.

On a political even if opportunistic level, critics of the use of TK to amplify the development imperative (under the auspices of IGC and other fora) question the legitimacy of state actors as champions of TK in relation to indigenous peoples and local communities, given the historic absence of unity of purpose between the latter and states. However, at the same time that there are concerns about so-called 'state biopiracy' (see Chapter 22 by Dutfield, this volume), there is perhaps no indigenous or local community constituency where legitimacy of representation is not at issue. In most indigenous and local communities, the question of representative legitimacy assumes greater complication than in the state context. Some of the objections to state role, albeit well-founded, tend to ignore the fact that the capacity of states, especially postcolonial states, to champion the protection of traditional knowledge is a factor of their constitutive ethnicity and politico-historical antecedents and factors. For example, comparing the legitimacy of the government of India as a state sponsor or champion of the traditional knowledge claims of indigenous and local communities to the legitimacy of the governments of Canada may be inappropriate.

In addition, critics of state roles seem oblivious to the reality that international processes have historically been suspicious of indigenous peoples' rights and have regularly constrained indigenous peoples' participation in those processes vis-à-vis sovereign states (see Chapter 11 by Solomon, this volume). For example, indigenous peoples' attendance at, and their contributions to, the IGC process is constrained both by funding and rules of procedure that place them at the mercy of sovereign states. As a caucus, their attendance has progressively declined in the course of the long, drawn-out IGC process.[16] Nevertheless, there is already a good, albeit not robust, emergent international regime on indigenous peoples' rights. In specific regard to genetic resources, for example, the Nagoya Protocol has further strengthened indigenous and local communities as legitimate beneficiaries of TK-related innovations in a manner that constrains the proclivity of states to exclude the interests of indigenous peoples and local communities. In essence, depending on context, it does not follow that state championship of rights over traditional knowledge amounts to state biopiracy in a manner akin to corporate biopiracy.

In developing countries, both the state and its constitutive indigenous and local communities may be the direct or indirect beneficiaries of TK. Either way, such an outcome is more likely to yield grassroots transformation through direct or indirect impacts on individuals, peoples and communities in contra distinction to the phenomenon of corporate biopiracy, which, for the most part, is externally driven corporate exploitation and exclusion of indigenous and local community rights. When developing countries and indigenous and local communities are empowered as stakeholders in TK through intellectual property or *sui generis* formulations, contracts and other responsive models of knowledge governance, then the development imperative is practically translated. The IGC is a strategic platform for advancing this objective.

In knowledge governance frameworks, development as freedom and capacitation[17] entail inclusive accommodation of all contributors to the process of knowledge production and its transformation. In, for example, pharmaceuticals, biotechnology, medicines, agriculture, chemicals, cosmetology, and the life sciences, knowledge production and transformation reinforce the fluidity (or futility) of boundary-marking across knowledge systems. They also demonstrate that TK is an extremely valuable source of innovation and insight as is evident in the patterns of biopiracy globally. As such, the legitimate desire of TK's perennially marginalized curators, the world's indigenous and local communities, for the innovation system to recognize their contributions, and address their needs, contexts and concerns, is critical to the interrelated concepts of capacitation, freedom, development and self-determination – in contrast to the prevailing system that fuels mis-appropriation of TK through its de-legitimization.

Slow progress at the IGC

Deliberations at the IGC do not seem to be fully attuned to the nexus between the committee's work and the development imperative. Developed-world opposition to the push by developing countries and indigenous and local communities for a pragmatic recalibration of the patent system (to tackle prevalent misappropriation of TK and the glaring abuses of the patent system) only inflame existing mistrust by developing countries over the commitment of their Northern counterparts to the development agenda.

However, despite initial scepticism, some developing countries managed to develop a measure of (gradually eroding) faith in the process. The Africa Group, the representatives of ILCs (indigenous caucus) and other negotiating blocs in the IGC, have invested tremendous energy and tenacity in the process. According to the Iranian delegation to the IGC's 28th session, the IGC process is symbolic of 'development-oriented IP norm-setting in WIPO'.[18] Whereas its success (through substantive developed-oriented treaty text(s)) would reassure developing countries of WIPO's commitment to taking development concerns in intellectual property rights seriously, its failure (by way of potential stalemate) would spread a sense of disillusionment and exclusion, among developing countries and ILCs, in respect of the international intellectual property system. Accordingly, failure at the IGC could yield instances of unpredictable backlash in the WIPO committee process and in relation to other international TK-oriented regimes. Failure could, further, fuel geopolitical tension, widen inequity and escalate the trust deficit in the international intellectual property regime complex.

Conclusion: IGC glass half full?

Notwithstanding the concerns raised above, the IGC's metaphorical glass can arguably still be seen as half full. For developing countries, the benefits of the decade and a half of exchanges at the IGC can still be redeemed and leveraged. With the exception of the US, Japan, Canada, South Korea,[19] and to a lesser extent the EU, most negotiating blocs in the IGC have achieved or are close to achieving concrete consensus on core issues regarding international instruments for effective protection of GR, TK and TCEs. This consensus could constitute the basis for national regimes in willing states. The majority of these willing countries and regions constitute the world's mega-genetically-diverse sites and are home to many of the world's ILC custodians of traditional knowledge. All they require is the political will to push forward on their shared consensus via an international instrument, even if that happens on a new multilateral level and development forum outside WIPO.

In addition to the possibility of adapting their shared consensus within the IGC negotiations into a multilateral instrument, many 'demandeur' countries, regions and global South negotiating blocs (e.g. the Africa Group, Like Minded Countries, Latin American and Caribbean Group of Countries, the Asia-Pacific Group, etc.) also have the option of engaging in the aforementioned regime game, through deliberate and collaborative patronage of friendlier regimes – such as the Food and Agriculture Organization (FAO) Plant Treaty, the CBD Nagoya Protocol, and the broader CBD framework including cognate frameworks such as the United Nations Declaration on the Rights of Indigenous Peoples – in order to foster a development-oriented intellectual property system at the intersection of traditional knowledge and genetic resources.

Finally, at national levels, many countries are seizing the initiative of ensuring effective protection of traditional knowledge by fine-tuning domestic patent laws and related statutes to accommodate some of the contentious issues at the IGC, including, for example, disclosure, in patent applications, of source and origin of GR and associated TK. As well, initially inspired by India, these countries are embracing documentation schemes for traditional knowledge, as both a complementary and defensive strategy against biopiracy.

Thus, the IGC's potential for shaping policy, and for advancing the development imperative, may not necessarily be a function of the ultimate outcome of its overly prolonged deliberations. So far, IGC's impact is being felt in the progressive attempts in various fora and regimes, national and international, to pressure global knowledge governance to tackle head-on the injustice of global biopiracy and the abuse of the intellectual property system at the expense of the world's impoverished peoples and marginalized regions.

Notes

1 This takes into account the 2015 hiatus from the IGC negotiations.

2 See Paul Kuruk, 'Regulating Access to Traditional Knowledge and Genetic Resources: The Disclosure Requirement as a Strategy to Combat Biopiracy' (2015) 17:1 *San Diego International Law Journal* 1–74 [Kuruk].

3 This is evident through various work programmes of the CBD, the Bonn Guidelines, the Nagoya Protocol on ABS and developments in cognate regimes such as the FAO Plant Treaty, some of which partially integrate or are inspired by the elements and principles of the United Nations Declaration on the Rights of Indigenous Peoples.

4 Kuruk, *supra* note 2 at 15 nn. 22, 347–355.

5 See www.wipo.int/export/sites/www/tk/en/igc/pdf/igc_mandate_1617.pdf (accessed 10/01/2017).

6 For example, see Meunier, Sophie, and Karen Alter, 'The Politics of International Regime Complex' (2009) 7:1 *Perspectives on Politics* 13–24; Laurence R. Helfer, 'Regime Shifting: The TRIPs Agreement and New Dynamics of International Intellectual Property Lawmaking' (2004) 29 *Yale Journal of International Law* 1–83; 'Regime Shifting in the International Intellectual Property System' (2009) 7:1 *Perspectives on Politics* 39–44.

7 See Pater K. Yu, 'A Tale of Two Development Agendas' (2009) 35 *Ohio Northern University Law Review* 465–573 at 467.

8 In 2007, the WIPO General Assembly adopted 45 recommendations under the Development Agenda. For text of the recommendations, see www.wipo.int/ip-development/en/agenda/recommendations.html (accessed 10/01/2017).

9 Ibid. TRIPS mandate on discussions on TK/GR from 2001 Doha Declaration urges stakeholders to take 'into account the development dimension', this element which is often omitted when the mandate is quoted.

10 Specific reference here is to biotechnology and applications of recombinant DNA insights across different industrial sectors: pharmaceutical, chemical, agriculture, etc. In addition, the use of digital and information communication technologies to mine, fixate, adapt, and transform mainly oral forms of traditional cultural expressions is also part of the new innovation landscape often captured as 'bio' and digital revolutions.

11 It is important to indicate that biopiracy is not necessarily limited to the exploitation of TK associated with biological resources or broader biocultural knowledge forms. In some adaptations, it serves as a metaphor for the exploitation of the actual and potential creativity of indigenous peoples and local communities including in the realm of music, choreography, and various forms of traditional oral or non-oral cultural expressions especially as facilitated by the use of information and communication technologies. See Olufunmilayo B. Arewa, 'Piracy, Biopiracy and Borrowing: Culture, Cultural Heritage and the Globalization of Intellectual Property' (March 2006), *Case Legal Studies Research Paper* No. 04–19. Available at SSRN: http://ssrn.com/abstract=596921 or http://dx.doi.org/10.2139/ssrn.596921 (accessed 10/01/2017).

12 In the context of biopiracy and the intersection of traditional knowledge and genetic resources, the assumption is that patent is the regime of intellectual property most engaged. However, that assumption is not quite accurate. Other regimes of intellectual property such as trade secrets may also be involved in the biopiracy dynamic.

13 The terms 'demandeurs' and 'non-demandeurs' are used in negotiating circles to designate countries that favour disclosure of source and/or origin of genetic resources in patent application and those that are opposed to it, corresponding loosely to developing and developed country blocs.

14 The majority of non-demandeurs favour the establishment of traditional knowledge databases at national levels to serve as reservoirs of prior knowledge that can be leveraged as a defensive measure to avoid questionable claims for novelty required to grant patents. However, while not opposed, in principle, to traditional knowledge databases, many developing countries do not consider it as an alternative to expected treaty text(s) to the IGC. Moreover, while such databases account for traditional knowledge in the public domain, it is not yet clear how to account for those that are in the public domain.

15 See, generally, Chidi Oguamanam, 'Pressuring "Suspect Orthodoxy:" Traditional Knowledge and the Patent System', in Matthew Rimmer (ed., 2015), *Indigenous Intellectual Property: A Handbook of Contemporary Research* (Cheltenham: Edward Elgar) p. 313.

16 See Chris Koziol, 'Indigenous voices being marginalized at WIPO?' (3 June 2016) *ABS Canada* (blog), online at: www.abs-canada.org/news/indigenous-voices-being-marginalized-at-wipo/ (accessed 10/01/2017).

17 See, generally, Amartya Sen (1999) *Development as Freedom* (Oxford: Oxford University Press).

18 See WIPO/GRTKF/IGC/28/11 (Feb. 15, 2016) Report of 28th Session of IGC July 7–9, 2014, para. 114.

19 They are referred to as the 'Group B' negotiating bloc. But it is not clear if South Korea is formally part of this group even though it often aligns with the position of Group B.

Glossary

access and benefit-sharing The *Convention on Biological Diversity* (1992) prescribes 'the fair and equitable sharing of the benefits arising out of the utilization of genetic resources, including by appropriate access to genetic resources and by appropriate transfer of relevant technologies, taking into account all rights over those resources and to technologies, and by appropriate funding' (Article 1). This is often abbreviated to 'access and benefit-sharing'.

See also The *Nagoya Protocol on Access to Genetic Resources and the Fair and Equitable Sharing of Benefits Arising from their Utilization to the Convention on Biological Diversity* (2010) and the *International Treaty on Plant Genetic Resources for Food and Agriculture* of the Food and Agriculture Organization (FAO)

cultural expressions The United Nations Education, Science and Culture Organization (UNESCO) *Convention on the Protection and Promotion of the Diversity of Cultural Expressions* (2005) defines cultural expressions as 'those expressions that result from the creativity of individuals, groups and societies, and that have cultural content'. Also, see Chapter 9 by Marc Perlman in this volume for further explanation.

demandeurs Although regularly used by delegates in informal discussions, the term 'demandeurs' does not have any official definition. Developing countries and megadiverse countries are often considered to be demandeurs seeking further protections for traditional cultural expressions, traditional knowledge and genetic resources. Indigenous peoples and local communities are also often seeking protections in accordance with their own objectives, customary laws and protocols.

derivative Article 2(e) of the Nagoya Protocol on Access to Genetic Resources and the Fair and Equitable Sharing of Benefits Arising from their Utilization to the Convention on Biological Diversity (2010) provides the following definition: 'a naturally occurring biochemical compound resulting from the genetic expression or metabolism of biological or genetic resources, even if it does not contain functional units of heredity.'

disclosure of origin requirement A proposed patent requirement that would require the applicant to disclose the origin, source and/or legal provenance of a genetic resource (and associated traditional knowledge where relevant) used in the research and development that led to the claimed invention.

folklore As defined in the United Nations Education, Science And Culture Organization (UNESCO) *Recommendation on the Safeguarding of Traditional Culture and Folklore* (1989), 'folklore (or traditional and popular culture) is the totality of tradition-based creations, of a cultural community, expressed by a group or individuals and recognized as reflecting the expectations of a community in so far as they reflect its cultural and social identity; its standards and values are transmitted orally, by imitation or by other means. Its forms are, among others, language, literature, music, dance, games, mythology, rituals, customs, handicrafts, architecture and other arts'. In the context of the IGC, the terms 'traditional cultural expressions' and 'expressions of folklore' are often used alternatively.

genetic material Article 2 of the *Convention on Biological Diversity* (1992) defines 'genetic material' as 'any material of plant, animal, microbial or other origin containing functional units of heredity'. It has also been suggested that genetic material can be understood 'as material from any biological source where units of heredity are operating or having a function'.[1]

genetic resources Article 2 of the *Convention on Biological Diversity* (1992) defines 'genetic resources' as 'genetic material of actual or potential value'.

Other legal instruments make reference to genetic resources using different terms. Article 2 of the FAO International Treaty on Plant Genetic Resources for Food and Agriculture (2001) defines 'plant genetic resources' as 'any material of plant origin, including reproductive and vegetative propagating material, containing functional units of heredity'.

Group B countries A group of mostly 'developed' or 'high income countries' that often negotiates together as a grouping in the WIPO IGC as well as in other UN forums. The origin of this denomination goes back to the establishment of UNCTAD by resolution 1995 of the UN General Assembly. An Annex to the resolution lists countries in four major groups: A (African and Asian developing countries); B (developed countries in Europe plus Japan, Australia, New Zealand, Canada and the United States); C Latin American and Caribbean countries); and D (Russian Federation and Eastern European countries).

International Patent Classification The International Patent Classification (IPC) is 'a hierarchical system in which the whole area of technology is divided into a range of sections, classes, subclasses and groups. The Classification is a language independent tool indispensable for the retrieval of patent documents in the search for "prior art".'[2]

inventive step Inventive step (also referred to as 'non-obviousness') is one of the criteria of patentability and relates to the question of whether the invention would have been obvious to a person skilled in the art.[3]

According to Article 33 of the PCT, a claimed invention shall be considered to involve an inventive step 'if, having regard to the prior art as defined in the Regulations, it is not, at the prescribed relevant date, obvious to a person skilled in the art'.

Article 56 of the *European Patent Convention* and Section 35 U.S.C. 103 provide for similar definitions. Section 35 U.S.C. 103 uses the equivalent term 'non-obvious subject matter'.

licensing agreements Licensing agreements are described as agreements setting out certain permitted use of materials or rights that the provider is entitled to grant, such as agreements to license the use of genetic resources as research tools, or to license the use of associated traditional knowledge or other intellectual property rights.[4]

misappropriation In the field of intellectual property, *Black's Law Dictionary* defines 'misappropriation' as 'the common-law tort of using the non-copyrightable information or ideas that an organization collects and disseminates for a profit to compete unfairly against that organization, or copying a work whose creator has not yet claimed or been granted exclusive rights in the work. [. . .] The elements of misappropriation are: (1) the plaintiff must have invested time, money, or effort to extract the information, (2) the defendant must have taken the information with no similar investment, and (3) the plaintiff must have suffered a competitive injury because of the taking.'

The tort of misappropriation is part of unfair competition law in the common law system. Misappropriation thus entails the wrongful or dishonest use or borrowing of someone's property, and is often used to found action in cases where no property right as such has been infringed. Misappropriation may refer to wrongful borrowing or to the fraudulent appropriation of funds or property entrusted to someone's care but actually owned by someone else. In relation to the IGC the term has been used by some countries to refer to any access or use of genetic resources/traditional knowledge without prior informed consent or approval and involvement and, where applicable, without mutual agreed terms, for whatever purpose (commercial, research, academic and technology transfer). The inclusion of the term misappropriation has been debated in the IGC, with some countries preferring to exclude it.

novelty Novelty is one of the criteria of patentability in any examination as to substance. An invention is new if it is not anticipated by prior art.[5]

According to Article 33 of the *Patent Cooperation Treaty* (PCT), novelty is defined as follows: '[f]or the purposes of the international preliminary examination, a claimed invention shall be considered novel if it is not anticipated by the prior art as defined in the Regulations.' Rule 64.1(a) of the Regulations under the PCT defines 'prior art' as 'everything made available to the public anywhere in the world by means of written disclosure (including drawings and other illustrations) shall be considered prior art provided that such making available occurred prior to the relevant date'.

patent A patent is defined as 'a document which describes an invention which can be manufactured, used, and sold with the authorization of the owner of the patent. An invention is a solution to a specific technical problem. A patent document normally contains at least one claim, the full text of the description of the invention, and bibliographic information such as the applicant's name. The protection given by a patent is limited in time (generally 20 years from filing or grant). It is also limited territorially to the country or countries concerned. A patent is an agreement between an inventor and a country. The agreement permits the owner to exclude others from making, using or selling the claimed invention.'[6]

Article 27(1) of the World Trade Organization (WTO) *Trade-Related Aspects of Intellectual Property Rights Agreement* (TRIPS Agreement) states that '[. . .] patents shall be available for any inventions, whether products or processes, in all fields of technology, provided that they are new, involve an inventive step and are capable of industrial application. [. . .] patents shall be available and patent rights enjoyable without discrimination as to the place of invention, the field of technology and whether products are imported or locally produced'.

prior art Prior art is, in general, all the knowledge that existed prior to the relevant filing or priority date of a patent application, whether it existed by way of written and oral disclosure. In some legal instruments there is a differentiation between printed publications, oral disclosures and prior use and where the publications or disclosure occurred.[7]

For the purposes of the PCT, prior art is defined by Rule 33.1 of the PCT Regulations as 'everything which has been made available to the public anywhere in the world by means of written disclosure (including drawings and other illustrations) and which is capable of being of assistance in determining that the claimed invention is or is not new and that it does or does not involve an inventive step (i.e. that it is or is not obvious), provided that the making available to the public occurred prior to the international filing date.'

prior informed consent A right or principle of 'prior informed consent' (PIC) or sometimes 'free, prior and informed consent' (FPIC) is referred to or implied in several international instruments, particularly in the environmental field, such as Article 6(4) of the *Basel Convention on the Transboundary Movement of Hazardous Wastes* (1989), and the *Convention on Biological Diversity* (1992).

In respect to access to genetic resources, the *Convention on Biological Diversity* (1992) states in Article 15(5) that it 'shall be subject to prior informed consent of the Contracting Party providing such resources, unless otherwise determined by that Party'.

Article 16(1) of the *Nagoya Protocol on Access to Genetic Resources and the Fair and Equitable Sharing of Benefits Arising from their Utilization to the Convention on Biological Diversity* (2010) provides that '[e]ach Party shall take appropriate, effective and proportionate legislative, administrative or policy measures, as appropriate, to provide that traditional knowledge associated with genetic resources utilized within their jurisdiction has been accessed in accordance with prior informed consent or approval and involvement of indigenous and local communities and that mutually agreed terms have been established, as required by domestic access and benefit sharing legislation or regulatory'.

protection 'Protection' in the work of the IGC has tended to refer to when considering the protection of traditional knowledge and traditional cultural expressions against some form of unauthorized use by third parties.[8] Two forms of protection have been developed and applied.

positive protection Two aspects of positive protection of traditional knowledge and traditional cultural expressions by intellectual property rights are relevant; one is concerned with preventing unauthorized use and the other is concerned with active exploitation of the traditional knowledge and traditional cultural expressions by the originating community itself. The use of non-intellectual property approaches for the positive protection of traditional knowledge and traditional cultural expressions can be complementary and used in conjunction with intellectual property protection.[9] For instance, positive protection of traditional knowledge and traditional cultural expressions may prevent others from gaining illegitimate access to traditional knowledge and traditional cultural expressions or using them for commercial gain without equitably sharing the benefits, but it may also be used by traditional knowledge and traditional cultural expressions holders to build up their own enterprises based on their traditional knowledge and traditional cultural expressions.[10]

defensive protection Defensive protection refers to a set of strategies to ensure that third parties do not gain illegitimate or unfounded IP rights over TK, such as a disclosure of origin requirement or databases of TK. These measures include the amendment of WIPO-administered patent systems (the International Patent Classification system and the Patent Cooperation Treaty Minimum Documentation). Some countries and communities are also developing TK databases that may be used as evidence of prior art to defeat a claim to a patent on such TK.

public domain In general, a work is considered to be in the public domain if there is no legal restriction for its use by the public.[11]

Black's Law Dictionary defines the public domain as '[t]he universe of inventions and creative works that are not protected by intellectual-property rights and are therefore available for anyone to use without charge. When copyright, trademark, patent, or trade-secret rights are lost or expire, the intellectual property they had protected becomes part of the public domain and can be appropriated by anyone without liability for infringement.'[12]

registers of traditional knowledge Registers can be analysed from many different perspectives. According to their legal nature, registers can be termed either declarative or constitutive, depending upon the system under which they are established.[13]

A declaratory regime relating to traditional knowledge recognizes that the rights over traditional knowledge do not arise due to any act of government but rather are based upon pre-existing rights, including ancestral, customary, moral and human rights. In the case of declarative registers, although registration does not affect the existence of such rights, it may be used to assist patent officials in analysing prior art, and to support challenges to patents granted which may have directly or indirectly made use of traditional knowledge. In circumstances where these registers are organized in an electronic form and available through the Internet, it is important to establish a mechanism that ensures that entry dates of traditional knowledge are valid when carrying out searches related to novelty and inventiveness. A third function that these registers may have is to facilitate benefit-sharing between users and providers.[14]

Constitutive registers form part of a legal regime which seeks to grant rights over traditional knowledge. Constitutive registers will record the granting of rights (i.e. exclusive property rights) to the traditional knowledge holder as a means to ensure their moral, economic and legal interests are protected and recognized. Most model constitutive registers are conceived as public in nature, run by a national entity and under a law or regulation which clearly determines how valid registration of traditional knowledge can take place and be formally recognized and accepted. As such they may be more controversial and difficult to design and face some critical challenges and questions in moving from concept to practice.[15]

sui generis *Black's Law Dictionary* defines *sui generis* as '*[Latin "of its own kind"]* of its own kind or class; unique or peculiar. The term is used in intellectual property law to describe a regime designed to protect rights that fall outside the traditional patent, trademark, copyright, and trade-secret doctrines. For example, a database may not be protected by copyright law if its content is not original, but it could be protected by a *sui generis* statute designed for that purpose.' A *sui generis* system is a system specifically designed to address the needs and concerns of a particular issue. There are already several examples of *sui generis* intellectual property rights such as plant breeders' rights – as reflected in the *International Convention on the Protection of New Varieties of Plants*, 1991 ('the UPOV Convention').

traditional cultural expressions WIPO uses the terms 'traditional cultural expressions' and 'expressions of folklore' to refer to tangible and intangible forms in which traditional knowledge and cultures are expressed, communicated or manifested. Examples include traditional music, performances, narratives, names and symbols, designs and architectural forms. The terms 'traditional cultural expressions' and 'expressions of folklore' are used as interchangeable synonyms, and may be referred to simply as 'traditional cultural expressions', often in its abbreviated form 'TCEs'.

traditional knowledge Traditional knowledge (TK) is knowledge, know-how, skills and practices that are developed, sustained and passed on from generation to generation within a community, often forming part of its cultural or spiritual identity. The CBD, Article 8(j), describes it as the: 'knowledge, innovations and practices of indigenous and local communities embodying traditional lifestyles relevant for the conservation and sustainable use of biological diversity.'

utilization The *Nagoya Protocol* Article 2(c) describes utilization as: 'to conduct research and development on the genetic and/or biochemical composition of genetic resources, including through the application of biotechnology as defined in Article 2 of the Convention [on Biological Diversity]'.

Glossary references

Information derived partly from the WIPO Secretariat website and The WIPO Intergovernmental Committee on Intellectual Property and Genetic Resources, Traditional Knowledge and Folklore Background Brief Publication No. 2 (www.wipo.int/edocs/pubdocs/en/wipo_pub_tk_2.pdf), from author's chapters in this book, and as adapted from the 'Glossary of key terms related to intellectual property and genetic resources, traditional knowledge and traditional cultural expressions' which is WIPO/GRTKF/IC/29/INF/7, November 30, 2015.

Notes

1 Document UNEP/CBD/WG-ABS/9/INF/1 (The Concept of 'Genetic Resources' in the Convention on Biological Diversity and how it relates to a functional international regime on access and benefit-sharing), p. 8.
2 WIPO PATENTSCOPE Glossary.
3 *WIPO Intellectual Property Handbook*, WIPO Publication No. 489 (E), 2008, p. 20.
4 See document WIPO/GRTKF/IC/17/INF/12.
5 WIPO Intellectual Property Handbook, WIPO Publication No. 489 (E), 2008, p. 19.
6 WIPO PATENTSCOPE Glossary.
7 *WIPO Intellectual Property Handbook*, WIPO Publication No. 489 (E), 2008, p. 19.
8 Overview of Activities and Outcomes of the Intergovernmental Committee (WIPO/GRTKF/IC/5/12), para. 20.
9 Ibid., paras. 21 and 22.
10 Ibid., para. 21.
11 See document SCP/13/5.
12 *Black's Law Dictionary*, p. 1027 (8th ed., 2005).
13 The Role of Registers & Databases in the Protection of Traditional Knowledge: A Comparative Analysis. UNU-IAS Report, January 2004, p. 32.
14 Ibid.
15 Ibid.

Index

Page numbers in bold refer to glossary entries

For Product Safety Concerns and Information please contact our EU representative GPSR@taylorandfrancis.com
Taylor & Francis Verlag GmbH, Kaufingerstraße 24, 80331 München, Germany

www.ingramcontent.com/pod-product-compliance
Ingram Content Group UK Ltd.
Pitfield, Milton Keynes, MK11 3LW, UK
UKHW022008180926
475701UK00003B/250
9780367335236